The Technology of Large-Scale Zinc Production in Chongqing in Ming and Qing China

Wenli Zhou

BAR International Series 2835

2016

Published in 2016 by
BAR Publishing, Oxford

BAR International Series 2835

The Technology of Large-Scale Zinc Production in Chongqing in Ming and Qing China

ISBN 978 1 4073 1551 5

COVER IMAGE *'Reconstruction of a retort from Miaobeihou.' Illustration by the author.*

Printed in England

BAR
PUBLISHING

BAR titles are available from:

BAR Publishing
122 Banbury Rd, Oxford, OX2 7BP, UK
EMAIL info@barpublishing.com
PHONE +44 (0)1865 310431
FAX +44 (0)1865 316916
www.barpublishing.com

To My Family

Acknowledgements

This research was based on my Ph.D. thesis submitted to University College London (UCL) in 2012. This research was funded by the Sun Hung Kai Properties Kwok Foundation in Hong Kong (2008-2011). The International Centre for Chinese Heritage and Archaeology founded by UCL and Peking University provided invaluable support for the research. Financial assistance was also kindly provided by the Institute for Archaeo-Metallurgical Studies, UCL Institute of Archaeology (IoA) and UCL Graduate School.

I would like to acknowledge my UCL supervisors Prof. Marcos Martinón-Torres and Prof. Thilo Rehren, and external supervisors Prof. Chen Jianli and Prof. Paul T. Craddock, for their continuous academic guidance and supervision. I want to express my sincere gratitude to Marcos, who shared with me his expertise and knowledge, supported me to attend conferences and publish my research, patiently read my drafts and made invaluable comments, and kept me well on track towards the end of this research. I owe special thanks to Thilo for his help in the interpretation of the samples, his comments on my drafts and his care for my life in London. Prof. Chen (Peking University) has always been a great support: he kindly permitted me to continue his research, made samples available for my analyses, accompanied me to fieldwork, and encouraged and stimulated this research. I also thank Paul (British Museum), who provided me a lot of literature, shared his experiences in the history of zinc metallurgy, and provided helpful comments on my drafts.

This research is a continuation of the project *Multidisciplinary Study of Zinc Smelting Sites in Fengdu, Chongqing (2005-2008)* of the State Administration of Cultural Heritage of China, carried out by Peking University, University of Science and Technology Beijing (USTB), Henan Provincial Cultural Heritage and Archaeology Institute (HPCHAI) and Chongqing Municipal Bureau of Cultural Heritage. I am very grateful to Liu Haiwang (HPCHAI), Director Zou Houxi, Vice Director Yuan Dongshan, Bai Jiujiang and Ye Lin (Chongqing Municipal Cultural Heritage Institute), Director Li Guohong, He Hairong, Fu Kunming, Qin Jin, Liu Yuzhen (Fengdu County Bureau of Cultural Heritage), Director He Fenhua, Jiang Ping, Zeng Qihua and Jiang Yanfei (Shizhu County Bureau of Cultural Heritage), for providing samples and useful materials, and their cooperation. Special thanks are given to the State Administration of Cultural Heritage of China for the support of this research.

Many other individuals have contributed to various aspects of this research. I would like to thank: Vice Director Wu Xiaohong, Prof. Zhao Hui, Cui Jianfeng, Zhang Hai and Zhang Jianwei (Peking University), Prof. Li Yanxiang and Prof. Han Rubin (USTB), Prof. Mei Jianjun and John Moffett (Needham Research Institute, UK), Director Zhou Weirong (Chinese Numismatic Museum), Hu Guoqiang, Li Mijia and Lei Yong (Palace Museum, Beijing), Prof. Huang Quansheng (Guangxi Radio and TV University), Prof. Jin Zhengyao (University Science and Technology of China), Chen Hailian (University of Tübingen, Germany), Prof. George B. Souza (University of Texas, USA), Wang Quanyu and Eleanor Blakelock (British Museum, UK), Prof. Nigel Wood and Prof. Mark Pollard (University of Oxford, UK), Vincent Pigott (University of Pennsylvania, USA), Donald B. Wagner (Nordic Institute of Asian Studies, Denmark), and Justine Bayley (English Heritage, UK).

At UCL IoA, I enjoyed high-quality courses and lectures and also benefited from communications with a number of staff. I would like to thank John Merkel, Bill Sillar, Wang Tao, Lucas Nickel, Ian Freestone and Patrick Quinn for their advice and help, and thank Kevin Reeves, Philip Connolly, Simon Groom and Harriet White for their technical support and assistance at the Wolfson Archaeological Science Laboratories. I also enjoyed the friendship and support from my fellow students, especially my B53 mates: Janice, Kuang-Jen, Yixian, Qiyan, Min, Siran, Thomas, Edwinus, Alice, Miljana, Kristina, Carmen, Ruth, Loïc and Jane.

The publication of this research was encouraged by my present bosses, Director Zhang Baichun and Prof. Guan Xiaowu (Institute for the History of Natural Sciences, Chinese Academy of Sciences) and my supervisors. The publication was funded by the National Natural Science Foundation of China (Project No. 51404242) and the Chinese Academy of Sciences (Project No. Y522021007 and Y621011003). Many thanks to Jonathan Wood (UCL IoA), Huang Chao (Sun Yat-Sen University) and Li Chenyuan (USTB) for their professional proofreading, and my colleagues Wu Shilei and Bao Ning for patient copy editing. Any errors and misunderstandings that still remain in this work are entirely my own responsibility.

My final appreciation and thanks go to my parents, my parents-in-law, my husband Kang Baoqiang (Palace Museum, Beijing) and my beloved daughter Xiaoyan who have always supported me.

Table of Contents

List of Figures

List of Tables

Abbreviations

AAS, Atomic Absorption Spectroscopy 原子吸收光谱
AMS, Accelerator Mass Spectrometry 加速器质谱
CCCHAI, Changsha Civic Cultural Heritage and Archaeology Institute 长沙市文物考古研究所
CCLGSC, Compilation Committee of Local Gazetteer of Shizhu County 石柱县志编纂委员会
CMBCH, Chongqing Municipal Bureau of Cultural Heritage 重庆市文物局
CMBI, Chongqing Municipal Bureau of Immigration 重庆市移民局
CMCHAI, Chongqing Municipal Cultural Heritage and Archaeology Institute 重庆市文物考古所
CMCHI, Chongqing Municipal Cultural Heritage Institute 重庆市文化遗产研究院
CRMs, Certified Reference Materials 标准参考物质
EPMA-WDS, Electron Probe Micro-Analyser - Wavelength Dispersive Spectrometry 电子探针及波长色散光谱
FCBCH, Fendgu County Bureau of Cultural Heritage 丰都县文物管理所
HCZLW, Hezhang County Zinc and Lead Works 赫章锌铅矿
HPCHAI, Henan Provincial Cultural Heritage and Archaeology Institute 河南省文物考古研究所
IACASS, Institute of Archaeology Chinese Academy of Social Sciences 中国社会科学院考古研究所
ICP-AES, Inductively Coupled Plasma - Atomic Emission Spectrometry 电感耦合等离子体原子发射光谱
IMRCAGS, Institute of Mineral Resources Chinese Academy of Geological Sciences 中国地质科学院矿产资源研究所
MHGBSIC, Metallurgy History Group of Beijing Steel and Iron College 北京钢铁学院冶金史组
MIP, Metallurgical Industry Press 冶金工业出版社
OM, Optical Microscopy 光学显微镜
RGALHS, Research Group of Ancient Local Histories of Shizhu 石柱古代地方文献整理课题组
SEM-EDS, Scanning Electron Microscopy - Energy Dispersive Spectrometry 扫描电子显微镜及能量色散光谱
SPAI, Shanxi Provincial Archaeology Institute 山西省考古研究所
SPCHAI, Sichuan Provincial Cultural Heritage and Archaeology Institute 四川省文物考古研究所
UCL IoA, University College London Institute of Archaeology 伦敦大学学院考古研究所
USTB, University of Science and Technology Beijing 北京科技大学
XRD, X-Ray Diffraction X 射线衍射

Abstract

Zinc was one of the important metals in the ancient world, primarily as an essential component of the alloy brass. As a volatile metal, its production required sophisticated distillation installations. The production of this metal played a special role in both the technological and economic history of Ming and Qing China: zinc was employed for coinage and also exported to the world via maritime trade. Since the early twentieth century, many scholars have studied the history of Chinese zinc smelting technology based on evidence from historical records, ancient brass coins and traditional zinc smelting processes. A major gap in our understanding of ancient Chinese zinc production has been the lack of detailed technological reconstructions based on the study of ancient production remains. Recently excavated zinc production sites in Chongqing provide an excellent opportunity to address this issue.

This research presents the analyses, technical interpretation and socio-economic contextualisation of the production remains from three Ming sites in Fengdu (Miaobeihou, Puzihe and Muxiexi) and one Qing site in Shizhu (Dafengmen). Zinc ore, zinc metal, retorts and slag from these sites were analysed by OM, SEM-EDS, EPMA-WDS and XRD. The analytical results reveal that these sites used similar zinc distillation technology: the retorts were made of pots, condensers, pockets and lids, all well designed to meet specific performance characteristics; the pots were charged with iron-rich oxidic zinc ores, coal and charcoal; a high temperature of around 1200 °C and highly reducing atmosphere were achieved inside the pots to reduce the zinc ores; the zinc vapour formed within the pots ascended through the hole of pockets and was cooled and collected in the condensers. Some differences were found between two groups of sites. The Qing site used lower grade oxidic zinc ores, a lower proportion of reducing agents, and elongated retorts of inferior performance, leading to a lower technical efficiency, in spite of its ideal location near the necessary resources of zinc ores and coal. These differences can be explained as adaptation of the zinc production for coinage to the different social, political and economic constraints affecting these sites.

This research thus offers a first contribution towards a detailed reconstruction of Chinese zinc distillation technology that considers both variations in time and space as well as common elements that characterise the Chinese technological style. Finally, the significance of Chinese zinc production is contextualised and discussed with reference to coinage in Ming and Qing China, but also by comparing it to other brass and zinc making technologies in China, India and Europe, and by assessing the influence of Chinese zinc in global maritime trade.

锌是古代世界最重要的金属之一，主要作为黄铜合金的重要组分。由于其挥发性，锌的生产需要复杂的蒸馏设施。金属锌的生产在明清中国的技术史和经济史上有着特殊的作用：锌不但用于铸造铜钱，还通过海上贸易出口到世界各地。二十世纪初以来，不少学者从历史记载、古代黄铜钱币和传统炼锌技术等方面研究中国古代炼锌技术的历史。我们对中国古代炼锌技术的认识的局限，主要在于缺少基于古代冶炼遗物研究的详细的技术重建。重庆近年来发掘的炼锌遗址为解决这个问题提供了很好的机会。

本研究对重庆丰都庙背后、铺子河和木屑溪三处明代炼锌遗址和石柱大风门清代炼锌遗址的冶炼遗物进行了科技分析、技术解读以及社会和经济背景研究。采用光学显微镜、扫描电子显微镜及能量色散光谱、电子探针及波长色散光谱和 X 射线衍射等方法分析了这些遗址出土的锌矿石、金属锌、冷凝罐和炉渣等遗物。分析结果显示这些遗址所采用的炼锌技术相似，蒸馏罐由罐、冷凝器、兜和盖组成，每个部分均经过良好的设计以满足其特定的使用性能；罐内加入富铁的氧化锌矿石、木炭和煤炭；罐内达到大约 1200 °C 的高温，并保持较强的还原气氛，将锌矿石还原；罐内生成的锌蒸汽上升并通过兜上的孔，冷却并收集在冷凝器内。两组遗址的技术也有一些差别。清代遗址所用的锌矿石含有较多杂质、所加还原剂比例较小、蒸馏罐质量较差，导致生产效率较低，但是遗址位置较好，离锌矿和煤矿产地近。这些差别可以解释为这些遗址对其所处的不同的社会、政治、经济等环境所做的适应。

本研究第一次详细重建了中国古代炼锌技术，不但考虑时空的变化，还考虑到中国炼锌技术特征的共性。最后，通过考察锌用于明清时期钱币铸造，比较中国、印度和欧洲的锌和黄铜生产技术，以及评价中国锌在全球海上贸易的影响，阐述了中国古代炼锌技术的重要性。

Chapter 1

Introduction

1.1 A brief history of the metallurgy of zinc and brass

Zinc was one of the most important metals in the ancient world, primarily as an essential component of brass [1] (Figure 1.1). But zinc is very difficult to extract from ores because its smelting temperature (around 1000 °C) is higher than its boiling point (907 °C), with the metal vaporising immediately and re-oxidising as it comes in contact with air. The invention of zinc distillation technology, requiring complex reduction and condensation installations rather than ordinary smelting furnaces, essentially explains why the isolation of metallic zinc took place millennia later than the metals of antiquity - gold, silver, copper, lead, tin, iron and mercury.

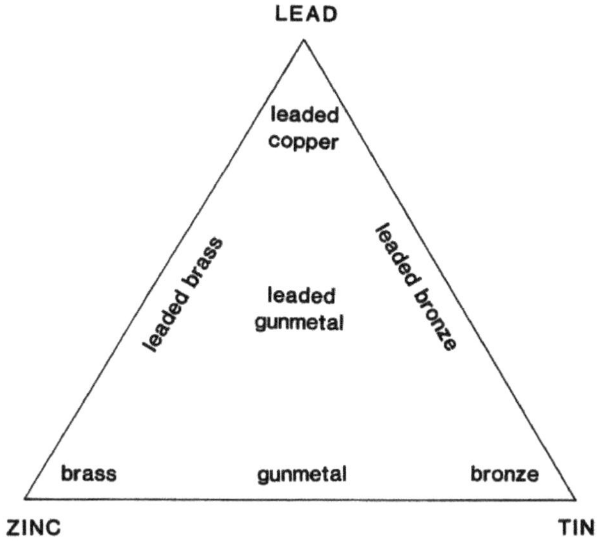

Figure 1.1 Ternary diagram showing the relationship between the alloy name and composition for copper alloys (Bayley 1998, 8) (© The Trustees of the British Museum).

Until recently, the main use of metallic zinc was to alloy it with copper to produce brass, in the so-called speltering process. However, brass was made for a long time before metallic zinc came into regular production. This has been intensively studied, especially in Europe, through a combination of historical evidence and analyses of brass artefacts and archaeometallurgical remains (Maréchal 1954; Caley 1964; Bayley 1984; Bayley 1998; Craddock 1978; Craddock and Eckstein 2003; Day 1998; Dungworth 1995; Dungworth 1996; Dungworth 1997; Rehren 1999a; Rehren 1999b; Martinón-Torres and Rehren 2002; Rehren and Martinón-Torres 2008). The earliest brass artefacts, mostly ornaments and small jewellery, found in the Near and Middle East, and dated to the Bronze Age and Early Iron Age, were probably produced by smelting zinc-rich copper ores. From the mid-first century BC, brass was manufactured more systematically using cementation. It was a process in which calamine (oxidic zinc ores), metallic copper and charcoal are heated together in a closed crucible, where brass forms through the direct absorption of zinc vapour by the copper. In the Roman World, brass was first used mainly for military hardware and copper-based coins, and later for general domestic use. After an apparent decline in the late Roman period, brass was not used extensively until the High Middle Ages, mainly for everyday implements and also secular and ecclesiastical works of art (Rehren and Martinón-Torres 2008).

The direct production of brass via cementation meant that zinc was never seen or even recognised as a metal by brass-makers – all they could observe was 'copper coloured yellow' by the action of calamine stone. Pure metallic zinc is thus very rare in early archaeological contexts. Although several zinc artefacts have been reported in Europe, dated to the Iron Age, Roman and Medieval periods, yet most have neither archaeological contextual information nor credible dates (Farnsworth *et al.* 1949; Brownsword 1988; Rehren 1996; Craddock 1998). Nevertheless, a few of them have been identified as authentic, possibly dated to as early as the Greek and Roman periods. However, rather than the products of deliberate distillation, it is likely that these exceptional objects were made by collecting the small amounts of zinc that condensed in the flues of lead and silver smelting furnaces while processing zinc-rich ores (Craddock 1998). This hypothesis was strengthened by the discovery of small lumps of zinc metal, dated to the Roman period in Poland and Britain, in association with lead smelting operations (Stos-Gale 1993; Todd 1993). There is also historical evidence that some zinc metal was collected from the furnace flues at Rammelsberg, Germany, in the seventeenth and eighteenth centuries (Craddock 1998, 4). However, the small quantities of zinc metal obtained this way were insufficient to meet the needs of large-scale brass making. Although this was achieved by the cementation process, a tradition which continued in Europe throughout the Middle Ages and the Renaissance to the mid-nineteenth century, it was not until the 1740s, when William Champion patented a zinc smelting process, that unalloyed zinc began to be produced in Europe (Day 1998). Some metallurgical remains related to zinc distillation in Bristol have been investigated (Day 1988; Dungworth and White 2007).

[1] In a strict sense, brass refers to an alloy of copper and zinc, just as bronze is an alloy of copper and tin. Copper alloyed with significant levels of both tin and zinc is known as gunmetal. If brass, bronze and gunmetal also contain lead, they are known as leaded brass, leaded bronze and leaded gunmetal, respectively. However, no clear-cut division exists between these alloys (Figure 1.1). In this research, all alloys containing significant amounts of zinc, such as brass, gunmetal, leaded brass and leaded gunmetal, are termed brass.

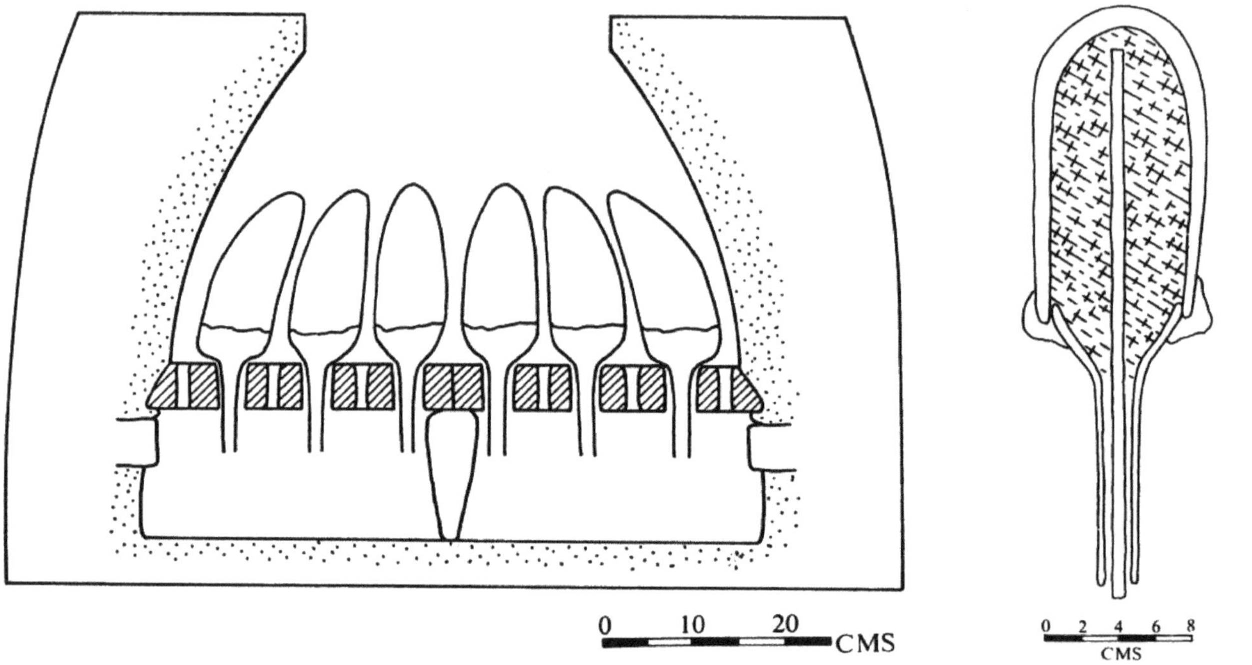

Figure 1.2 Zinc distillation installations at Zawar (Craddock *et al.* 1998). Left: a furnace with retorts loaded *in situ* at Zawar; right: a small retort with a condenser luted on with clay, the charge loaded and a stick inserted (© The Trustees of the British Museum).

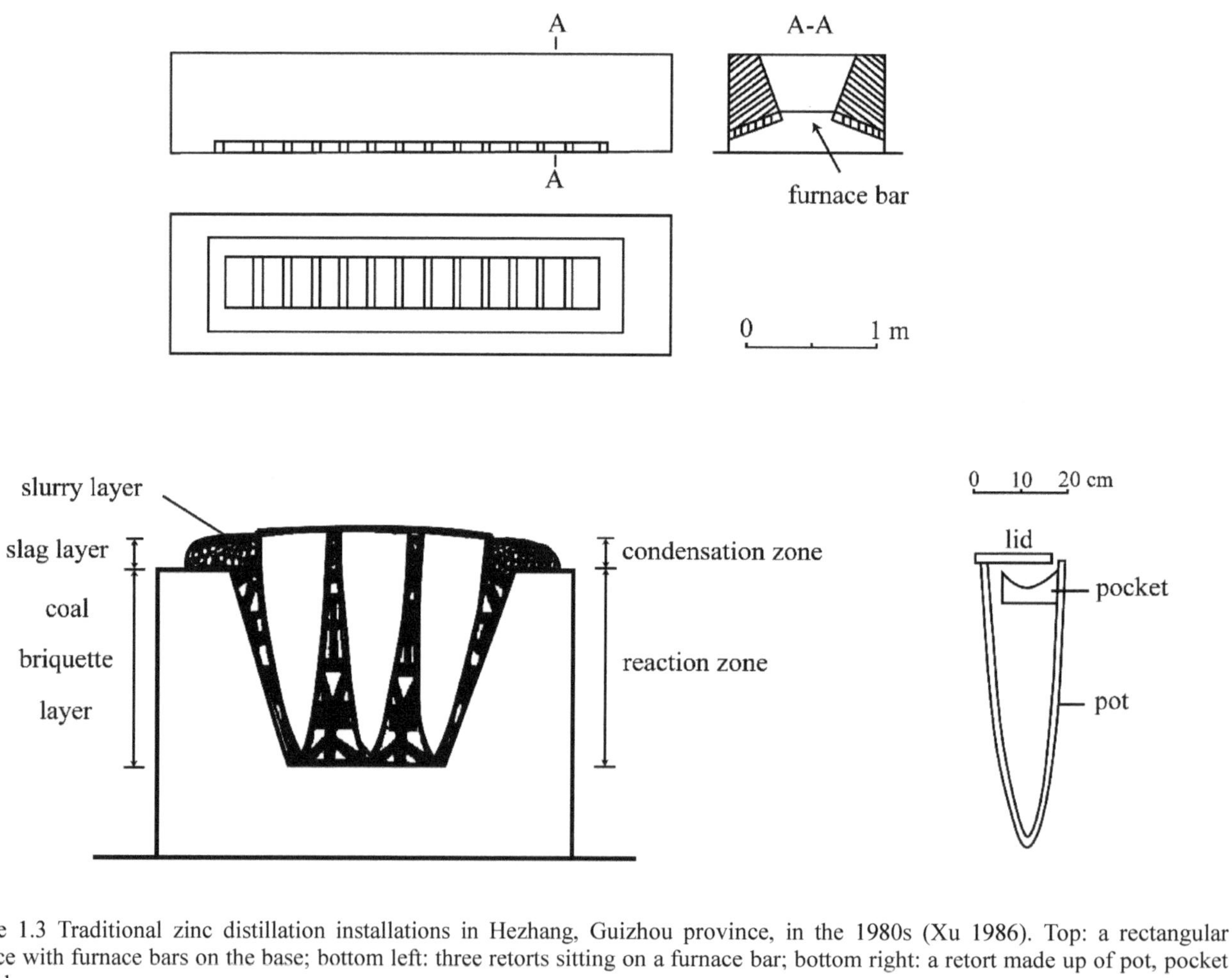

Figure 1.3 Traditional zinc distillation installations in Hezhang, Guizhou province, in the 1980s (Xu 1986). Top: a rectangular furnace with furnace bars on the base; bottom left: three retorts sitting on a furnace bar; bottom right: a retort made up of pot, pocket and lid.

Before the eighteenth century, only India and China produced zinc on an industrial scale. According to the available evidence, Indian zinc production appeared about 1000 years ago at Zawar, northwest India. The technology at Zawar boomed during the fourteenth to sixteenth centuries, declined in the seventeenth and eighteenth centuries and finally ceased in 1812. Archaeological and archaeometallurgical studies of production remains at Zawar reveal that the Indian zinc smelting process was based on the principle of distillation by descending (Figure 1.2). In this process, the retorts were composed of cylindrical aubergine-shaped pots and conical condensers with long tubes. The pots were filled with zinc ores and reducing agents and then luted onto the condensers. The charged retorts were then turned upside down and loaded into the upper chamber of the furnaces with their condensers sitting in the holes of the perforated platforms, with their tubular ends protruding into the lower chamber below. The pots were heated with fuel in the upper chamber and the zinc vapour, which formed subsequently, descended into the condenser tubes. The temperature was lower in the lower chamber, resulting in condensation of the zinc vapour that later dripped down into the collecting vessels beneath. No vessels have been recovered at Zawar (Craddock *et al.* 1983; Craddock *et al.* 1984; Freestone *et al.* 1985a; Freestone *et al.* 1985b; Hegde *et al.* 1986; Craddock *et al.* 1989; Freestone *et al.* 1991; Craddock *et al.* 1998).

China, which also contributed to the history of zinc distillation, has been producing metallic zinc since at least the early seventeenth century (Zhou 2007a). Chinese zinc was mainly used for making brass coins in the late Ming (AD 1368-1644) and Qing (AD 1644-1911) Dynasties. Some zinc was traded by European merchants to the wider world (see Section 2.2.4). Traditional zinc smelting processes were still practiced in southwest China until the late twentieth century (see Section 2.3). Through the investigation of these traditional processes, it has been shown that Chinese zinc distillation was based on the principle of distillation by ascending (Figure 1.3), which is fundamentally different from the Indian process. The Chinese processes generally used lidded cylindrical retorts with internal condensers at the top. Ceramic partitions, called pockets, were added to separate the upper condensation zone from the lower reaction zone. The retorts charged with zinc ores and coal were heated in rectangular furnaces fuelled with coal briquettes. The zinc vapour formed in the reaction zone, ascended via the hole made on one side of the pocket, and condensed in the upper zone (Xu 1986).

1.2 Problems and research gaps in the study of Chinese zinc distillation technology

Given the technological and economic importance of Chinese zinc distillation, many geologists and historians of chemistry/metallurgy have investigated the technology and its early history, especially its origins, since the early twentieth century (Hommel 1912). Significant progress has until recently been made on this issue through the interpretation of terminology and entries related to zinc in historical records (Zhang 1923; Zhang 1925; Needham 1974; Zhao 1984b; Zhou 1993; Zhou 2001), chemical analyses of brass coins (Wang *et al.* 2005; Dai and Zhou 1992; Zhou and Fan 1994) and field surveys of traditional zinc production in southwest China (Hu and Han 1984; Xu 1986; Mei 1990; Zhou 1996; Craddock and Zhou 2003). However, these sources of evidence have their limitations.

Table 1.1 Debates on the origins of zinc smelting in China before the 1990s.

Proposed date	Arguments for the view	Arguments against the view
The Han Dynasty (202 BC-AD 220)	-*Lian* 连/鏈 denoted zinc or zinc ore. -Some Han coins were found to contain zinc (Zhang 1923; Zhang 1925).	-*Lian* 连/鏈 denoted lead or lead ore (Zhou 1993). -No other Han coins with zinc have been found (Zhou 2004).
The Five Dynasties (AD 907-960)	-*Woyuan* 倭铅 denoted zinc; it can be traced back to an entry dated to the period: *woyuan kegoujin* 倭铅可勾金 (Wang 1955, 92-93; Liu 1991). -Local gazetteers in Guizhou record that zinc smelting began in the period (Xu 1986).	-This entry was a later addition (Zhao 1984b); *woyuan* 倭铅 denoted lead (Zhou 2007a). -It was silver smelting that started from the period according to the local gazetteers (Zhou 2007a).
The Song Dynasty (AD 960-1279)	-*Baixi* 白锡 denoted zinc; it was used for making coins. -Some Song coins were found to contain zinc (Zhang 1923; Zhang 1925; Needham 1974).	-*Baixi* denoted tin (Zhou 1992). -No other Song coins with zinc have been found (Zhou 2004).
The early Ming Dynasty (AD 1368-1644)	-*Woyuan baishuiyuan* 倭源白水铅 denoted zinc, which was used for making Xuande incense burners (Zhang 1964). -Brass coins were made of speltering brass from the Jiajing period (Bowman *et al.* 1989). -Zinc coins dated to the early Ming period were found (Leeds 1954).	-*Woyuan baishuiyuan* was not used for Xuande incense burners and most likely was lead (Zhou 2007a). -The Jiajing and Wanli brass coins were made of cementation brass (Zhou 2007a). -Zinc coins actually dated to the late Ming and Qing periods (Bowman *et al.* 1989).

1. Historical records

One of the main problems when working with historical records lies in the difficulties in the unequivocal identification of the terms employed for zinc and other associated materials and processes, which often have different meanings in different contexts. A number of terms related to zinc have been recognised, such as *lian* 连/鍊, *baixi* 白锡, *woyuan* 倭铅/窝铅 and *woyuan baishuiyuan* 倭源白水铅. The interpretation of the different terms has fed heated debates on the origins of the technology, which, coupled with limited analyses of coins, has resulted in different dates for the origins of zinc distillation technology being proposed, including the Han Dynasty, the Five Dynasties, the Song Dynasty and the early Ming Dynasty. Each has been challenged (Table 1.1). Recent research by Zhou Weirong and his colleagues has suggested that zinc was not produced on a large scale for minting until the late Ming Dynasty (see Section 2.1).

Among the various terms, *woyuan* 倭铅 has been widely accepted as the key term for zinc. *Woyuan* 倭铅 and its production are recorded in the renowned technical compendium *Tian Gong Kai Wu* 天工开物 (The Exploitation of the Works of Nature)[2], written by Song Yingxing and published in 1637:

> *Woyuan* 倭铅, a term of recent origin, does not appear in ancient books. … Fill each earthen jar with ten catties of *luganshi* 炉甘石, then seal it tightly with mud, and let it dry slowly so as to prevent cracking when heated. Then pile a number of these jars in alternate layers with coal briquettes, with kindling on the bottom layer for starting the fire. When the jars become red-hot, the *luganshi* melts into a mass. When cooled, the jars are broken open and the substance thus obtained is *woyuan* 倭铅, with a twenty percent loss. This metal is easily burnt off by fire if not mixed with copper. Because it is similar to lead, yet more fierce in nature, it is called *woyuan* 倭铅. (Sun and Sun 1966, 247) (Figure 1.4)

As it was produced from *luganshi*, a term employed to refer to oxidic zinc ores (see Section 6.2.1), and it is described as a volatile metal, *woyuan* 倭铅 was generally believed to denote zinc. *Woyuan* 倭铅, literally 'Japanese lead', was so named probably because it is similar to lead but as fierce, due to its volatility, as the Japanese pirates invading the southeast coast of Ming China. A potentially earlier, yet controversial, reference to zinc appears in the medical compendium *Ben Cao Gang Mu* 本草纲目 (Compendium of Materia Medica), written by Li Shizhen in 1593 (Li 1975); here one entry refers to *woyuan kegoujin* 倭铅可勾金, quoting from *Bao Zang Lun* 宝藏论 (Discourse on the Precious Treasury of the Earth) of the Five Dynasties (AD 907-960). This reference was used as the main evidence to support the view that zinc production started in the Five Dynasties – a proposal also supported by a British scholar Joseph Needham (Needham 1974). However, the entry has been argued to originate from a later addition by Li Shizhen; furthermore, according to the literal interpretation of the entry[3], *woyuan* 倭铅 is more likely to have referred to lead. Recent textual research on this term has claimed that *woyuan* 倭铅 was a term originally denoting lead, but was incorrectly applied to zinc; the true name for zinc was *woyuan* 窝铅 (note the same pronunciation but different Chinese spellings), literally 'pocket lead', i.e. lead from the pockets of retorts (Zhou 2007a). Thus, even *woyuan* 倭铅 may not be a reliable term for tracing the use of zinc.

Figure 1.4 The illustration of zinc smelting in *Tian Gong Kai Wu* (Song 1637).

With doubts surrounding epigraphic interpretations, this leaves only the text and illustration in *Tian Gong Kai Wu* as early evidence of the zinc smelting process in China. This also presents problems. Although the text suggests that zinc formed inside the jars, and that the jars had to be broken open to collect the metal, no details describe how

[2] *Tian Gong Kai Wu* is highly valued as the most important Chinese technical compendium of the early seventeenth century and a unique piece of work among Chinese writings on technology before the twentieth century. This lavishly illustrated book covers a wide range of manufacturing and technological activities, including agriculture, textiles, ceramics, metallurgy, papermaking, weapons and transport.

[3] Needham interpreted the entry as stating that: '*woyuan* can by alloyed with other metals' (Needham 1974, 214); whereas Zhou interpreted the text as saying that: '*woyuan* can be used to prompt some metals to be extracted' (Zhou 2007a).

the zinc smelting process took place inside the jars. Judging by the illustration, the jars seem to have been fired in a stacked heap, but no furnaces are mentioned in the text. The text also fails to mention the use of reducing agents, either charcoal or coal, without which zinc ores could not be reduced to the metal. All of these elements are at odds with a realistic description of a method of distillation for the production of zinc.

2. Brass coins

Analyses of copper-based coins have been another prominent approach to trace the use of zinc. It is rather difficult to differentiate brass coins made by speltering from those made by cementation on basis of their chemical compositions. Although zinc and cadmium levels of brass coins can be indicative of the manufacturing method, these are not absolute criteria. It is generally accepted that brass produced via cementation has an upper zinc limit of 28%, as commonly encountered in Roman and Medieval brass coins and artefacts and in several simulation experiments (Haedecke 1973; Welter 2003). However, recent experiments have shown that it is possible to produce cementation brass with zinc contents of over 40% (Newbury *et al.* 2005; Bourgarit and Bauchau 2010). Evidently, the effectiveness of the upper zinc limit as a threshold to differentiate brass coins made by either process is questionable, while brass coins with lower levels of zinc could certainly be manufactured by either process. Turning to cadmium, this element usually coexists with zinc ores, but it is more volatile than zinc, with a boiling point of 767 °C. Therefore, cementation brass tends to contain little cadmium as most of it would vaporise, while speltering brass can have detectable traces of cadmium, which would have been collected via condensation together with the zinc (Zhou and Fan 1994). Nevertheless, it should be noted that the cadmium levels recorded in brass coins not only depend on the different behaviour of cadmium in the two processes, but also on different zinc ore sources, ore preparation methods and smelting technology (Cowell *et al.* 1993). Clearly, both these criteria need to be considered carefully (Further discussion see Section 7.1.1).

3. The traditional processes

Traditional zinc smelting processes in southwest China in the twentieth century have been described many times since it was first documented by foreigners in the 1910s and 1920s (Moore-Bennet 1915; Wheler 1922). Further traditional processes have been recorded by Chinese geologists, metallurgists and historians of metallurgy (see Section 2.3). Although these sources provide us with valuable information on Chinese zinc distillation technology, caution is required before assuming that the twentieth-century technology corresponds to the ancient technology. There may be continuities between the ancient and the modern traditional processes, but there is a risk of anachronism in interpreting the ancient technology based on its modern counterpart.

The problems and limitations of the traditional approaches to the study of the history of zinc distillation in China will be discussed in more detail in the relevant chapters, especially as new data and a reinterpretation of past studies prompt new ideas and research strategies. However, this brief outline suffices to highlight the need for fresh approaches. When previous studies are reviewed, a major gap in our understanding of ancient Chinese zinc production becomes apparent: this is the lack of detailed technological reconstruction based on the studies of production remains. The opportunity to rectify this situation has arisen with the discovery and excavation of zinc smelting sites in Chongqing, which have generated much needed archaeological evidence to reconstruct the technology and organisation of production.

Since 2002, about 20 zinc smelting sites, dated to the fifteenth to seventeenth centuries, have been found along the Yangtze River in Fengdu county, Chongqing municipality, southwest China (Figure 1.5). In nearby Shizhu county, many retort fragments and slag heaps have been found in the Laochangping region, broadly dated to the Qing Dynasty. A preliminary technological reconstruction was made on raw materials, structures of furnaces and retorts, and the smelting process from Miaobeihou, one of the Fengdu sites (Liu *et al.* 2007). The main parts of retorts were identified and their function discussed: pots containing reagents, top condensers, perforated pockets and lids (Figure 1.6). Although this research provided a useful starting point for our understanding of Chinese zinc distillation technology, further analyses of different types of production remains are necessary to reconstruct the technology, especially with regard to the mapping of temporal and regional variants and the specific questions outlined in Section 1.3. The analytical studies of production remains from three sites in Fengdu and one site in Shizhu constitute the core of this research. These case studies will enable, for the first time, a high-resolution reconstruction of the technology based on production remains, which will, in turn, help provide much more insight than the confusing description and illustration in *Tian Gong Kai Wu*. These case studies will also be employed as a foundation to offer a broader picture of the production, transport and consumption of the metal in Ming and Qing China.

A final problem existing in most previous studies is that zinc distillation technology has not been placed within relevant social contexts, and the broader significance of zinc production in Ming and Qing China has been ignored. To reconstruct and explain the technology in a historically meaningful way, it is necessary to go beyond the technicalities. As a minting material, zinc and its mining and smelting had significant impact on the domestic economy. Furthermore, Chinese zinc not only played an important role in the world trade, but it also inspired the scientific experiments with the metal that, in due course, would lead to its industrial production in Europe. The history of zinc distillation technology in Ming and Qing China cannot be detached from the broader, changing context in which it was embedded. The interplay between geography, economy and politics will help contextualise the archaeological evidence, while the technological reconstruction will also add to our understanding of the complex dynamics. Therefore, this study of large-scale zinc production in Chongqing is presented as an innovative strategy to fill the gap in our knowledge of the history of the technology, and also, more broadly, as an archaeological contribution to important aspects of the socio-economic history of Ming and Qing China.

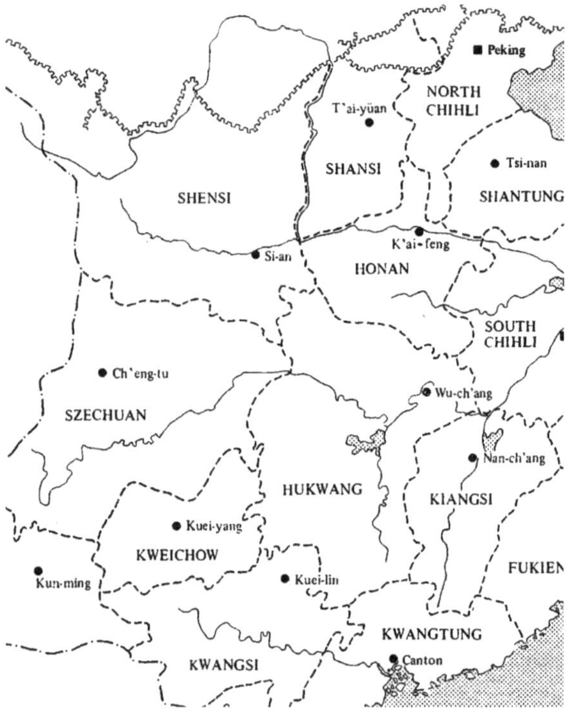

Figure 1.5 Map of the thirteen provinces and two metropolitan regions (South Zhili and North Zhili) of the Ming empire (AD 1368-1644), reproduced from (Huang 1974, xvi) (© The Trustees of Cambridge University Press). Chongqing was then a prefecture in Sichuan province.

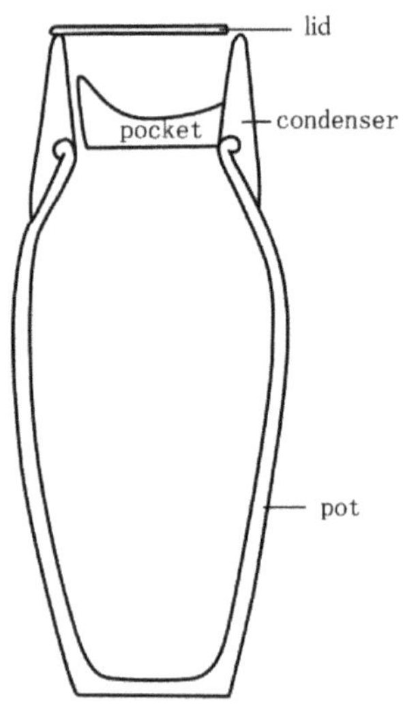

Figure 1.6 Reconstruction of a retort from Miaobeihou. The retort is made of pot, condenser, pocket and lid.

1.3 Research aims and objectives

This research sets out to provide a substantial contribution towards our knowledge of zinc smelting technology, the organisation of production and its broader socio-economic implications in Ming and Qing China.

The first key aim of this research is to understand the zinc smelting technology practiced in Fengdu and Shizhu. Miaobeihou, the most extensively excavated site in Fengdu, was selected as the main case study of this research, providing the initial reference point. Production remains of two contemporary sites in Fengdu, Puzihe and Muxiexi, were also examined for comparison. In addition, Dafengmen, a site in Shizhu dated to the Qing period, was also included as a further case study in order to address temporal and regional variability. To fulfil this aim, each site was investigated in a similar way:

1. Comparative characterisation of different fabrics used to make the different parts of the retorts (pots, condensers, pockets and lids), identification of raw materials used and assessment of how their formal and material properties were adapted to optimise their performance characteristics.

2. Analyses of raw materials, zinc metal and metallurgical residues to establish details about the original charge, zinc products and the metallurgical process taking place within the retorts.

3. Overall reconstruction of the zinc smelting technology and the whole process of production based on archaeological evidence, analytical results, historical documents and traditional zinc smelting, including a trial of estimation on the production scale.

The second aim of the research is to juxtapose the technology and organisation of production at these sites in Fengdu and Shizhu. Factors such as resource procurement (zinc ores, reducing agents, fuel, clay and water), transport of raw materials and products and choice of location are assessed in order to explain spatial organisation of production at the two groups of sites. The different choices documented are explored by considering the relevant social contexts affecting each group of sites. As a result, the defining characteristics of the Chinese zinc distillation tradition are summarised by integrating historical, ethnographic, archaeological and archaeometric evidence, thereby allowing detailed comparisons between the zinc distillation traditions in China and India.

The last and broader objective is to investigate the historical impact of Chinese zinc both at home and abroad. Given that zinc was mainly produced for state minting, the significance of zinc production in China is explored by considering the reasons behind the transition from the cementation process to the speltering process for making brass coins; this transition is explained from a technical perspective, but also by considering the changing historical settings. Chinese zinc was also exported via the international maritime trade. The influence of Chinese zinc abroad is assessed by considering the different situations that Chinese zinc met in different parts of the world, in particular, India and Europe.

1.4 Structure of the research

This general introduction is followed by chapter two, which provides a more detailed historical and technological background to this research. A general overview of the history of brass and zinc in China is first presented. The chapter then introduces the uses of brass and zinc in Ming and Qing China. In addition, various ethnographic records of traditional zinc smelting processes in the twentieth century are summarised in terms of their operational sequences. These are analysed synthetically for the first time.

Chapter three outlines the theoretical framework underpinning this archaeological study, as well as the methodology employed in the scientific analyses of the zinc smelting remains in Chongqing. It first addresses several theoretical concepts that are of use in defining the research questions, shaping the approaches and explaining the data, as well as some other sources of evidence whose critical exploitation could help gain a better understanding of archaeological materials. This chapter then details and justifies the choices made with regards to the sampling strategy and the analytical methods.

Chapter four presents the analytical results of production remains from three river sites in Fengdu. It starts with fundamental aspects of the geography, archaeology and history in the region studied, followed by an introduction of the discovery and excavation of zinc smelting sites in Fengdu. The zinc distillation technology at Miaobeihou, the main case study site, is investigated by analysing a variety of production remains, such as minerals, coal, zinc, different parts of retorts and metallurgical residues. Retort design, the charge and conditions of reduction and condensation are discussed. Production remains from two other sites, Puzihe and Muxiexi, are also analysed and discussed. Finally, the technology and organisation of zinc production in Fengdu are reconstructed.

Chapter five then turns to the Laochangping region, Shizhu, where possible zinc mines supplying zinc ores for the river sites and zinc smelting sites, dated to the Qing period, were found. It first introduces the geology, historical records and archaeological finds of this area, followed by a description of the field investigation and the analyses of geological zinc minerals at one large old mine working, Yushi Cave. The main part of the chapter focuses on the analytical results of retorts and slag from Dafengmen, with discussions on retort design and the charge.

Chapter six presents a comparison of the technology and organisation of zinc production at the two groups of sites. Possible reasons for the differences are explored by making use of relevant theoretical models which take into consideration the social, political and economic constraints affecting zinc production in the region in the different periods. The second section of this chapter attempts a synthesis by outlining the four key technical features that define the Chinese zinc distillation tradition – these are identified as the commonalities running through the case studies presented as well as the historical and ethnographic sources examined. This is followed by a detailed comparison of the zinc distillation technologies in China and India.

Chapter seven investigates the role of Chinese zinc at home and abroad. The transition from cementation brass to speltering brass is discussed based on historical records, a reinterpretation of compiled chemical data on brass coins and from simulation experiments. It is argued that the speltering process shows advantages over the cementation process in terms of raw materials, efficiency of the process, control of alloy compositions and costs of minting. The reasons for this transition and the eventual change from bronze coins to brass coins are explained by considering the wider historical contexts. The influence of Chinese zinc exported to the world is explored in two regions: India and Europe.

Chapter eight summarises the main finds of this research and outlines some avenues for future research.

The appendices include the chronology of the Chinese dynasties, notes on weights and measures, and formulae of minerals and phases mentioned in this research, for the convenience of readers. Appendix four lists some historical texts relevant to this research in both Chinese and English.

Chapter 2

Historical and Technological Background to Zinc Production in China

This chapter introduces the historical and technological background to this research, starting with an overview of the history of brass and zinc in China. It then turns to the uses of brass and zinc in Ming and Qing China. Finally, various ethnographic records of traditional zinc smelting persisting throughout the twentieth century are reviewed and summarised in terms of their operational sequences, in the hope that this information will aid the interpretation of the archaeological remains.

2.1 The history of brass and zinc in China

Since the 1980s, Zhou Weirong and his colleagues have reviewed relevant textual evidence, offered new interpretations for some terminology and documents on which previous discussions had been based, and carried out systematic analyses on copper-based coins (Zhou 1990; Zhou 1992; Zhou 1993; Zhou and Fan 1994; Zhou 1996; Zhou 2000; Zhou 2001; Zhou 2004; Zhou 2007a). With substantial historical and numismatic evidence, they argued that China did not begin to smelt zinc on a large scale before the Wanli period (AD 1573-1620) of the Ming Dynasty. As a result of this evidence, the history of brass and zinc in China was divided into four stages: accidental brass, imported brass, cementation brass and speltering brass.

2.1.1 Accidental brass

The earliest brass artefacts found in China consist of only a few scattered pieces. The two earliest were discovered at the Jiangzhai site (Yangshao culture I: 4700-4000 BC), in Xi'an, Shaanxi province. A thin round piece was analysed and found to contain 25.6% zinc, while a tube contained 32% zinc (Han and Ke 1988). Later, a long *ji* 笄 (hairpin) with 27-32% zinc was found at the Late Yangshao culture site (around 3000 BC) in Weinan, Shaanxi province (Han and Ke 2007, 182). A brass awl with 20.2-26.4% zinc was found at the Shanlihe site (Longshan culture: 2300-1800 BC), in Jiaoxian, Shandong province (IACASS 1988, 196-199). These brass artefacts are extraordinary in their early dates and high levels of zinc, with all over 20% zinc. Some scholars have questioned their dates (An 1981) and the accuracy of the analyses (Zhou 2001). However, simulation experiments have shown that such brass artefacts could have been produced by smelting mixed copper ores and zinc ores under rudimentary conditions (MHGBSIC 1981; Fan *et al.* 2010; Fan *et al.* 2012). As bronze was the dominant copper-based alloy mainly used for ritual vessels from the Shang Dynasty (sixteenth to eleventh century BC) to the Han Dynasty (202 BC-AD 220), it is reasonable to define these early brass artefacts as accidental products predating the development of intentional brass smelting. Zinc levels of early Chinese bronzes rarely exceed 1% (Gettens 1969; Chase 1999).

2.1.2 Imported brass

With the introduction of Buddhism, cementation brass appears to have been imported via the Silk Road from the Western Regions, mainly Persia and India, in the Eastern Han period (AD 25-220) (Zhou 2000). The imported brass was called *toushi* 鍮石, a name derived from the ancient Persian language (Lin 1999). Contemporary historical documents show that *toushi* was known to resemble gold, but it was recognised as a different metal; it was always listed after gold and silver, but before copper and iron, indicating its relative value. In addition to Buddhist sculptures and incense burners, only small-sized objects, such as needles, belts and seals, are reported to have been made of *toushi* (Zhou 2000).

Turning to the archaeological evidence, a few brass artefacts have been identified at some cemeteries excavated on the Silk Road. At the Yingpan cemetery (fourth century AD), Xinjiang province, most copper-based artefacts were ornaments, three of which analysed were made of brass with 20-22% zinc (Li and Zhou 1998; Han *et al.* 2002). A ring and a metallic fragment with zinc contents of 6.3% and 18.3%, respectively, were found at a Xianbei cemetery (late fifth century AD), Inner Mongolia (Li 2004). Five leaded brass artefacts were identified at a Tubo cemetery (ninth century AD), Qinghai province. Their zinc contents varied between 18% and 29% (Li and Han 1992). Some brass vessels dated to the ninth to thirteenth centuries were found at Talgar in Kazakhstan, just to the west of Xinjiang province (Park and Voyakin 2009). These brass artefacts are most likely imports, as there are neither historical documents nor archaeological evidence indicating domestic brass making before the tenth century.

2.1.3 Cementation brass

It was not until the Five Dynasties (AD 907-960) that a *toushi* making method with copper and *luganshi* 炉甘石, oxidic zinc ores, was first described by Chinese alchemists. The method was first described in *Rihuazi Dian Geng Fa* 日华子点庚法 (Methods of Metal Manufacture of Rihuazi): one catty (c. 0.6 kg) of red copper, one catty of *luganshi* and eight catties of charcoal produced cockscomb-coloured pellets. In *Wai Dan Ben Cao* 外丹本草 (Materia Medica for External Elixirs) of the Song Dynasty (AD 960-1279), Cui Fang offered a simple record of this method without mentioning the necessary charcoal: smelting three catties of copper and one catty of *luganshi* produced one and a half catties of *toushi*. During the Song Dynasty, knowledge of this method spread from the alchemists to the commoners. Because of the high value of *toushi*, common people even tried to make it by remelting copper-based artefacts or coins – a practice soon forbidden by the government. Meanwhile, the import of *toushi* ceased due to the decline of the trade on the Silk Road. In this period, brass was called *huangtong* 黄铜, literally 'yellow copper', instead of *toushi*, and was still deemed precious and often used in palaces and temples. It was not until the mid-Ming period that brass began to be produced on a large scale for making coins (Zhou 2004). Before the Ming Dynasty, relatively few copper-based artefacts containing significant levels of zinc have been reported (Jett and Douglas 1992; Cowell *et al*. 2003).

2.1.4 Speltering brass

Tian Gong Kai Wu [1637], the earliest document clearly describing zinc, records the use of *luganshi* for making cementation brass, and for smelting *woyuan* 倭铅, zinc, for the first time (Sun and Sun 1966, 247). The author, Song Yingxing, stated that *woyuan* could not be found in old books, but was a recent term (Sun and Sun 1966, 247). Previously, in the medical compendium *Ben Cao Gang Mu* [1593], Li Shizhen had written that all the brass objects were made with copper and *luganshi* at his time (Needham 1974, 208-209). The two texts thus seem to suggest that zinc was first produced in the early seventeenth century. This is in agreement with historical records relevant to minting materials, in which brass and tin were mentioned before this period, but copper and zinc were referred to from this period. It has also been reported that brass coins made after 1621 contain higher cadmium levels than those made earlier, suggesting a transition from cementation brass to speltering brass (Zhou and Fan 1994). On this basis, Zhou argued that brass coins were first made of cementation brass from the Jiajing to Wanli periods (AD 1522-1620), while from the Tianqi period (AD 1621-1627) to the Qing Dynasty brass coins were made of speltering brass (Zhou 2007a). This transition will be discussed in more details in Section 7.1.

Another source describing zinc smelting technology is *Dian Nan Kuang Chang Tu Lue* 滇南矿厂图略 (An Illustration Account of the Mines and Smelters of Yunnan), an administrative handbook on metal mining and smelting in Yunnan province, which was compiled by the provincial governor Wu Qijun in the 1840s:

> *Baiyuan* 白铅 is usually called *woyuan* 倭铅. Smelting *yuan* utilises earthen jars in four-wall furnaces. The mixed ores and coal are charged into the jars and pockets are inserted afterwards. The jars are arranged inside the furnaces and surrounded with coal. The furnaces are powered with leather-made bag bellows. Every two or four jars are placed on each furnace bar. The size of furnaces depends on the number of furnace bars. (Wu 2002a, 184)

This passage records a Qing process of zinc smelting similar to the traditional ones with rectangular furnaces, but it fails to describe the principle of distillation necessary for the process.

In late Ming documents, zinc was named *woyuan* 倭铅, *woyuan* 窝铅 or *yuan* 铅 for short. In the Qing Dynasty, zinc was called *baiyuan* 白铅, literally 'white lead', as opposed to *heiyuan* 黑铅, literally 'black lead'; *baiyuan* and *heiyuan* were generally called *yuan*. The close association of zinc and lead is probably because their ores often exist together in lead-zinc deposits and are both greyish in colour. A possible reason for why zinc was called white lead and lead was called black lead is that zinc is silvery grey and lead is dull grey.

During the late Ming and Qing periods, Chinese zinc was used for making brass coins and copper-based artefacts, and was also exported via the maritime trade routes. This is further discussed in the following section.

2.2 The uses of brass and zinc in Ming and Qing China

The Ming (AD 1368-1644) and Qing (AD 1644-1911) Dynasties were the last two dynasties of imperial China, covering a vast, diversified territory primarily based on an agricultural economy. The Ming Dynasty was founded by the Han people, the dominant ethnic group native to China; it was overthrown by the Manchu people, an ethnic minority originating from Manchuria (present northeast China), who founded the Qing Dynasty. The Qing Dynasty inherited the complex institutions of bureaucratic government that characterised the preceding Ming Dynasty. Both dynasties were governed by the emperors in a hierarchy of central, provincial and local governments. (Twitchett and Fairbank 1998; Fairbank 1978).

Brass had been a luxury metal only used in palaces and temples until the late Ming Dynasty, when it became the main alloy employed for the minting of copper-based coins. In Ming and Qing China, brass was also used for manufacturing vessels, figures, architectural fittings and astronomical instruments. Zinc was also used to make Paktong, an alloy of copper, nickel and zinc. In addition, zinc was exported via maritime trade to many regions of the world.

2.2.1 Brass coins

Copper-based coins generally constituted the fundamental form of currency in China from the Eastern Zhou period (770-221 BC) to the early twentieth century. Typical Chinese copper-based coins were round with square holes in the middle, inscribed with their values and issuing periods, and made by mould casting, stack casting and sand casting (Zhou 2002). Throughout history, minting in China was generally under the monopoly of the state, who controlled the mint metal procurement, the production and the circulation of the coins (Vogel 1987, 9). In the Ming period, copper-based coins were manufactured in official mints: the *Baoyuanju*

宝源局 under the Ministry of Works in the capital and *Baoquanju* 宝泉局 in most provinces. In the Qing Dynasty, coins were cast in two capital mints: *Baoquanju* under the Ministry of Revenue and *Baoyuanju* under the Ministry of Works, and many provincial mints (Peng 1958). The Ming and Qing coins were made by sand casting, as documented and illustrated in *Tian Gong Kai Wu* (Sun and Sun 1966, 165-169).

Since the 1980s, scholars from both the British Museum (Bowman *et al.* 1989; Cowell *et al.* 1993; Wang *et al.* 2005) and the Chinese Numismatic Museum (Dai and Zhou 1992; Zhou 2004) have analysed a large number of Chinese coins, producing similar results. These studies have shown that all early copper-based coins were made of bronze. It was not until the beginning of the sixteenth century that coins made of a quaternary alloy containing considerable amounts of tin, lead and zinc (Zn≤16%) appeared. The coins issued in the Jiajing period (AD 1522-1566) were all made of quaternary alloys, most of which contained 12-20% zinc. Later the zinc contents of coins increased, reaching 30% on average during the Wanli period (AD 1573-1620) and 34% at the end of the Ming Dynasty (Chongzhen period: AD 1628-1644). The manufacture of brass coins continued into the Qing Dynasty, with increasing levels of zinc, from 25% to 40%. Technological and economic reasons for the fluctuations in coin compositions will be discussed in more detail in Section 7.1.

2.2.2 Brass artefacts

Copper-based artefacts from the Song to Qing Dynasties, from about AD 1000 to 1900, are often named 'later Chinese bronzes' in order to distinguish them from the earlier bronzes, especially those of the Shang and Zhou Dynasties. These later Chinese bronzes were mostly vessels and figures. Domestic vessels, such as wash-basins, incense burners, flower vases and mirrors, were used in everyday life for functional or decorative purposes. Ceremonial vessels, in the forms of incense burners, *jue* 爵 and vases, were used for tombs, temples and domestic altars. Figures were mostly Buddhist and Daoist statues used for worship in temple altars; some were burial items made in animal and human forms (Kerr 1990).

These 'later Chinese bronzes' were not all made of traditional bronze, a copper-tin alloy, as shown by some technical studies (Kerr 1990; Jett and Douglas 1992; Cowell *et al.* 2003). From the Ming period, zinc levels in bronzes increased and zinc even became the dominant alloying component. Such bronzes are named 'brasses' in this research. Some typical Ming and Qing brass artefacts have been identified, such as Xuande incense burners, Buddhist statues, architectures and instruments.

1. Vessels

Among various domestic and ceremonial vessels, Xuande incense burners stand out as a group of copper-based incense burners of exceptional quality, often inscribed with the Xuande reign (AD 1426-1435) marks on the bases (Figure 2.1). They were first recorded in *Xuande Ding Yi Pu* 宣德鼎彝谱 (Atlas of Xuande Ding and Yi), written by Lv Zhen in the Xuande period. The document describes that the Xuande Emperor ordered the Ministry of Works to make ritual vessels, such as *ding* 鼎, *yi* 彝 and incense burners, to be displayed in the court and temples for offerings to the ancestors, the heaven and the earth. It is also recorded that these vessels were made of a high quality copper named *fengmo* 风磨, literally 'wind polishing', which was interpreted as *toushi* (Zhou 1990). However, numerous copies of Xuande incense burners were made and even inscribed with the Xuande reign marks throughout the late Ming and Qing Dynasties. Some scholars have suggested that *Xuande Ding Yi Pu* is a fake book written in the late Ming period rather than the Xuande period, and that therefore there are no real Xuande incense burners at all (Lu 2008). Therefore, 'Xuande incense burners' refer generally to incense burners with the Xuande reign marks or incense burners of similar style, but no conclusive date may be ascribed based solely on their style.

Figure 2.1 A Xuande incense burner from the Palace Museum (Palace Museum 2012, 259).

As early as 1925, Wang Jin, a historian of chemistry, analysed two Xuande incense burners from his own collections and found that they were made of brass with 20.4% and 36.4% zinc, respectively (Zhang 1964, 110). Most of the Xuande incense burners in the Victoria and Albert Museum's collections are made of brass with 29-35% zinc (Kerr 1990, 36). The Xuande incense burners collected at the Palace Museum were identified by portable XRF to be made of brass (Figure 2.1). Their zinc levels changed throughout the Ming and Qing Dynasties, from 12-14% in the Xuande period (AD 1426-1435) to 2-7% at the end of the Ming Dynasty, to 6-7% in the Kangxi period (AD 1662-1722), 10.4-14.6% in the Yongzheng period (AD 1723-1735), and 16.6-21.9% in the Qianlong period (AD 1736-1795). Li identified only one incense burner dated to the Xuande period based on its inscriptions, patina, manufacturing techniques and also chemical composition (Li Mijia 2013). A less controversial brass vessel of the Xuande period is a gilded vase, now at the Qinghai Museum, which was one of the gifts from the Xuande Emperor in 1427 to the Qutan Temple in Qinghai province. The vase was identified to be brass with 11.8-14.7% zinc (Li Mijia 2013).

2. Figures

The most famous brass figures are the gilded Buddhist statues manufactured in the Yongle and Xuande courts. Since the Yuan Dynasty, the introduction of Tibetan Buddhism into central China had brought the tradition of brass Buddhist statues into the imperial court (Luo 2010). The Ming court cast Buddhist statues mainly in the Yongle (AD 1403-1424) and Xuande periods (AD 1426-1435). These Yongle-Xuande Buddhist statues were well-known for their delicate craftsmanship and consistent artistic styles. They were usually sent as religious gifts to monks who came from Tibet to the court to present themselves to the Emperor (von Schroeder 2001; Huang 2003a; Huang 2003b). He Xing and Hu Guoqiang from the Palace Museum found that 30 gilded Buddhist statues with Yongle or Xuande reign marks (Figure 2.2) were all made of brass. These statues contain 6-20% zinc, with most around 14% (He and Hu 2012). Similar alloy compositions were found in Yongle-Xuande gilded Buddhist statues from two private collections (Xing 2011 ; Zhou and Huang 2012).

3. Architectural fittings

Brass was sometimes used to make architectural components which had traditionally been made of wood in China. An encyclopaedia compiled by Chen Renxi, published in the Chongzhen period, mentions that the top of the Huangji Hall of the Forbidden City (i.e. the Palace Museum) was made of brass (Zhou 1990). A few structures, such as halls and towers, were assembled with individually cast parts of copper-based alloys. Among the nine copper-based halls and towers of the Ming and Qing Dynasties that have survived, six have been identified (with portable XRF) to be made of brass, two made of bronze, one made of bronze and brass and one made of bronze, brass and copper. The first brass one is the well-known gilded copper-based hall on the Wudang Mountain, Hubei Province, built during the Yongle period (Figure 2.3). Surface analyses of the hall (on areas where the gilding layer was worn) showed that the substrate alloy was brass with 10-17% zinc and 1-2% lead. The more recent one is the Baoyun Pavilion at the Summer Palace, Peking, built in the Qianlong period (AD 1736-1795). It was made of brass with zinc as high as 23-27% and lead as low as 1-2% (Zhang 2013, 237-269). The increasing zinc levels of these brass halls and towers from the Wanli period are similar to those of brass coins.

4. Astronomical instruments

China has a long history of astronomical observations using copper-based instruments, which can be traced back to the Western Han period (202 BC-AD 9). Astronomical instruments were mostly made for the imperial observatories and they were often assembled with many accessories (Lin 1987). Two large instruments built in 1437, an armillary sphere and an abridged armilla, have been analysed. Their complex supporting parts are leaded bronze, while measuring parts are largely leaded brass (Li and Wu 1994). Eight large instruments from the Peking Ancient Observatory, dated to the Qing Dynasty, were found to contain 27.6% zinc and 2.5% lead on average (Lin 1987).

Figure 2.2 A gilded brass Buddhist statue from the Palace Museum (Palace Museum 2012, 243).

Figure 2.3 Wudang Mountain Metal House, 5.54 m in height (photo by Zhang Jianwei)

2.2.3 Paktong

Another traditional use of zinc in China was to make *baitong* 白铜, literally 'white copper', a copper-nickel-zinc ternary alloy. It is known in the West as Paktong, the romanisation of the Cantonese pronunciation of the Chinese characters for *baitong* because it was exported to Europe from the late sixteenth to nineteenth century mainly from Canton (present Guangzhou in Guangdong province) (Bonnin 1924; Gilmour and Worrall 1995; Pinn 1999) in the forms of tableware, candlesticks and other domestic vessels (Figure 2.4).

Figure 2.4 A selection of Paktong artefacts (photo by Huang Chao).

According to historical documents, Paktong may have been produced as early as the fourth century AD. However, its large-scale production did not begin in the provinces of Sichuan and Yunnan, southwest China, until the Ming Dynasty. The Huili county in Sichuan province was the main centre of Paktong production. Analyses of the ingots and slag from ancient smelting sites in Huili indicate that the Paktong in Huili was produced from a mixture of copper ores and nickel ores to form a copper-nickel binary alloy, which was only later alloyed with zinc in Yunnan province to improve malleability and produce a white appearance. Chinese Paktong artefacts of the eighteenth to nineteenth centuries in both Chinese and British collections have been reported to contain 35-45% zinc, which should have been introduced into Paktong as metallic zinc rather than brass. There are also a few earlier Paktong artefacts containing considerable levels of zinc. A Paktong artefact dated to the twelfth century was found to contain about 30% zinc; a padlock and a key in the Victoria and Albert Museum collection, dated to the Jiajing period, are also made of copper-nickel-zinc alloys (Mei and Ke 1989; Gilmour and Worrall 1995; Mei 1995; Pinn 1999; Huang 2016). It is worth further investigating how the zinc in these earlier Paktong artefacts was introduced, considering that zinc was not produced regularly until the late Ming Dynasty. If these earlier Paktong artefacts were made by adding cementation brass, it is unlikely that the final artefacts would contain zinc as high as 30%.

2.2.4 Exports

In addition to the primary usage of zinc for making copper-based alloys, zinc produced in southwest China was also exported via the maritime trade routes as ballast goods under the name of *tutenag* from the seventeenth to nineteenth centuries. The earliest detailed research on this topic is Alfred Bonnin's *Tutenag and Paktong*, published in 1924. This treatise is a series of references, mostly English, on Chinese zinc by a few of the great chemists, mineralogists, and authorities of the age (Bonnin 1924, 6). The export of zinc is also mentioned, although not extensively explored, by some historians of metallurgy and archaeometallurgists (Tylecote 1976; de Ruette 1995; Craddock and Hook 1997; Craddock and Zhou 2003; Craddock 2013). The role of Chinese zinc in the international trading networks of the seventeenth to eighteenth centuries was later investigated by George B. Souza and Chen Hailian (Souza 1986; Souza 1991; Chen 2013).

The earliest export of Chinese zinc appears to date to the late sixteenth century and would have involved Portuguese merchants (Maréchal 1954; Souza 1991). However, this cannot be fully confirmed from the few detailed surviving Portuguese records (George B. Souza pers. comm. 2010). Tylecote stated that zinc slabs with date inscriptions (AD 1585) were being exported (Tylecote 1976, 77), but these were actually found in a hill near Lianzhou in northern Guangdong province (Browne 1916) and were not necessarily meant to be exported.

More convincing evidence of the export of zinc from the very early seventeenth century onwards has emerged. Souza's studies have shown that Portuguese merchants at Macao and Chinese merchants at Canton were the first to carry Chinese zinc to Japan, as well as starting to export Chinese zinc to Southeast Asia and South Asia, India in particular, in the early seventeenth century. According to Souza's estimates, by 1637, nearly 3,000 picols of zinc were shipped to Japan annually - the total annual exports did not exceed 5,000 picols in the first half of the seventeenth century (Souza 1991, Table 2.1). Zinc ingots have also been found in some Dutch shipwrecks, such as the Mauritius, en route from China to Amsterdam via Batavia, wrecked off Gabon, West Africa in 1607, and the *Witte Leeuw*, en route from Bantam in Java to Amsterdam, which sank off the island of St Helena in the South Atlantic in 1613 (Craddock and Hook 1997).

During the Qing Dynasty, Dutch, English and Swedish entrepreneurs were frequently involved in the maritime trade of Chinese zinc. The first decades of the Ming-Qing transition saw the decline of zinc exports, with a maximum amount of 1,000 picols being traded annually. Subsequently the Portuguese, Chinese and English developed the market in the South China Sea, the Indian Ocean and Europe, which led to a boom of zinc exports in the late seventeenth century (8,000 picols in 1686). This was followed by a temporary reduction in the early eighteenth century (3,000 picols). From the 1730s, zinc exports increased significantly to the point of exceeding over 50,000 picols in 1817. The principal market of Chinese zinc was in Asia, India being the largest market (Souza 1991, Table 2.1). Chinese zinc ingots have been found in two later shipwrecks: the SOIC *Göteborg*, en route from Canton and wrecked near the city Gothenburg, Sweden, in 1745 (Carsus 1960) and the EIC *Diana*, en route from Canton to Calcutta, India, which sank off Malacca in the Malacca Straits in 1816 (Ball 1995).

Table 2.1 The weights of zinc exported annually, estimated by Souza (1991).

Period	Zinc exported in picols/tonnes
First half of the seventeenth century	< 5,000/302.5
First decades of Qing period	< 1,000/60.5
Late 1670s	3,000-4,000/181.5-242
1682	6,000/363
1686	8,000/484
1693	6,000/363
First fifteen years of the eighteenth century	3,000/181.5
1730s	11,000/665.5
1790s	>40,000/2420
1817	>50,000/3025

In summary, at least since the early Ming Dynasty, brass was used for different types of copper-based artefacts, including Buddhist statues, ceremonial vessels and even metal halls. Brass artefacts of the early Ming Dynasty were quite valued, with their use restricted to palaces and temples. They were mostly manufactured for ritual or religious purposes commissioned by the Emperor. These brass artefacts contain relatively low levels of zinc, up to ~20%, as compared with the later ones, dated to the very late Ming and Qing Dynasties, with 20% to over 30% zinc. The change in zinc levels is particularly evident in brass coins. The mass production of zinc was strongly associated to minting. Zinc, as a minting metal, was produced and distributed under the control of the state. It was also used for making brass and Paktong artefacts as well as for export, which was sometimes prohibited by the state when the metal stock was insufficient for minting (Souza 1991).

In addition, zinc was used for making zinc oxide by burning the metal, with the resulting zinc oxide being very white, fine-grained and free of lead and other impurities, as opposed to that made directly from zinc ores. This type of pure zinc oxide had been produced from metallic zinc in India since the tenth century, probably for medicinal purposes (Craddock *et al.* 1998, 30). In 1834, zinc oxide was introduced to London as a white pigment under the name of 'Chinese white' by the Winsor & Newton Ltd (Jettens and Stout 1966, 177; Kühn 2012, 171), implying that its origins might have been related to Chinese zinc. Although the brittleness of metallic zinc makes it unsuitable for use in its own right, after hot working at temperatures of between 100-150 °C the brittleness decreases and unalloyed zinc can be used, mainly as a roofing material. A final use of zinc that should be mentioned is its application to galvanise iron by the formation of a layer of zinc-iron intermetallic phases which can prevent the iron from corroding. Both the use of zinc for making zinc sheets and galvanising iron became popular in Europe in the nineteenth century, when the mass production of zinc began (Day 1998). However, no evidence of the use of zinc for either purpose before the twentieth century has been found in China.

2.3 Traditional zinc smelting in modern China

Traditional zinc smelting technology survived and evolved in southwest China throughout the twentieth century (Craddock and Zhou 2003). These processes are called *tufa* 土法, literally 'domestic method' and translated in English as 'traditional', as opposed to the modern methods utilising pyrometallurgy and hydrometallurgy (Peng 2005). These traditional processes were in operation in Hunan province and the area where the provinces of Sichuan, Guizhou and Yunnan meet (Figure 2.5), where raw materials for zinc smelting were abundant, including zinc ores, coal and fireclay. There are over ten ethnographic descriptions of these processes, written by people from different backgrounds and different purposes (Table 2.2).

In the late nineteenth and early twentieth centuries, Western mining engineers and geologists visited China to prospect for mineral resources. The first descriptions of zinc production were given in the 1910s and 1920s and zinc smelting furnaces and retorts were also illustrated (Moore-Bennet 1915; Wheler 1922) (Figure 2.6). W. R. Ingalls reported seeing photographs of zinc smelting furnaces used in remote places of China (Ingalls 1936).

Chinese geologists have investigated zinc mines and smelters since the early twentieth century. A geological survey in 1914 described the traditional zinc smelting process conducted at the Songbai zinc factory in Changning, Hunan, where a similar process was still in operation in the 1930s. Further traditional zinc smelting processes were documented in geological surveys in Huize, Yunnan, in 1939, Weining, Guizhou, in 1940, and Huili, Sichuan, in 1941 (Mei 1990). All of these traditional processes from different areas were based on the principle of distillation by ascending, but some technological variations were outlined. However, the descriptions of these processes were very basic and neither photographs nor illustrations are found in these geological reports.

Figure 2.5 Locations of traditional zinc smelting processes in the provinces of Hunan, Sichuan, Guizhou and Yunnan and zinc smelting sites in Fengdu, Chongqing. From Google Earth.

Table 2.2 Traditional zinc smelting practices recorded in various sources.

Location	Year smelts observed	Scale	Observers	Reference
Yunnan, Guizhou, Sichuan	1915	unknown	foreigner	(Moore-Bennet 1915)
Central and South China	1922-23	unknown	foreigner	(Wheler 1922)
Changning, Hunan	1914; 1932-34	factory	geologist	(Mei 1990)
Huize, Yunnan	1939	factory	geologist	
Weining, Guizhou	1940	factory	geologist	
Huili, Sichuan	1941	factory	geologist	(Mei 1990)
	1944		metallurgist	(Li 1944)
	1945		metallurgist	(Sun 1945)
Huili, Sichuan	1958	factory	metallurgist	(MIP 1958)
Hezhang, Guizhou	1958	factory	metallurgist	
Hezhang, Guizhou	1984	workshop	historian of metallurgy	(Hu and Han 1984)
Hezhang, Guizhou	1986	workshop	historian of metallurgy	(Xu 1986)
Huize, Yunnan	1998	workshop	historian of metallurgy	(Xu 1998, 123)
Huizhe, Yunnan; Bijie, Guizhou	1994-95	workshop	historian of metallurgy	(Craddock and Zhou 2003)
Huize, Yunnan	1996; 2000	workshop	historian of metallurgy	(Li 2008)

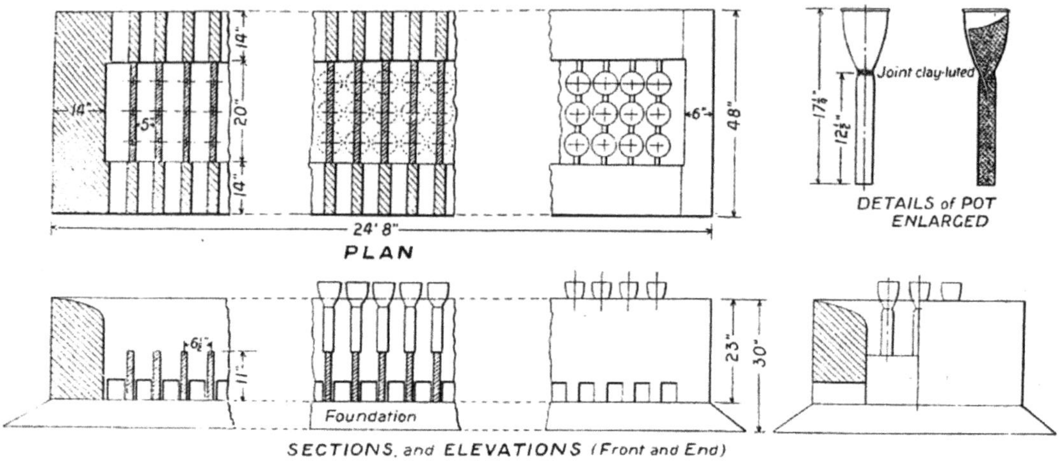

Figure 2.6 A trough furnace (Wheler 1922).

In the 1940s and 1950s, detailed reports on traditional zinc smelting were written for industrial purposes. During the Sino-Japanese War (AD 1937-1945), the production of zinc by the traditional process was encouraged in isolated areas of southwest China with limited resources, as seen in Huili (Li 1944; Sun 1945). In 1958, a handbook *Tu Fa Lian Xin* 土法炼锌 (Traditional Zinc Smelting) was published to give guidance on how to set up small zinc factories, followed by two examples: Hezhang County Zinc and Lead Works (HCZLW) in Guizhou, and a factory in Huili, Sichuan (MIP 1958).

In the 1980s, some Chinese historians of metallurgy started to investigate traditional zinc smelting processes in order to gain a better understanding of the ancient technology, in Hezhang, Guizhou (Hu and Han 1984; Xu 1986; Xu 1998). In 1994-95, Craddock and Zhou made two visits to Bijie, southwest Guizhou, and Huize, northeast Yunnan, where zinc mining and smelting activities were recorded and filmed (Craddock 1997; Craddock and Zhou 2003). In 1996 and 2000, the traditional process at Huizhe was investigated (Li 2008b).

Based on these various ethnographic sources, the operational sequences (i.e. the *chaîne opératoire*: see Section 3.1.1) of the traditional zinc smelting processes can be described, including retort production, mining and ore preparation, reducing agent and fuel preparation, furnace construction, tool making and zinc smelting.

1. Retort production

Typical retorts were conical or cylindrical pots. With the exception of the flat-bottomed cylindrical pots described by Wheler, most were conical with pointed or small flat bases, which could leave more space between them, improving ventilation and promoting even-distribution of combustion within the furnaces (Hu and Han 1984, 37). Pots varied in size, with the heights ranging from 30 to 90 cm and the rim diameters from 6 to 24 cm.

The production of the pots involved a series of steps: preparing and mixing raw materials, forming, drying and firing. The early sources did not report the clay type, or they simply stated that the pots were made of fireclay, as in Huili in the 1940s, or fireclay and steatite, as in Huili in 1958. The main materials used in Hezhang in 1986 for making pots were three types of clay: 53% *zhuni* 主泥, 26% *huangni* 黄泥 and 21% *shahuang* 砂荒 (Table 2.3). *Zhuni*, literally 'main clay', also named sticky *huang* 荒, was found in the lower layers of coal measures and was refractory and adhesive. *Huangni*, literally 'yellow clay', was a type of common clay rich in fluxing oxides. *Shahuang*, from the upper layers of coal seams, was not adhesive but sandy, acted as a temper. A similar recipe was also used in Bijie in 1994-95, but refractory white clay replaced the yellow clay and the ratio of three clays was 2:1:4.5. The Huize process in 1996/2000 also used three raw materials: fireclay, yellow clay and sand. A small proportion of other materials were added in Hezhang in 1958, including *wupi* 五匹 clay, white sand and grog. These materials were ground and mixed wet by treading. The mixed clay was shaped into cylinders by throwing; the bottom parts were usually formed by moulding. After drying, the pots were fired in ceramic kilns fuelled by coal or coal briquettes. The kilns were similar to the trough-shaped zinc smelting furnaces in Huili in 1944-45 and 1958, while dragon kilns with several linked chambers were used in Hezhang in 1958 and in Bijie in 1994-95.

Some sources stated that there were strict requirements for the quality of the pots. They should be dense, without cracks, able to withstand high temperatures and sudden temperature changes, and of good conductivity and resistant to chemical attack (Li 1944, 84; MIP 1958, 21). In Hezhang in 1986 and Huize in 1996/2000, the pots used were glazed to prevent zinc vapour leakage during smelting; in Bijie in 1994-95, they were soaked in a muddy limewater before smelting. Pots were normally used from five to over ten times. If slightly cracked after use, they were repaired with fresh clay.

Table 2.3 Chemical compositions of three raw materials for making pots and calculated chemical compositions of their mixture (wt%), normalised to 100% (Xu 1986).

Raw materials	*Zhuni*	*Huangni*	*Shahuang*	Mixture
Ratio	53	26	21	100
SiO_2	54.42	69.17	59.64	59.35
Fe_2O_3	3.75	8.22	4.60	5.09
Al_2O_3	37.86	19.55	30.54	31.56
TiO_2	0.81	1.05	1.40	1.00
CaO	0.81	0.25	1.13	0.73
MgO	2.33	1.77	2.69	2.26

Condensers exhibited various structures that can be grouped into three main types:

Type A: Elongated condensers (Figure 2.7)

This type of condensers was common in Changning, Huize and Huili in the first half of the twentieth century. They were made by adding conical or cylindrical parts or building up from the tops of pots. In Wheler's case, cylindrical pots were capped with crucible-shaped extensions. In Changning in 1914, pots were connected with bell-mouthed condensers made of yellow fireclay. In Huize in 1939, pots were elongated by adding 'ash clay', probably clay mixed with ash. In Huili in 1941, pots were joined with cylinders made of clay and cinders. Similar condensers were still used in Huili in 1944, when clay cylinders made of fireclay, yellow clay and furnace ash were shaped onto the rims of pots by wheeling. These add-on condensers were soon replaced by connecting pots and separately-made condensers in Huili in 1945.

Inside elongated condensers, pockets made of clay and furnace/coal ash were added onto the loaded charge and usually placed several centimetres below the rims of the condensers. The top surfaces of pockets were shaped into a concave depression to collect the metal. In Huili in 1944-45, wooden rods, called 'jar needles', were inserted down one side of the pockets into the charge, creating a hole within the pockets and a gas passage in the charge.

The tops of retorts were covered by iron or clay round discs which acted as lids. A small hole was left in each lid opposite the hole in the pockets.

Type B: Internal condensers (Figure 2.8)

In addition to the earlier example in Weining in 1940, most retorts described in the second half of the twentieth century had internal condensers: no extensions were added and zinc was collected in the pockets placed within the pots. It is surprising that in Huizhe in 1996/2000 no holes were left on the pockets for the ascent of zinc vapour, but the pockets were made porous enough for zinc vapour to pass through but dense enough to hold molten zinc.

Type C: External condensers (Figure 2.9)

Only two examples of retorts with external condensers, one in Huili in 1958 and another in Huizhe in 1998, have been reported. The separate external condensers were aubergine-shaped, with their bellies acting as pockets. The condensers used in Huili in 1958 were made of the same materials (steatite and fireclay) as the pots. The processes utilising retorts with such external condensers were not really successful and stopped being used in the 1990s (Paul T. Craddock pers. comm., 2010).

2. Mining and ore preparation

Traditional underground mining was still carried out on a small scale by private enterprises with labour-intensive, simple technology in Huize in 1994. Small but rich veins of smithsonite (ZnCO3), were exploited, and a preliminary beneficiation of the ores took place on flat platforms near the mine entrances (Craddock 1997; Craddock and Zhou 2003). Most of the traditional processes used oxidic zinc ores, mostly smithsonite. The oxidic ores were further beneficiated at the smelters, and sometimes were calcined by placing them by the furnaces, as practiced in Huize in 1939. However, several traditional processes also processed sulphidic ores, namely sphalerite (ZnS) by adding a stage of roasting prior to smelting. In the Changning processes in 1914 and 1932-34, ores were roasted in round furnaces three times, in a process which lasted over 20 days. In Huili in 1941, rectangular reverberatory furnaces were employed to roast sulphidic ores, whereas in 1958 reverberatory furnaces were used to remove water, carbon dioxide and a little sulphur from oxidic ores to improve their grades.

3. Reducing agent and fuel preparation

The reducing agent for the traditional processes was coal in most cases. Anthracite and high-quality bituminous coal were the most common types. They contain high levels of carbon but little ash and sulphur. As exceptions, it should be noted that the Changning process in 1914 used mutan 木炭, charcoal, and the Huize process in 1939 used huitan 灰炭, a term that may refer to either charcoal or coal. The Huili process in 1945 first used bituminous coal and later used jiaotan 焦炭, coke.

The most widely used fuel was powdered coal briquetted with clay and, sometimes ash. Coal briquettes were called *tanba* 炭巴 and their preparation involved a series of procedures including mixing, treading and cutting. Coal briquettes of various shapes and sizes were placed in different parts of the furnaces.

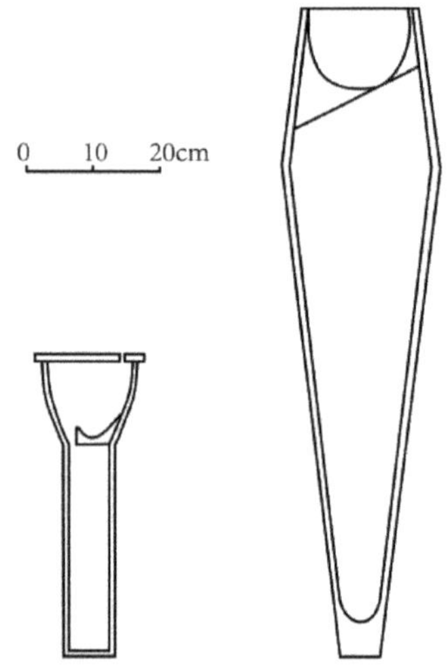

Figure 2.7 Retorts with elongated condensers. Left: Wheler's process (Wheler 1922). Right: in Huili (Li 1944).

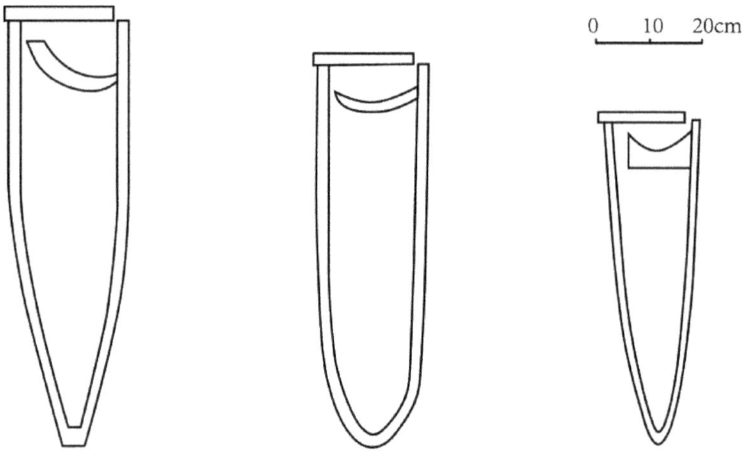

Figure 2.8 Retorts with internal condensers. Left: in Hezhang (MIP 1958); middle: in Hezhang (Hu and Han 1984); right: in Hezhang (Xu 1986).

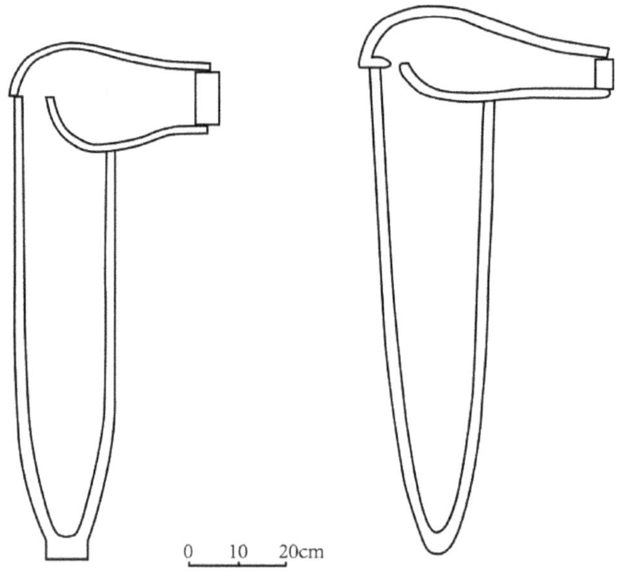

Figure 2.9 Retorts with external condensers. Left: in Huili (MIP 1958); middle and right: in Huizhe (Xu 1998) (© The Trustees of the British Museum).

4. Furnace construction

Most of the furnaces used in traditional zinc smelting were rectangular, morphologically resembling troughs for feeding animals (Figure 1.3 and 2.6). The four walls of the furnaces were built with bricks. A number of bricks were first arranged at regular intervals at the bottoms of the two long walls, so the space between them could serve as airholes. Then the furnace walls were built, and brick bars were built between each pair of airhole bricks. Generally, three retorts were placed on each furnace bar; as for retorts with external condensers, only two retorts were placed on each furnace bar due to their projected condensers and large sizes (Figure 2.9). The number of furnace bars varied from 10 to 40, so each furnace could accommodate between 30 and 120 retorts. Furnace sizes depended on the number and sizes of the retorts and were in the range of from 4 to 10 m in length, and about 1 m in width and height. In some processes, one furnace was separated into two or three units. In this arrangement, one unit could be recharged while another was in operation, which greatly improved the efficiency of the process.

These original trough furnaces, also called *yan* 铅 furnaces, were developed into 'high airhole' trough furnaces by Yang Kan, a worker of HCZLW (Xu 1986). Furnaces described as 'high airhole' and 'fixed airhole' furnaces were also used at this factory; furnaces described as 'ash base' and 'hard base' furnaces were used in Huili in 1958. Although none of these structures were described or illustrated in detail, the differences between these furnaces, as indicated by their names, were probably related to the way the draught was controlled.

Since the 1970s, larger furnaces, 'saddle' furnaces and 'climbing hillside' furnaces, have been developed (Craddock and Zhou 2003). Two-unit saddle furnaces were the most common ones used in the 1990s. Each unit could hold 60-90 retorts. Different from trough furnaces, their fireboxes were separated from the retort chambers and two units shared one chimney (Figure 2.10). The less common 'climbing hillside' furnaces resemble typical Chinese 'dragon' pottery kilns, as both consist of a series of linked chambers, each with its own fire. In the case of 'climbing hillside' furnaces the flames and gases are sucked continuously through all the chambers from the lower to the upper end (Craddock and Zhou 2003, 275).

5. Tools making

Tools were mostly made of iron or wood. To prepare ores, iron hammers were used to crush the rocks and bamboo baskets to sieve them. Iron or wooden cleavers were used to cut the coal briquettes. Wooden rods, iron spoon-shaped tools or iron curved sheets were used to perforate pockets. During the process, hooked tools were employed for raking out coal ash. Afterwards, iron tongs were used to lift ingots and retorts, and iron points and angled tools were used to clean out retorts. Iron ladles were used for ladling molten zinc during refining (descriptions and illustrations of these tools see Xu 1998, 126, 128).

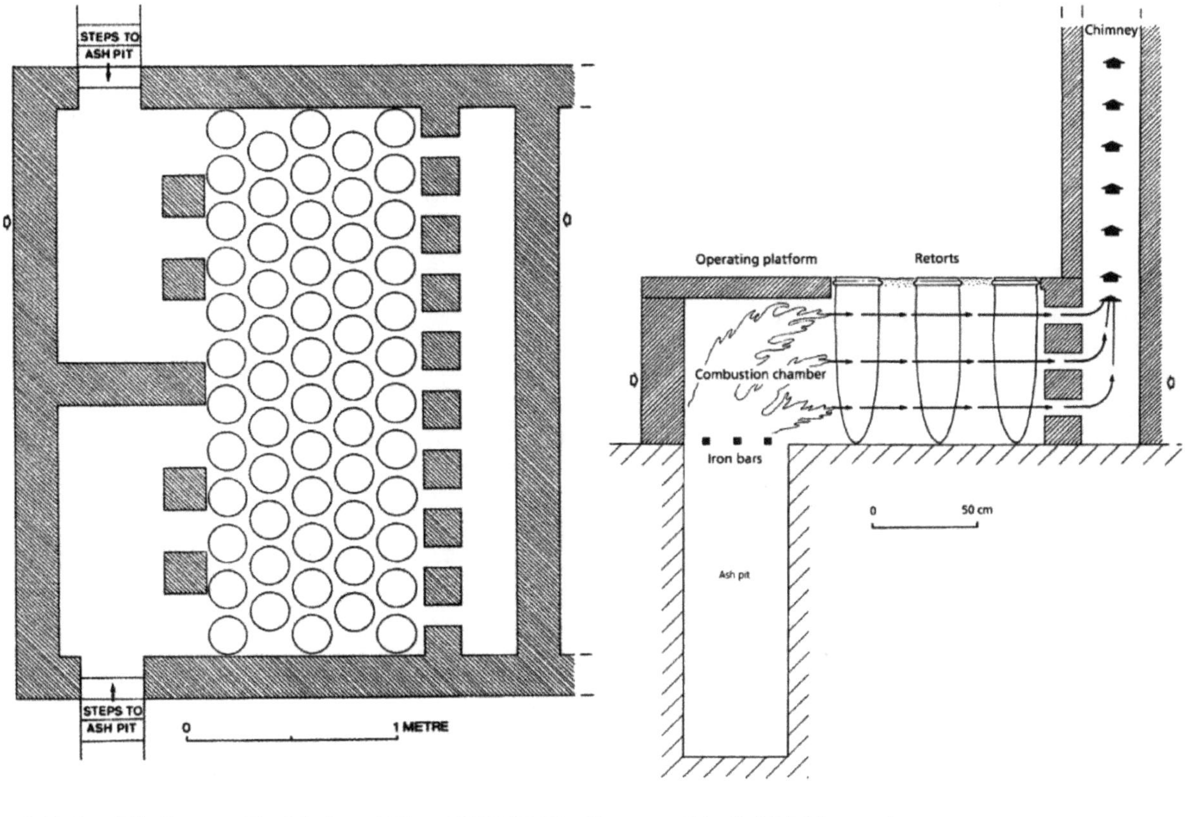

Figure 2.10 A saddle furnace (Craddock and Zhou 2003) (© The Trustees of the British Museum).

Table 2. 4 The smelting process in Huili in 1944-45 (Li 1944; Sun 1945).

Operation		Explanation
Loading furnace	出罐 Taking out retorts	After cooling, retorts from the previous smelt are taken out of the furnace.
	掏炉底 Drawing the furnace bottom	Remove large coal clinker and leave coal ash at the bottom of the furnace.
	过炉铲 Passing furnace spade	Flatten small coal clinker in airholes to connect each pair of airholes.
	打炉脚 Raking furnace foot	Put large coal clinker on the furnace bed to avoid the blockage of airholes.
	发火 Igniting	Ignite coal briquettes by the furnace and put them between each pair of furnace bars.
	下罐 Loading retorts	Place retorts onto furnace bars in the furnace.
	下井 Arranging coal briquettes	Put coal briquettes around the retorts, leaving some space between the upper parts of retorts.
	加盖 Adding lids	Cover the retorts with lids, leaving openings.
	码渣子 Adding clinker	Add coarse clinker and fine clinker onto the coal briquettes.
	放炉脚 Piecing furnace foot	Pierce airholes with thin bamboo sheets to adjust air.
	码边火 Piling side fire	Surround the furnace with bricks or ores.
Distilling	看火 Observing fire	When the fire has reached the coal at the top in 4-5 h, observe the flame from the opening of retorts. The flame turns yellow to white and blue, indicating some zinc escaping.
	打针 Piercing with needles	When the passages for zinc are blocked, remove the lids and pierce the pockets with needles.
	点眼炭 Covering coal	When zinc is ascending too fiercely, cover the lids with some coal.
	提火 Lifting fire	When the fire is not strong enough, pierce the coal briquettes and cover the pierced briquettes with more briquettes.
	收白灰 Collecting white ash	When the fire is too fierce, the zinc vapour could penetrate through the pockets and escape into air. Cover with retorts to collect the zinc oxide.
Collecting zinc	搂铅 Collecting zinc	After about 20 h firing, the blue flame faints. Ladle the zinc from the pockets, melt in iron bowls and cast into ingots.

6. Zinc smelting

The zinc smelting processes which operated in different areas and periods were generally similar. Taking the 1944-45 Huili process as a paradigmatic example, a typical smelting process included three main stages: loading the furnaces, distilling and collecting zinc. These stages can be further divided into a series of small operations (Table 2.4)

Some of the operations were of great importance to the success of the process. To regulate the draught of the furnaces, three steps (drawing furnace bottom, passing furnace spade and raking furnace foot) were conducted in Huili in 1945. This was also achieved by raking ash in different shapes on the bottoms of furnaces in Hezhang in 1986. To keep the top parts of retorts away from strong fire and maintain appropriate temperatures, coarse clinker and fine clinker/sand were piled onto the top of coal briquettes, as practiced in Huili in 1944-45, sometimes followed by a layer of slurry made of yellow mud and fine furnace ash to cover the fire, as in Hezhang in 1958 and 1986. During distilling, an excess of carbon monoxide came out of the openings of the lids and was burnt together with some zinc vapour, producing 'zinc flames'. Observing the zinc flames and taking corresponding measures to regulate the combustion in the furnaces were crucial.

After smelting, the raw zinc was sometimes refined and cast into ingots. The Huili zinc producers in 1958 refined raw zinc of 98.7% purity in reverberatory furnaces and obtained grades of zinc with levels up to 99.5-99.6%. Zinc products were divided into five grades, ranging from the fifth (98.7% Zn) to first grades (99.96% Zn). In 1958, the Hezhang craftspeople refined raw zinc with a second distillation using the same furnaces and retorts, followed by remelting it in iron woks or reverberatory furnaces and casting into ingots. If lead-rich zinc ores were processed, lead particles remained in the retort residues, which were sometimes washed to recover lead, for example in Hezhang in 1986.

During the process, some zinc would be oxidised to form zinc oxide which would deposit onto the internal surfaces of retort rims. The whitish substance, a mixture of zinc oxide and metallic zinc, was called 'blue powder'. In Hezhang in 1958, residues with 4-6% zinc deposited under condensers were called 'white ash', while most residues with only 0.2-0.5% zinc were termed 'secondary ash'. The blue powder, white ash, clay lids and pockets were resmelted to recover zinc, but secondary ash was usually discarded. In Huili in 1958 and Huize in 1998, external condensers were also crushed and recycled.

The smelting process usually lasted about one day, which, for example, included 8-9 h preheating, 8 h smelting, and 4 h cooling, as shown in Hezhang in 1984. The yield of each furnace per firing ranged from 15 to 50 kg, depending on ore grades, sizes of retorts and furnaces, and recovery rates. The recovery rates also varied widely from 20%, as in Hezhang in 1984, to 90%, as in Hezhang in 1958. In Huili in 1945, the recovery rates increased from 30% to 55-61% by taking steps to improve the technology, such as condensers, ore grades, amounts of reducing agents, savings of fuel and retorts, recycling of by-products etc.

Each firing involved 3-8 people, including one highly experienced craftsperson, called the 'furnace master', and several less skilled workers. Furnace masters were in charge of crucial operations during the process, while ordinary workers assisted masters and operated other tasks. For example, in Huili in 1945, each furnace was attended by one master, three workers and two helpers. The furnace master was in charge of important issues, such as attending the fire, drawing out furnace ash, clearing air passages, repairing furnaces, and also managing other important issues. The three workers were responsible for taking out retorts and clinker from furnaces, repairing retorts, making condensers, mixing the charge, loading the charge into retorts, adding pockets, etc. The two helpers assisted by mixing coal, crushing ores, removing slag from the retorts, etc.

Overall, the variations documented within each stage of the *chaîne opératoire* of zinc production demonstrate the room for 'technological choices' (see Section 3.1.1) made by the contemporary craftspeople, depending on their learning traditions as well as other material and cultural constraints. The detailed examination of these ethnographic accounts facilitates our understanding of some of the technical parameters involved and identification of archaeological remains, in addition to providing a wealth of data for the later discussion of spatial and temporal variations in the archaeological materials.

2.4 Summary

This chapter has introduced the technological and historical background to this research. There is a wealth of literature, both in China and the West, addressing many topics relevant to this research. In particular, a combination of studies of the textual evidence, the analyses of brass coins and artefacts, and the investigation of traditional zinc smelting processes in southwest China, has provided a general picture of the zinc distillation technology in China. The problems and limitations of these sources of evidence, as already clarified in Section 1.2, can be overcome by the studies of the archaeological remains from Chongqing. It is hoped that the case studies of the Fengdu and Shizhu sites presented in this research will contribute to a more comprehensive and historically accurate picture of zinc distillation technology in Ming and Qing China.

Chapter 3

Theoretical Framework and Methodology

This chapter outlines the theoretical framework and methodology employed for the scientific studies of the zinc production remains in Chongqing. The first section introduces the general theoretical framework, by addressing some theoretical concepts to the characterisation of technology, including the *chaîne opératoire*, technological choice/style and organisation of production. For this research, historical texts and ethnographic accounts are of great use and thus critically exploited in order to gain a better understanding of the archaeological and archaeometric data. The second section justifies the choices of samples and analytical techniques, and details sample preparation methods and the principles and operating parameters of the analytical techniques used.

3.1 Theoretical framework

3.1.1 Technology, the chaîne opératore*, technological choice/style and organisation of production*

Technology has a multiplicity of definitions and uses, in a wide variety of perspectives. Modern studies of technology tend to perceive it 'in the context of an outwardly expanding, nested set of actions and relationships: from production to the organisation of the production process to the entire culture system of processes and practices associated with production and consumption' (Miller 2007, 4). Crucially, technology not only involves *production*, i.e. 'creating objects from start to finish', but also *organisation of production*, i.e. 'organisational arrangement within which production takes place' (Miller 2007, 5). At the same time, studies of technology also consider *consumption* of objects, including distribution, as well as use to disposal. As such, technology examines 'the interconnections between people and objects' during the processes and practices of production and consumption of objects (Miller 2007, 5). Thus technology is not only about the physical actions, but also about their embeddedness in wider social contexts and the social meanings with which they are imbued.

A variety of theoretical concepts and approaches have been developed to investigate the social construction of technology, many engaging a variety of terms such as technological choice, technological style, practice, agency, materiality, ethnoarchaeology, material culture studies and archaeometry (Killick 2004). This research draws specifically from some well-established concepts such as the *chaîne opératoire*, technological choice/style and organisation of production.

A common first step in the study of ancient technologies is to reconstruct the *chaîne opératoire*, 'the range of processes by which naturally occurring raw materials are selected, shaped and transformed into usable culture products' (Schlanger 2005). Importantly, the characterisation of those processes can only be accomplished if one considers the contexts where they take place as well as the impact of the technology studied in that sociocultural environment.

The remit of the *chaîne opératoire* approach has progressively expanded from production to incorporate processes of distribution, use, reuse and final discard (Sillar and Tite 2000). This integration is of great importance for the present study of zinc production remains, as zinc in China was only an intermediate material, mostly for making brass. Only by considering its further use in brass making and the use of brass can the technology of zinc production be fully interpreted. Each element of the technological sequence, such as the manufacture of retorts or the procurement of zinc ores and fuel, can be regarded as an individual sequence. All of them, however, are interdependent.

The reconstruction of a *chaîne opératoire* is a foundation to identify the human choices made regarding raw materials, tools, energy, techniques and their combination in a sequence. 'Technological choices' are influenced not only by material constraints, such as properties and performance characteristics, but also by social, economic, political and ideological factors. Why particular techniques were chosen and potential functionally equivalent alternatives were rejected can only be explained by investigating the overall situational contexts of the production and the use of products (Sillar and Tite 2000). In practice, the culturally determined factors can be identified by assessing the extent to which technical constraints have influenced technological choices (Rehren *et al.* 2007). The sum of the many choices that take place in a technological sequence can be understood as the 'technological style', i.e. 'the many elements that make up particular technological activities' (Lechtman 1977, 6), or, in other words, 'the formal integration of the behaviors performed during manufacture and use of material culture which, in its entirety, expresses social information' (Childs 1991, 332).

In order to approach to the social aspects of the technology, the organisation of production should also be examined. The organisation of craft production can be characterised by four parameters: context, concentration, scale and intensity. Based on these parameters, eight generic categories have been proposed to classify different models of production organisation, ranging from individual specialisation to retainer workshop. Production loci and debris can provide direct evidence for the reconstruction of the organisation of production (Costin 1991).

This research will attempt to reconstruct the *chaîne opératoire* related to zinc production in different contexts. By obtaining high-resolution characterisations of the many choices defining each technology, it is hoped that the comparison of regional and temporal variants will be elucidated. Furthermore, it will be possible to identify common elements present in various contexts, thus establishing the broad technological tradition that has emerged from the social and material constraints shaping Chinese technology over time. At the same time, individual variants will reflect the specific engagement between the materials, the craftspeople, and the sociocultural environments in each case. In addition, by contributing to a contextualised history of technology, it is hoped that this research will provide a reference baseline for future comparative studies within China and beyond. This approach may therefore be a useful complement to research focused on technological origins, as it may allow comparisons of specific choices, for example between zinc smelting in India and China, but also between zinc distillation and crucible iron/lead smelting or other distillation technologies within China.

In order to obtain a comprehensive understanding of the technology, archaeological and archaeometric evidence of the production remains will be integrated with any other sources of evidence available. This research deals with the technology of zinc production in Ming and Qing China, AD 1500-1900, a period in which written sources are plentiful. In addition, the rich ethnographic accounts on the traditional zinc smelting processes in the twentieth century outlined in Section 2.3 will aid the interpretation of the archaeological remains in Chongqing. Therefore, as textual and ethnographic sources are frequently referred to in this research, it is necessary that the problems and limitations of these sources of evidence are recognised as well as their potential.

3.1.2 Historical texts

Most scholars now accept that archaeology should not be subordinated to history in text-informed periods, since both disciplines play different roles and complement each other in understanding the past. Technology is an area where archaeological data is relatively easy to reveal, compared to the usually limited textual evidence. This disparity is partly due to the fact that craftspeople were generally illiterate or otherwise unable, or unwilling, to convey their technology in written texts, while the elite or the literate did not regard crafts activities worthy of detailed description.

Only a limited number of technical texts or treatises have survived. These technical texts can assist archaeology in several ways for a better understanding of technology: they can help identify unclear archaeological assemblages, archaeometallurgical remains in particular; they may inform about different aspects of technology, such as the ideas and perceptions of craftspeople, which are not always easy to grasp from archaeological finds; they may sometimes fill the gaps where no or few archaeological data are available. However, these technical texts, varying in quality and detail, are sometimes vague, inaccurate, misleading, or even intentionally deceitful. Thus they must be assessed critically by considering the background and experience of the authors, the assumed readership and the accuracy and details of the technology described (Martinón-Torres 2008; Freestone 2008; Bayley 2008). There are few technical treatises referring to the technology of zinc distillation and brass making in Ming and Qing China. The most important one is *Tian Gong Kai Wu*, which not only describes zinc distillation and brass making, but also some other relevant aspects, such as coin casting, coal mining and mercury distilling.

Besides technical treatises, another type of written source of great importance to this research is *fangzhi* 方志, gazetteers for various administrative units. They usually presented on similar topics, such as geography, taxes and revenues, population, education, transport, products, customs, biographies of officials and celebrities etc. (Twitchett and Fairbank 1988, 778). These gazetteers were often based on local archives and field investigation, so they are generally quite reliable (Zhao and Zhou 1998, 22). Some of the topics they record, such as products and taxes, are informative for studies on mining and metal production. The gazetteers were usually compiled by local officials, and therefore tend to cover more administrative aspects of the technology from the viewpoint of the officials and gentry. As such, they provide a useful complement to the technical information revealed through archaeological studies. This research will make reference to both regional and local gazetteers, such as gazetteers of Sichuan, Fengdu and Shizhu (see Appendix 4).

Other useful textual sources include writings by alchemists and pharmacologists, as well as administrative and geological documents recording the procurement of mint metals and the making of coins.

3.1.3 Ethnoarchaeology

Ethnoarchaeology is the ethnographic study of living cultures from archaeological perspectives in order to aid the interpretation of archaeological remains by the use of analogy – a form of inference that holds that if something is like something else in some aspects it is likely to be similar in others (David and Kramer 2001, 1-2). Ethnoarchaeological approaches have long been applied to the study of craft production, especially ceramic production (Costin 2000; Underhill 2003) and African iron smelting (Childs 1991; Killick 1991; David 2001; David and Kramer 2001). Ethnoarchaeologists can more directly reconstruct technology by recording the operational sequences, investigating various technological choices, and distinguishing between those choices determined by material constraints and those affected by potential cultural choices. However, it is recognised that the material and cultural constraints are not always neatly divided (David 2001, 58).

Ethnoarchaeological studies can be classified in three broad types: 'informal ethnographic observation to inform the archaeologists', 'ethnographic study of a craft or technology for an archaeological purpose' and 'ethnographic studies of whole cultures, the full context for archaeological inference' (Thompson 1991, 233). The various records of traditional zinc smelting processes employed in this research fall into the first and second types. African archaeometallurgy provides many examples of how the observation of recent iron smelting practices greatly helped understand ancient processes. The ethnographic records not only point at potential processes that generate material patterns comparable to archaeological remains, thus helping process identification, but they also aid the reconstruction of the furnaces from incomplete remains and the elucidation of the *chaîne opératoire* of the smelting processes (Killick 1991).

The great potential of ethnoarchaeology notwithstanding, one should be cautious before drawing inferences based on observations of modern crafts. In particular, when using traditional zinc smelting literature to help interpret archaeological remains, some problems should be taken into consideration. First, these various ethnographic records, like the historical texts, are of greatly varying quality and detail, depending on the knowledge, experience and specific interest of the observers. Generally, reports by metallurgists and historians of metallurgy are more reliable for this research than those left by foreign visitors and geologists. Secondly, even when the processes may have been described in great detail, practices are not necessarily the same as the ancient ones. For example, the demand of products, the transport and the organisation of production were different. The types of ores, the furnace and retort design and the location of the workshops may have been also affected by different constraints. Therefore, ethnographic analogies must be applied carefully, and they do not replace the direct and detailed investigation of archaeological remains.

3.2 Methodology

3.2.1 Sample selection

To fulfil the aims and objectives of this research, a wide range of materials of metallurgical remains from four zinc smelting sites in Chongqing were selected for analysis, including possible raw materials (minerals and coal), products (zinc metal), retorts and metallurgical residues (slag). Retorts are the most abundant and important debris at zinc smelting sites, and thus form the core of the research. Analyses of different ceramic parts can identify the choices of clay and temper, the manufacturing methods and performance characteristics. Used retorts inform on the operating temperatures and redox conditions of the processes, especially the temperature control for distillation. Slag within the retorts, together with potential ores and zinc metal, can provide information on the original charge, the zinc products and the metallurgical process.

The first stage of sample selection was carried out by Prof. Chen Jianli in his several visits to the sites during excavation at Fengdu. His collection of samples at the School of Archaeology and Museology, Peking University, was complemented by the samples picked up from the surface by the author during fieldwork in August 2009, when the bulk of these sites had already been flooded by a dam construction. The samples from Dafengmen in Shizhu were collected during fieldwork in June 2010. The second stage involved photography, material classification and final sample selection for analysis.

Due to the diversity and complexity of the assemblages, several samples were chosen for each category of materials from each site in order to assess their internal variability (Table 3.1). Different parts of retorts and unused and used pots were selected. Where possible, the bottom, middle and top parts of used pots were also deliberately selected in order to assess the temperature differences. Since Miaobeihou is the main case study site, more samples for each category were selected from its assemblage. The number of samples selected also depended on the availability of each type of sample in the collection. For example, only one lid was available for the site at Dafengmen.

Table 3.1 Sample numbers of each category of production remains from four zinc smelting sites in Chongqing. The four mineral samples of Dafengmen come from Yushi Cave (Y).

Site	Mineral	Coal	Zinc	Pot	Condenser	Pocket	Lid	Slag	Sum
Miaobeihou (YM)	3	8	3	6	8	5	-	5	38
Puzihe (YP)	-	-	-	4	3	-	-	2	9
Muxiexi (CM)	-	-	-	4	3	-	-	3	10
Dafengmen (SD)	4	-	-	6	3	-	1	5	19
Sum	7	8	3	20	17	5	1	15	76

The underlying concern when sampling was to obtain as much information as possible with minimum damage to the samples. To achieve this, for example, small parts were cut from fractured edges of pot samples with the cross sections vertical to the surfaces of pots, so that the slag or vitrification layer could be analysed at the same time. In most cases, relatively small fragments of samples were taken for analysis (~1 cm^3). However, two larger samples were cut from the top parts of retorts to address specific questions.

All the samples were named with three capital letters followed by a number. The first one or two letters refer the site (YM for Miaobeihou, YP for Puzihe, CM for Muxiexi, Y for Yushi Cave, SD for Dafengmen); the last letter denotes the material type (O for ore, C for coal, Z for zinc, B, M and T for bottom, middle and top parts of retorts, respectively).

3.2.2 Optical Microscopy (OM)

There are two main types of optical microscopy: reflected light microscopy, i.e. when light is reflected by the polished surface of a thick sample; and transmitted light microscopy, i.e. when light passes through a thin section of the sample. Reflected light microscopy is frequently used in the examination of the microstructures of metal artefacts, to identify their compositions and manufacturing methods (Scott 1991; Han *et al.* 2002), and also used for metallurgical remains, such as crucibles and slag (Bachmann 1982; Rehren and Pernicka 2008, 237). It is considered an important precursor to further analyses of a variety of archaeological materials.

The samples selected for reflected light microscopic analysis were mounted in epoxy resin. Labels showing sample numbers, sample types, the preparation date and the initials of the person who prepared the blocks were mounted with the samples. Some porous samples, such as ceramics and slag, underwent a second filling of epoxy resin under vacuum. The blocks were then first ground using silicon carbide paper of several grades (Grit B120, 320, 600, 1200, 2500 and 4000) and then polished to 1 μm. The same polished blocks were subsequently used for further examination with SEM-EDS and EPMA-WDS.

The polished blocks were inspected under two modes of light, plain polarised reflected light (PPL) and cross polarised reflected light (XPL), to examine the general appearances and identify different phases in the samples. These two modes complement each other. PPL is useful in studying metallic materials: non-metallic inclusions can be easily recognised in zinc metal; bright metallic iron and zinc are clearly visible in the darker metallurgical residues. Moreover, it is found that carbon-rich coal is best observed under PPL because of its high reflectivity. Under XPL, the real colours of minerals/phases are revealed, which is extremely helpful in the studies of samples of minerals and metallurgical residues (Figure 3.1).

The microscope used was a Leica DM LM. Pictures were taken under PPL and XPL at magnifications of 50x, 100x, 200x and 500x, corresponding to widths of pictures of 2 mm, 1 mm, 0.5 mm and 0.2 mm respectively. Careful examination of polished blocks allowed a general understanding of the samples and to locate phases or areas of special interest, which facilitated the work done in the SEM directly using these blocks.

Figure 3.1 PPL (top) and XPL (bottom) images of the same area in slag YMB2-1. The width of picture is 0.5 mm. The metallic iron (white) is better illustrated under PPL, while the zinc sulphide (yellow) is better identified under XPL.

Transmitted light microscopy of thin sections is extensively used for the mineralogical analyses of ancient stones, ceramics and other inorganic materials (Reedy 2008). The minerals in samples mounted in thin sections with a standard thickness of 30 μm, can be identified by observing a variety of optical features. Thin section petrography is particularly suitable for the studies of paste preparation techniques, forming methods, firing parameters and provenance of ancient ceramics (Freestone 1995). Some of the retort samples were prepared as thin sections following standard sample preparation procedures (Reedy 2008, 2-3) and examined with transmitted light microscopy under PPL and XPL. It was found, however, that thin section petrography was not particularly informative for the retort samples: the pots from river sites were very high-fired, so few types of minerals survived except quartz grains and quartz-rich rock fragments; the condenser and pocket samples from all the sites were low-fired but heavily contaminated by zinc oxide, so the original mineralogical compositions are difficult to identify; the Dafengmen pot fabrics were also heavily altered by zinc oxide. Therefore, it was decided that reflected light microscopy of polished blocks coupled with SEM-EDS would be employed as the main analytical techniques.

3.2.3 Scanning Electron Microscopy - Energy Dispersive Spectrometry (SEM-EDS)

Scanning electron microscopy with micro-analytical facilities is widely used in the examination of various archaeological materials (Olsen 1988; Pollard *et al.* 2007; Artioli 2010). It combines high resolution imaging of the electron microscope with quantitative analyses, allowing for analysing small areas and detecting spatial variations in compositions (Pollard *et al.* 2007, 109). It is especially suitable for complex and multi-phase materials, such as crucibles and slag (Rehren and Pernicka 2008, 237).

The basic principle of SEM is that the sample bombarded by a beam of electrons emits several signals, including secondary electrons (SE), backscattered electrons (BSE) and characteristic X-rays. The low energy secondary electrons (less than 50 eV) emitted from the surface generate images showing the topography of the sample. The signals of higher energy backscattering electrons (above 50 eV) emitted from deeper within the sample reflect both the topographic and qualitative compositional information. The characteristic X-rays of higher energy are detected from further deep into the sample by an energy dispersive spectrometer (EDS). The energy of X-rays is characteristic of the elements present in the sample, thus generating quantitative compositions (Pollard *et al.* 2007, 109-111).

The SEM used for this research was a Philips XL30 environmental SEM with an Oxford Instruments INCA spectrometer package. It was equipped with both SE and BSE detectors and an X-ray micro-analyser (EDS). The acceleration voltage applied to all analyses was 20 kV, the working distance 10 mm, the spot size 5.0-5.3, the beam current adjusted to a deadtime of 35-40% and the livetime 50 s. The polished blocks were carbon coated to be conductive, and were observed under SE and BSE modes at regular magnifications (50x, 100x, 200x, 500x and 800x) as well as other magnifications when needed. When necessary, SE images were taken in order to reveal specific morphological information or differentiate phases which were difficult to identify under BSE mode.

The EDS micro-analysis was performed under BSE mode. Given the heterogeneous nature of most of the samples studied, the bulk compositions of the samples were obtained by averaging five measurements of large areas of ~2 by ~2.5 mm taken at 50x. In addition, the ceramic matrices of the pots were analysed at areas of ~100 by ~150 μm, avoiding large inclusions; the glassy phases of slag were studied at areas of ~400 by ~600 μm; individual phases were probed by measuring spots of a few micrometres in diameter.

The results were combined with oxygen by stoichiometry where appropriate. All the chemical compositions presented in this research are given as weight percentage (wt%). The results were usually normalised to 100%; however, the low totals of mineral and coal samples were mostly due to the presence of light elements such as carbon, so their results were not normalised in order to estimate the proportions of light elements. Sodium was omitted in the analyses of samples rich in zinc (minerals, pots, condensers, pockets and slag), because both the Na Kα and Na Kβ peaks overlap with the Zn Lα peak.

Data quality was measured by testing accuracy and precision of three basalt certified reference materials (CRMs) from the United States Geological Survey (Table 3.2). The powdered CRMs were made into pellets by mixing with epoxy resins; the pellets were cut, mounted into blocks, and then ground and polished. These standard samples were analysed with SEM-EDS. In the accuracy test, the analytical results were compared with the certified values; the relative errors of major element oxides were mostly below 5%. Coefficients of variation in precision tests were below 5% for major elements. The relative errors and coefficients of variation of minor element oxides in CRMs were mostly over 20%. Lower confidence limits were established at about 0.5%, but analytical results below the threshold are reported for indicative purposes.

3.2.4 Electron Probe Micro-analyser - Wavelength Dispersive Spectrometry (EPMA-WDS)

While SEM-EDS is a sufficient instrument to analyse major elements, it has relatively high detection limits, preventing an accurate quantification of minor and trace elements. EPMA is a similar instrument to SEM, but differs in that it is equipped with a range of crystal spectrometers that enable quantitative chemical analyses by measuring the wavelength of the characteristic X-rays diffracted by crystals (WDS) with higher spectral resolution and lower detection limits (Pollard *et al.* 2007, 111). Thus EPMA-WDS was employed to detect minor and trace elements in zinc samples.

The EPMA used was a JEOL JXA 8600 with an integrated operating system for the WDS analysis. The acceleration voltage applied was 20 kV, the beam current was $5X10^{-8}$ A, with an acquisition time of 50 s. Three CRMs (41X4380Zn2, 41X0336Zn2, 41X0336Zn6) and three zinc samples were analysed at areas of ~80 by ~120 μm ten times. Three crystals were used: TAP (thallium acid phthalate) for the Lα line of As; PET (pentaerythritol) for the Lα lines of Sb, Sn, Ag and Cd, the Mα lines of Pb and Bi, and the Lβ line of In; LIF (lithium fluoride) for the Kα lines of Zn, Fe and Cu. The detection limits of these elements are about 0.03%, but again analytical results below the limits are presented.

Table 3.2 Comparisons of measured mean values and certified values for three CRMs (wt%), normalised to 100%. Analyses on polished sections at areas of ~200 by ~300 µm by SEM-EDS. BIR-1: Icelandic basalt; BHVO-2: Hawaiian Volcanic Observatory basalt; BCR-2: Colombia River basalt.

CRM		Na_2O	MgO	Al_2O_3	SiO_2	P_2O_5	K_2O	CaO	TiO_2	MnO	FeO
BIR-1	Mean value (n=3)	1.6	9.1	14.7	49.4	0.1	-	13.4	1.0	0.3	10.5
	Certified value	1.6	9.0	14.6	49.5	0.3	0.02	13.4	1.1	0.2	10.4
	Relative error %	1.4	1.3	0.6	0.1	81.1		0.4	5.8	20.0	1.1
	Coefficient of variation %	2.8	1.1	0.5	0.2	411.6	186.3	0.8	11.8	21.5	1.1
BHVO-2	Mean value (n=3)	2.1	6.8	12.8	51.8	0.3	0.5	11.5	2.9	0.2	11.3
	Certified value	2.2	6.6	13.0	51.3	0.3	0.5	11.5	3.0	0.1	11.5
	Relative error %	3.5	2.5	1.8	0.9	10.8	0.7	0.3	5.6	36.9	1.8
	Coefficient of variation %	3.2	2.7	1.7	0.8	8.7	7.6	0.5	5.5	33.2	1.7
BCR-2	Mean value (n=3)	2.8	3.4	12.6	56.2	0.3	1.8	7.1	2.5	0.2	13.0
	Certified value	2.8	3.4	12.6	56.3	0.5	1.8	7.0	2.4	0.2	12.9
	Relative error %	0.1	0.8	0.1	0.2	28.5	0.4	1.2	3.7	9.7	0.5
	Coefficient of variation %	3.0	1.5	0.3	0.2	42.5	2.1	1.5	5.8	21.4	0.5

3.2.5 X-Ray Diffraction (XRD)

X-Ray Diffraction is a technique traditionally employed for the identification of crystalline phases in geological minerals and rocks. It is a cheap, rapid and reliable technique for studying various archaeological materials, including metals, slag, ceramics, soils, building stones, pigments, plasters, etc. (Aritioli *et al.* 2010, 51).

When a beam of monochromatic X-rays of known wavelength is incident upon a crystalline sample, the X-rays are reflected, in a manner similar to the reflection of light from a mirror, from successive atomic layers in the crystal lattice of the sample, producing a series of reflected X-ray intensity maxima at angles determined by the spacings between crystal planes. Each mineral has a unique atomic lattice arrangement and can generate a unique X-ray diffraction pattern; the mineral can be identified by referencing to sets of standard X-ray powder diffraction patterns (Pollard *et al.* 2007, 113-116).

In this research, XRD is a supplementary technique to OM and SEM in the studies of mineral and pot samples. With SEM-EDS, minerals could only be roughly identified to be zinc carbonates or zinc silicates, while XRD could ascribe them to specific minerals, i.e. smithsonite and hemimorphite. The pot samples were analysed by XRD in order to confirm the presence of mullite, a high temperature phase which is not easily identified with SEM.

The XRD used was a Rigaku D/max 2550PC Diffractometer fitted with Cu Kα radiation, generating X-rays with a wavelength of 1.54056Å, and graphite monochromator. A small portion of samples was ground with an agate pestle and mortar and the powder was placed in a glass sample holder. The operating conditions were 40 kV and 150 mA, with a 1° divergent slit, a 1° scattering slit and a 0.3 mm receiving slit. Diffraction data were collected in the range of 3° and 90° 2θ in steps of 0.02°, at a scanning speed of 8° 2θ per minute. The data were processed and interpreted using MDI Jade software version 6.5.

3.2.6 Refiring experiments

One of the methods to estimate the firing temperatures of ceramics is to refire them at a series of temperatures and observe the changes in the degree of vitrification with SEM. When the refiring temperatures exceed the original firing temperatures, the degree of vitrification starts to develop (Table 3.3). By comparing the degree of vitrification of the ceramics in the as-found state with that of the refired ceramics, the original firing temperatures could be estimated. Refiring experiments have been carried out on ancient pottery (Maniatis and Tite 1975; Maniatis and Tite 1981; Tite *et al.* 1982), bricks (Wolf 2002), and also technical ceramics, such as crucibles, furnaces and tuyères (Tite *et al.* 1985; Freestone and Tite 1986; Hein *et al.* 2007). It should be noted that the development of vitrification of ceramics depends on various factors, including not only the type of clay from which the ceramics were manufactured, but also the atmospheres in which they were fired, the duration of firing and the size.

The refiring experiments were carried out in a Lenton electrical furnace with Eurotherm temperature programmer at a heating rate of 200 °C/h with a soaking time of 1 h at peak temperatures of 1100 °C and 1200 °C. As the furnace cannot achieve 1300 °C, higher temperatures could not be tested. Two pot samples YMM13 and YMM8 were cut into small fragments, ~1 cm thick, and refired under oxidising and reducing atmospheres, respectively. The reducing atmosphere was achieved by placing the samples inside one small graphite crucible covered by another. In order to examine the effect of long exposure to high temperatures, two pieces were refired at a peak temperature of 1200 °C with a soaking time of 4 h in oxidising and reducing atmospheres, respectively. The fresh fractured surfaces of the refired samples were gold coated and examined under the SE mode of SEM.

Table 3.3 The stages of vitrification and their characteristics (Maniatis and Tite 1981).

Stage of vitrification	**Characteristics**
NV: no vitrification	The aggregates of flaky clay particles are the same as the raw clay.
IV: initial vitrification	Isolated smooth-surfaced areas or filaments of glass appear.
V: extensive vitrification	A network of smooth-surfaced glass filaments forms an open or cellular structure.
CV: continuous vitrification	A continuous vitrified surface containing isolated pores develops from the fine bloating pores (FB), 0.2-4 μm in diameter, to medial bloating pores (MB), 4-10 μm in diameter, to coarse bloating pores (CB), 10-50 μm in diameter.
TD: terminal distortion	Total collapse or large-scale bloating to produce a friable and highly porous body.

Chapter 4
Analyses of Zinc Smelting Remains in Fengdu

Analyses of the production remains from the Fengdu sites constitute one of the core research contributions of this research. This chapter first introduces the archaeological contexts and main finds of these sites. The analytical results of various types of production remains from Miaobeihou are first presented and discussed in terms of retort design, the charge and parameters of reduction and condensation. This is followed by the analytical results of production remains from two further sites in the region, Puzihe and Muxiexi, and the interpretation of the results. Based on these studies, the overall technology of zinc distillation in Fengdu is summarised and the organisation and scale of production are discussed.

4.1 Zinc smelting sites in Fengdu

4.1.1 Three Gorges archaeology

About 20 zinc smelting sites have been found in the Three Gorges reservoir region in Fengdu county, Chongqing municipality, southwest China. The Three Gorges, namely Qutang Gorge, Wu Gorge and Xiling Gorge, are located along the Yangtze River between Fengjie county, Chongqing municipality and Yichang city, Hubei province. Since 1992, a hydropower-complex project, the Three Gorges project, has been set up to construct a grand dam near the easternmost gorge, Xiling Gorge, in Yichang. The dam, finished in 2008, not only generates electricity, but also increases the river's shipping capacity and reduces the potential for floods downstream. This project has led to the formation of a large reservoir region, stretching from Chongqing to Yichang for about 600 km. The water level of the reservoir region has gradually risen to 175 m, and a great number of archaeological sites have been submerged. To rescue the cultural heritage in the region, archaeological surveys and subsequent intensive excavations have been carried out since 1992. It was under these circumstances that a great number of zinc smelting sites were found along the Yangtze River in Fengdu.

The Yangtze River, the longest River in China, flows for over 6,000 km from the glaciers on the Tibetan plateau in Qinghai province eastward across southwest, central and east China before entering the East China Sea at Shanghai. The Three Gorges region is a relatively independent and complex geographical unit along the upper and middle reaches of the Yangtze River. It connects the Sichuan basin to the west and the Yunnan-Guizhou plateau to the southwest, with the middle and lower reaches of the Yangtze River to the east. In this region, the valley of the Yangtze River is wide when the river passes through syncline mountains of soft rocks (mudstone, shale and sandstone), while it is narrow and surrounded by precipitous cliffs when the river flows through anticline mountains of dense and hard rocks, mainly composed of limestone. The Three Gorges are three sections of narrow valleys with rapid currents and submerged reefs. The Yangtze River and its tributaries are surrounded by mountainous areas. The climate is mild subtropical with distinct seasons, abundant sunlight and monsoon rainfall. Thus the region has been suitable for the living of a variety of animals and plants, and also for human settlement.

The region within Chongqing is one of the birthplaces of Chinese civilisation. A wealth of archaeological remains dated to the Paleolithic period has been found, including stone tools, mammal and human fossils. The famous 'Wushan people' are the earliest *homo erectus* in Asia, living about two million years ago in karst caves in Wushan county. The Neolithic period witnessed the development of several archaeological cultures, from Yuxi Lower culture (before 5000 BC), Yuxi Upper culture (5000-4000 BC), Laoguanmiao Lower culture (4000-3000 BC) to Weijialiangzi culture (3000-2000 BC). These cultures have close links with other cultures to the east in provinces of Hunan and Hubei. In the Shang (sixteenth to eleventh century BC) and Zhou (eleventh century to 256 BC) Dynasties, the cultures of the Ba state developed in the regions, and later the culture of the Chu state diffused from the east. After the imperial unification in 221 BC, the region entered the reign of the Qin Dynasty (221-206 BC) and the subsequent Han Dynasty (202 BC-AD 220). A strong funerary culture emerged in the Eastern Han period (AD 25-220), when tombs built with brick and stone were carved with pictorial decoration. The funerary culture developed in the Six Dynasties (AD 220-581), a period of disunity, instability and warfare; the funerary objects in brick tombs of this period are mainly celadon porcelains. After the Sui Dynasty (AD 581-618) reunified China, the region saw fast social, economic and cultural development from the Tang Dynasty (AD 618-907) onwards (CMBCH and CMBI 2001; 2003; 2006; 2007a; 2007b; 2010).

Along the Yangtze River in Fengdu, dozens of archaeological sites have been found and excavated, including settlements, tombs and workshops. In addition to the zinc workshops, several ceramic kilns have also been found. For example, at Yuxi (Figure 4.1, no.9), five pottery kilns, dating from 5000-4000 BC, were found. At Yuxiping, another site occupied since the Neolithic period, a well-preserved brick kiln of the Eastern Han period was excavated. In addition, over 2000 gilded copper-based statues of the Tang Dynasty were found at Yuxiping in 2001, which probably belonged to a temple destroyed by flood. At Puzihe (Figure 4.1, no.3), one of the zinc smelting sites studied in this research, two ceramic kilns of the Song Dynasty were excavated (SPAI and CMBCH 2007).

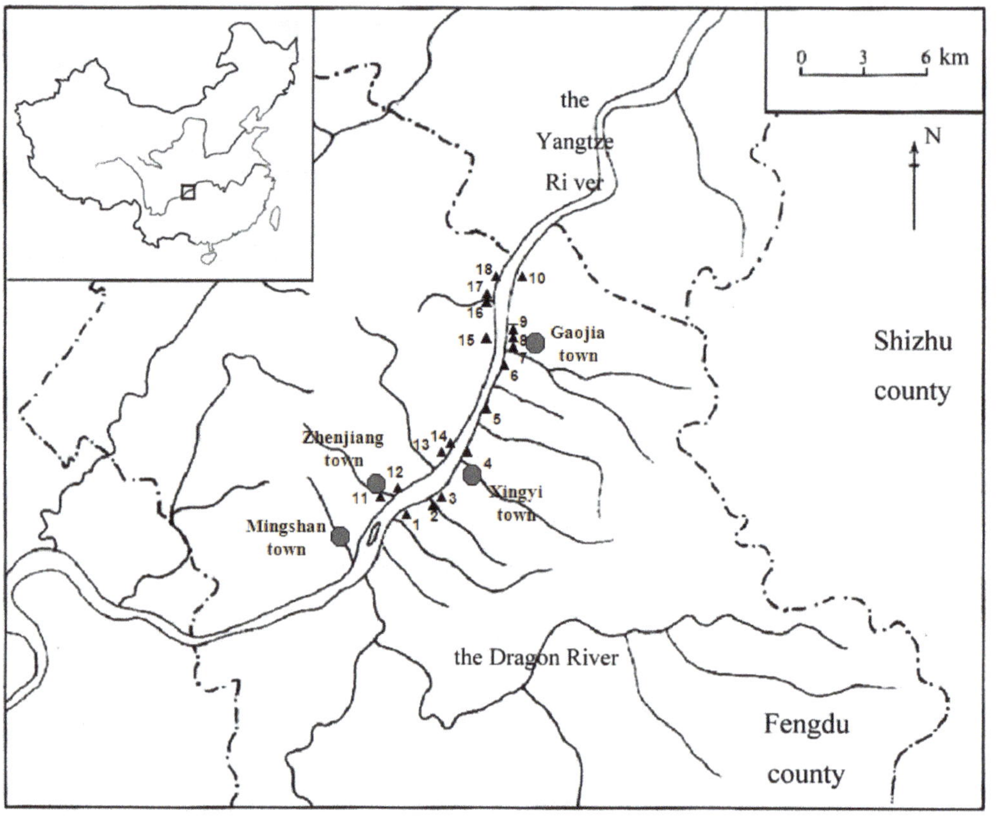

Figure 4.1 Distribution map of zinc smelting sites and four towns along the river in Fengdu, Chongqing. 1. Daoliuzi, 2. Miaobeihou, 3. Puzihe, 4. Muxiexi, 5. Shaxizui, 6. Qingjiayuan, 7 Yuanjiayan, 8. Shidiba, 9. Yuxi, 10. Hejiaba, 11. Jiudaoguai, 12. Langxi, 13. Chixi, 14. Qingyuan, 15. Tingxi, 16. Hezuishang, 17. Tangfang, 18. Xiaoshuangxi.

Figure 4.2 Left: the landscape of Miaobeihou with three areas of smelting debris marked; right: retorts excavated from Miaobeihou (photos by Liu Haiwang).

Historically, Fengdu became a county in the Eastern Han period. Its seat was Mingshan, a town at the foot of the Mingshan hill, situated on the northern bank of the Yangtze River. Mingshan town in Fengdu is now well-known as a 'ghost town' with many legendary stories about the Chinese Hell and related temples built from the Six Dynasties on the Mingshan hill. The county seat has now been moved to the opposite bank of the river because of the Three Gorges project. Along the Yangtze River, Gaojia town, built in the Ming Dynasty, was also an important town for land and water transport. In modern Fengdu, the towns of Xingyi and Zhenjiang are also located along the river (Figure 4.1). A road runs along the southern bank of the river, serving the towns of Xingyi and Gaojia and numerous villages.

4.1.2 Zinc smelting sites

During the field survey and test excavations in 1992-94, zinc smelting remains were found at several sites

(SPCHAI 1998). A peculiar type of sand-tempered pots, with clear evidence of high temperature exposure, was regarded as a diagnostic feature to identify these as smelting sites (Figure 4.2 right). These pots are flat-bottomed jars, around 25 to 30 cm tall. Slag and coal clinker were found adhering to the surfaces of the pots.

Intensive excavations in the region since 2001 have revealed further smelting remains and furnace foundations (Li *et al.* 2013). In 2001, Changsha Civic Cultural Heritage and Archaeology Institute (CCCHAI) and SPAI excavated the sites of Zhangjiahe and Puzihe. Similar smelting pots were found in a large quantity at both sites; in addition, cylindrical accessories which appeared to have been attached onto the pot rims were recognised. However, these sites were originally reported to be iron smelting sites due to the abundant rust within the pots (HPCHAI *et al.* 2007; SPAI and CMBCH 2007). Between 2002 and 2005 Henan Provincial Cultural Heritage and Archaeology Institute (HPCHAI) excavated the sites of Miaobeihou, Puzihe and Muxiexi and first time identified them as zinc smelting sites. In 2004-05, several field surveys were made in Fengdu and neighbouring Shizhu county to investigate zinc smelting sites, ore sources and transport route (see Section 5.1); and Chongqing Municipal Cultural Heritage and Archaeology Institute (CMCHAI) excavated the sites of Qingjiayuanzi, Yuanjiayan, Shidiba and Jiudaoguai in Fengdu (CMCHI and FCBCH 2013).

A research project, *Multidisciplinary Study of Zinc-smelting Sites in Fengdu,* was carried out by Peking University, USTB, HPCHAI and Chongqing Municipal Bureau of Cultural Heritage (CMBCH) from 2005 to 2008. Preliminary scientific studies on smelting remains from Miaobeihou have revealed that the ceramic pots were used as retorts for zinc distillation and the site was dated to the Ming Dynasty (AD 1368-1644) (Liu *et al.* 2007). In addition, the technology of zinc distillation was roughly reconstructed in terms of raw materials, structures of the furnaces and retorts, and the smelting process. This project laid the foundation for the present research.

Overall, nearly 20 zinc smelting sites have been found along both banks of the Yangtze River in Fengdu (Figure 4.1), all located on the first terraces of the banks at altitudes of 150-170 m (Figure 4.2 left). Dozens of rectangular furnace foundations have been found with surviving brick bars arranged at regular intervals, resembling those operated in traditional zinc smelting (see Section 2.3). The furnaces are between 1 and 2 m in width but vary in length, with the furnace at Puzihe being around 15 m long (see Section 4.3.1).

4.2 Miaobeihou (YM)

4.2.1 Site description

Miaobeihou is located at Yangliusi village, Xingyi town, Fengdu county. The site is situated on the first terrace of the southern bank of the Yangtze River, covering a length of more than 1000 m. It is divided by four small ditches into five areas, named Area I, II, III, IV and V from the west to the east. In 2002 and 2004, three excavation seasons were carried out and a total area of 4000 m^2 was excavated by HPCHAI. The stratigraphy can be divided into four main levels: the first stratum is modern earth; the second is from the Ming and Qing periods; the third is from the Northern Song period; the fourth is from the Eastern Zhou period to the Han Dynasty.

Abundant smelting debris and several furnace foundations were found in Areas II, IV and V. In Area II, several waste pits filled with retort fragments, furnace fragments and slag were excavated, including some relatively intact retorts. In Area IV, two pits for crushing coal and three slag pits were found. In Area V, several furnace foundations were identified but their shapes could not be reconstructed. Beside one of the furnace foundations there was a series of vestiges composing a relatively complete zinc smelting workshop, including a working shed, a working platform for crushing ores, a pit for crushing coal, a pit for charging pots and a waste pit. The charging pit was previously mistaken for a round furnace foundation (Liu *et al.* 2007, 176).

Three types of furnace fragments were excavated. The most common are quartz-tempered, highly vitrified fragments from the internal parts of furnace walls. The second type are large blocks of tomb bricks of the Han Dynasty, called 'Han bricks', used to construct the middle parts of furnace walls. The third type is composed of clayey fragments from the external parts of furnace walls. It is worth noting that the Han bricks used were usually 7 cm thick, some with losange-shaped decoration on the surfaces (Figure 4.3).

Accelerator Mass Spectrometry (AMS) radiocarbon dating on charcoal fragments taken from the various levels of the site and from the bottom of some retorts shows that the site dates to from the fifteenth century to the first half of the seventeenth century (Table 4.1 and Figure 4.5), corresponding with the Ming Dynasty (AD 1368-1644).

Figure 4.3 Two Han bricks found at Miaobeihou during fieldwork in 2009. Note the losange-shaped decoration on the surface of the brick on the left.

Table 4.1 AMS radiocarbon dates of charcoal samples from Miaobeihou by Peking University AMS lab, calibrated by OxCal 4.1 and IntCal 09 (Liu *et al.* 2007).

Lab code	**^{14}C date (BP, 1σ)**	**Calibrated date (AD)**	
		68.2% probability	**95.4% probability**
BA04196	400±40	1442 (57.4%) 1513, 1601 (10.8%) 1617	1432 (66.6%) 1527, 1556 (28.8%) 1633
BA04199	385±40	1447 (52.0%) 1519, 1594 (16.2%) 1619	1439 (58.2%) 1529, 1544 (37.2%) 1634
BA04200	345±40	1485 (24.1%) 1525, 1557 (44.1%)1632	1460 (95.4%) 1641
BA04201	325±40	1512 (53.9%) 1601, 1616 (14.3%) 1640	1468 (95.4%) 1647
BA04203	385±40	1447 (52.0%) 1519, 1594 (16.2%) 1619	1439 (58.2%) 1529, 1544 (37.2%) 1634
BA04204	330±40	1495 (19.5%) 1530, 1538 (37.2%) 1602, 1615 (11.4%) 1635	1465 (95.4%) 1645
BA04206	330±40	1495 (19.5%) 1530, 1538 (37.2%) 1602, 1615 (11.4%) 1635	1465 (95.4%) 1645

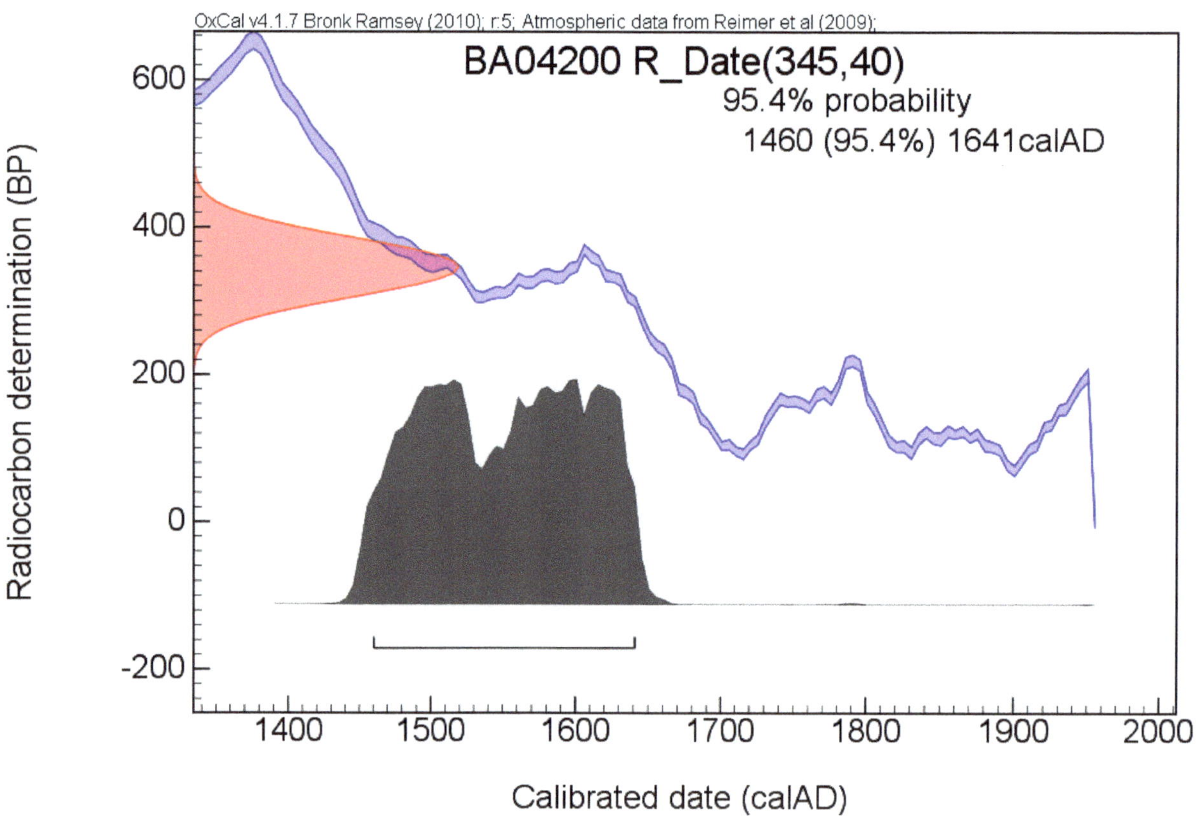

Figure 4.4 The calibration plot of BA04200, showing that the wider range of the calibrated date (AD 1460-1641, 95.4% probability) than the radiocarbon date (345±40 BP).

4.2.2 Results

This section presents the analytical results of the samples from Miaobeihou by category. The results of raw materials (minerals and coal) and products (metallic zinc) are given first, as they can assist in interpreting the results of retorts, slag and zinc-rich crusts in the later parts of this section.

4.2.2.1 Minerals

Some pieces of possible zinc ores were excavated at Miaobeihou. Two large pieces were recorded in the excavation report (Liu Haiwang pers. comm., 2009). One piece (2004CFXYMVT1:28) is an oxidic zinc ore, subspherical in shape, 18 cm long, 15 cm wide and 10 cm thick. It is dark greyish yellow on the outside, but the fracture appears brick red with some white and translucent minerals present in vesicles (Figure 4.5). The other one (2004CFXYMIT1307③:1) is 20 cm long, 15 cm wide and 8 cm thick. It is brown on the surface, but looks blackish with a metallic lustre inside (Figure 4.6). The excavators suspect that this might constitute a lead-zinc ore, but the supposition cannot be confirmed as this piece was not available for analysis. Three mineral samples from Miaobeihou were analysed (Figure 4.7).

Figure 4.5 One piece of zinc ore (2004CFXYMVT1:28) from Miaobeihou.

Figure 4.6 One piece of zinc ore (2004CFXYMIT1307③:1) from Miaobeihou. One corner is broken, showing the dark interior.

YMO1 and YMO2 have similar chemical and mineralogical compositions (Table 4.2 and 4.3). They are rich in ZnO (31-40%) and FeO (24-26%), and contain 5-6% SiO_2, around 0.5% of both CaO and MgO, and around 0.2% Al_2O_3 and CdO, with analytical totals of 64-71%. The XRD results show that the dominant mineral is smithsonite ($ZnCO_3$). YMO1 also contains hemimorphite [$Zn_4Si_2O_7(OH)_2{\cdot}H_2O$] and goethite [FeO(OH)]; YMO2 has hydrozincite [$Zn_5(CO_3)_2(OH)_6$], hemimorphite, dolomite [$CaMg(CO_3)_2$] and hematite (Fe_2O_3).

The matrices display finely intergrown oolithic structures of yellowish or reddish minerals, goethite or hematite, and white minerals, smithsonite (Figure 4.8). The matrices also contain significant concentrations of ZnO and FeO, together with a few percent of CaO, MgO and SiO_2.

Minerals with the appearance of two kinds of zinc carbonates seem to have grown in cavities. One kind is transparent and present in small pores (Figure 4.9 left). This phase is dominated by ZnO (60-65%) with a low analytical total (64-70%), indicating that it might be smithsonite. The other kind, only found in YMO2, is white and present in larger pores, where numerous tiny flaky minerals formed layered radial aggregations (Figure 4.9 right). The structure of this mineral is similar to that of hydrozincite, which is reported to be abundant in the Laochangping lead-zinc deposits (Chen and Zhang 1943).

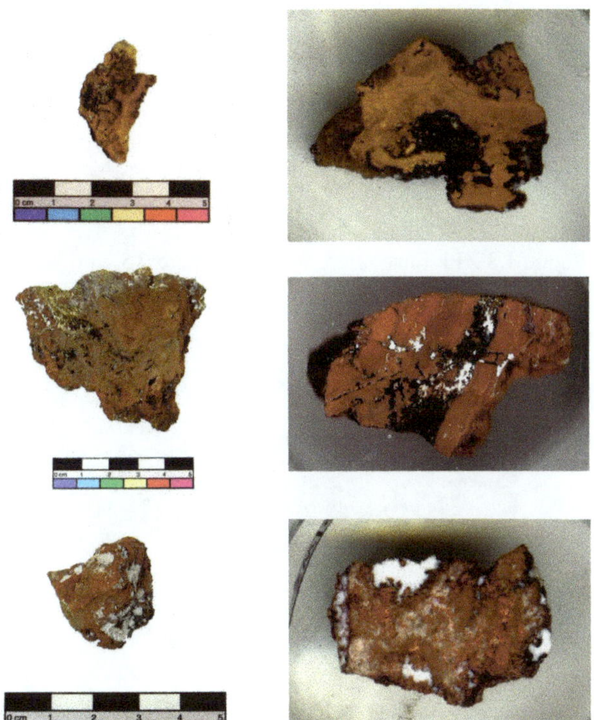

Figure 4.7 Mineral samples (left column) from Miaobeihou and their cross sections (right column). Top: YMO1; middle: YMO2; bottom: YMO3.

Table 4.2 Average bulk compositions of three mineral samples from Miaobeihou (wt%). Analyses on polished sections for areas of ~2 by ~2.5 mm by SEM-EDS. Both raw (top half) and normalised (bottom half) results are presented. '-' means 'not detected'.

Sample	MgO	Al_2O_3	SiO_2	CaO	FeO	ZnO	CdO	PbO	Total
YMO1	0.6	0.1	5.7	0.5	24.5	39.7	0.2	-	71.3
YMO2	0.3	0.2	5.1	0.6	26.3	31.3	0.2	-	64.1
YMO3	0.3	0.8	68.1	0.5	3.0	32.9	0.1	0.1	105.9
YMO1	0.8	0.1	7.6	0.7	34.5	56.0	0.3	-	100.0
YMO2	0.5	0.3	7.7	0.9	41.0	49.2	0.3	-	100.0
YMO3	0.3	0.8	63.5	0.5	3.1	31.5	0.1	0.1	100.0

Table 4.3 Mineralogical compositions of three mineral samples from Miaobeihou. Analyses by XRD.

Sample	Mineral
YMO1	smithsonite, hemimorphite, goethite
YMO2	smithsonite, hydrozincite, hemimorphite, dolomite, hematite
YMO3	hemimorphite, quartz, dolomite, hematite

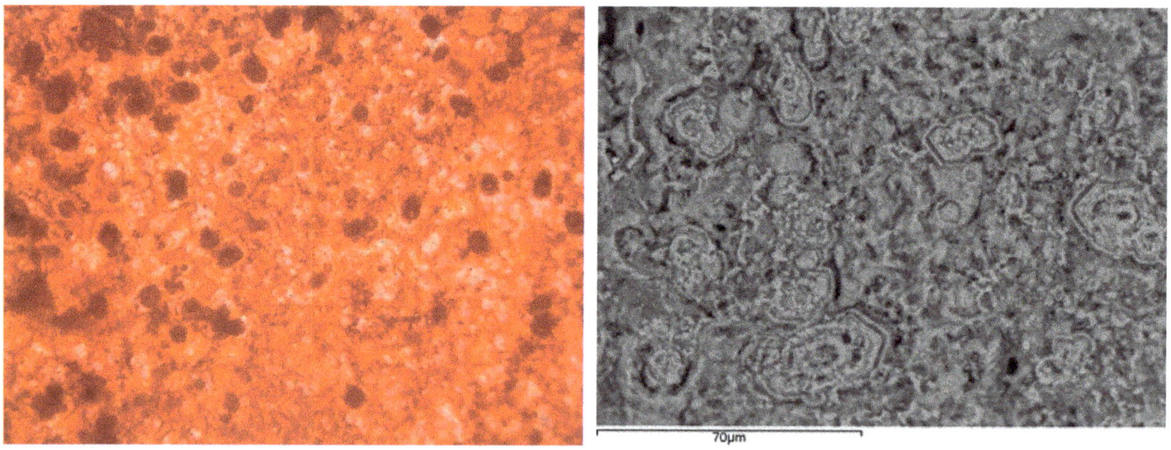

Figure 4.8 XPL image (left, width of picture 0.5 mm) and BSE image (right) of finely intergrown oolithic structures of the matrix of YMO2.

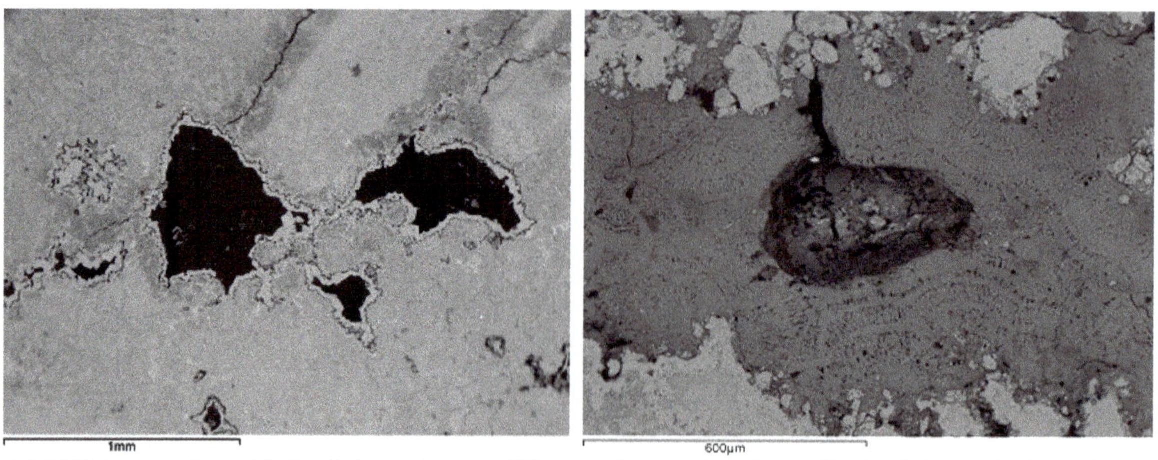

Figure 4.9 BSE images of two kinds of zinc carbonates filling cavities: smithsonite in YMO1 (left) and hydrozincite in YMO2 (right).

Besides zinc carbonates, there are white minerals with a zinc to silicon atomic ratio of around 2:1, which is likely to be hemimorphite. Occasionally, a layer of zinc carbonates can be seen between the matrices and the hemimorphite, indicating that the hemimorphite formed after the zinc carbonates (Figure 4.10).

Iron is enriched within some pores and at the interfaces of the matrices. This suggests that a few iron- and zinc-rich silicates with concentric structures were first deposited, sometimes followed by a layer of iron-rich minerals. It can sometimes be seen that the iron-rich minerals are embedded within hemimorphite minerals, again indicative of the later formation of hemimorphite.

Compared to YMO1 and YMO2, YMO3 shows similar concentrations of ZnO (33%), but far more SiO_2 (68%), less FeO (only 3%), more Al_2O_3 (0.8%) and a little PbO (0.1%), with an analytical total of about 100% (Table 4.2). Its XRD result shows that YMO3 is dominated by hemimorphite and quartz, and also contains dolomite and hematite (Table 4.3). Microscopically, hemimorphite is embedded among quartz grains, and in some areas large crystals develop (Figure 4.11). Within some quartz grains, there are a few small carbonate minerals containing 49-71% CdO, 20-34% CaO, 3-6% ZnO and less than 1% of SiO_2, PbO, As_2O_3 and SeO_2.

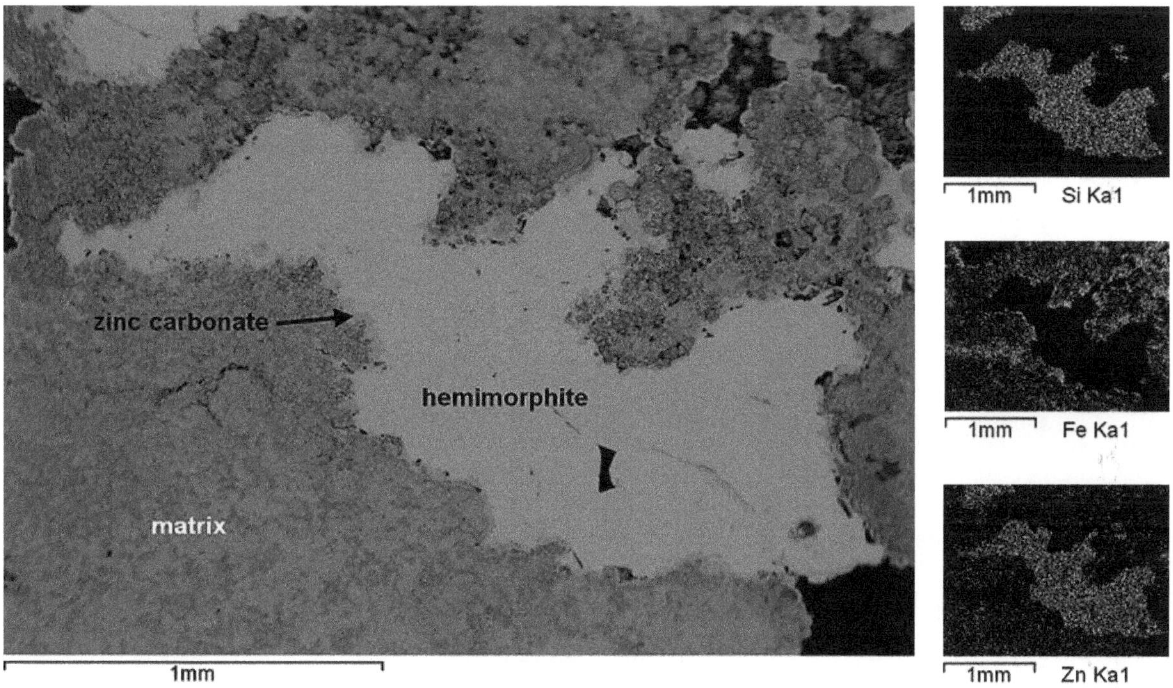

Figure 4.10 Hemimorphite embedded within the matrix of YMO1. Left: BSE image; right: X-ray maps of silicon, iron and zinc concentrations.

Figure 4.11 BSE images of YMO3. Left: the large crystals of hemimorphite; right: hemimorphite and zinc-rich silicate (light grey) among quartz grains, and tiny cadmium and calcium carbonates within quartz grains.

These mineral samples are likely to have derived from gossans of lead-zinc deposits. Primary sulphidic ore minerals of lead-zinc deposits are typically zinc sulphide (sphalerite, ZnS), lead sulphide (galena, PbS), and iron sulphides (pyrite and marcasite, FeS_2). In the upper parts of sulphidic ore deposits, secondary oxidic ore minerals form: iron sulphides are oxidised to iron oxides, which formed gossans, also known as an 'iron cap'; galena oxidises to lead carbonate (cerussite, $PbCO_3$); sphalerite oxidises to zinc carbonates (smithsonite and hydrozincite) and silicate (hemimorphite). As iron oxides are strong colorants, other oxidic ore minerals are usually concealed among gossans. This is the reason why these archaeological minerals from Miaobeihou look like iron ores. In actual fact, they also contain significant amounts of oxidic zinc minerals, such as smithsonite, hydrozincite and hemimorphite.

This raises to the question as to whether these archaeological minerals were used for smelting, or whether they were discarded by ancient craftspeople. Based on their high zinc contents, these minerals may be considered as ores that could have been employed for zinc smelting. By this rationale, they provide useful information on the nature of the minerals charged in the retorts, and thus their study may facilitate the understanding of the slag formed during smelting.

4.2.2.2 Coal

Coal is a combustible black or brownish-black sedimentary rock mainly consisting of carbonised plant matter. Plant bodies accumulating in swamps, protected from oxidisation and biodegradation, led to the formation of peat bogs. After peats were covered by sediments, the biochemical effects weakened and eventually ceased. This was followed by the coalification, transforming peat to lignite by diagenesis, which subsequently metamorphosed into bituminous coal and anthracite. Coal contains water, organic matter and inorganic minerals. Organic matter in coal is primarily composed of carbon, hydrogen, oxygen, nitrogen and sulphur. Inorganic minerals in coal include clay, sulphides, phosphates, oxides and carbonates (Diessel 1992).

Coal was used as reducing agents inside retorts and as fuel outside retorts, as documented in most traditional zinc smelting practices (see Section 2.3). The common types of coal used as reducing agents were anthracite and high-quality bituminous coal, which contain high levels of carbon but little ash or sulphur. The quality of the coal used for fuel was not strictly controlled.

At Miaobeihou, coal preparation pits, coal fragments, coal powders and coal gangue were reported by the excavators. The typical blackish colour and layered structure of coal are diagnostic features. Some coal fragments still remain inside some of the excavated retorts. Eight coal fragments were collected for analysis. A few of them are black and covered with soil, while others are yellowish or reddish (Figure 4.12).

Although these coal samples were identified as such by macroscopic observation, they were found to bear large quantities of inorganic constituents. To show the proportions of inorganic and organic components in these samples, raw bulk compositions are presented, with analytical totals ranging from 34% to 93%. The coal samples contain large proportions of SiO_2 and Al_2O_3, and a few percent of CaO, FeO and K_2O (Table 4.4). Based on their SiO_2 to Al_2O_3 ratios, the eight samples can be grouped into three types, which are characterised below.

Type A: YMC1, YMC3 and YMC4

These three samples have SiO_2 to Al_2O_3 ratios of 1.5-1.6. YMC1 is relatively rich in light elements, which amount to 66%. It can be seen from BSE imaging that black areas, rich in organic matter, and grey areas, rich in other minerals, form a layered structure (Figure 4.13 left). In some areas rich in organic matter, tiny potassium feldspar minerals ($KAlSi_3O_8$) appear interspersed, forming a fine structure. This is also seen in YMC3 and YMC7. In YMC1, a few particles of pyrite appear entrapped between sedimentary layers (Figure 4.13 right), resulting in a relatively high sulphur content.

YMC3 and YMC4 have similar SiO_2 to Al_2O_3 ratios to YMC1, but contain much less organic matter, which is only present as a few layers between more substantial mineral layers (Figure 4.14). YMC3 contains a large amount of iron oxides, which are either scattered or present along cracks. Part of the sample is abundant in minerals with a silicon to aluminium atomic ratio of about 1:1, and around 2% of both K_2O and TiO_2. In YMC4 small barite minerals ($BaSO_4$) were identified between carbon-rich and silica-rich layers.

Type B: YMC2, YMC5 and YMC6

These three samples have SiO_2 to Al_2O_3 ratios of 2.6-3.4. They are mostly composed of mineral layers with some iron-rich minerals clustered together or filling cracks and pores. YMC2 has a thin layer of fluorapatite [$Ca_5(PO_4)_3F$] and a small area rich in quartz grains. YMC5 has a vitrified surface with semi-molten quartz grains, pyroxene, spinel, iron sulphide and iron oxides. Compared to the body, this layer is richer in MgO, K_2O, CaO and FeO (Table 4.4, YMC5v), an enrichment probably derived from the reaction with coal ash.

Type C: YMC7 and YMC8

These two samples have the highest SiO_2 to Al_2O_3 ratios (4.9-5.1). They are rich in silt-sized quartz and potassium feldspar (Figure 4.15 left). Minerals such as zircon ($ZrSiO_4$), rutile (TiO_2) and iron oxides were also identified. They contain thin layers of organic matter, some of which appear to have burned off. YMC8 has a vitrified surface with semi-molten quartz and plagioclase feldspar [$(Ca, Na)Al_{1-2}Si_{3-2}O_8$]. This layer, similar to the vitrified surface of YMC5, has higher levels of MgO, K_2O, CaO and FeO than the body (Table 4.4, YMC8v), probably introduced by the reaction with coal ash. Iron oxides are present on the surface of and also within the vitrified layer (Figure 4.15 right).

Table 4.4 Average bulk compositions of eight coal samples and vitrified surfaces of two of the coal samples from Miaobeihou (wt%). Analyses on polished sections for areas of ~2 by ~2.5 mm by SEM-EDS. v=vitrified surface. The raw data of coal samples are presented with the original analytical totals given in the penultimate column, while those of the vitrified surfaces of two samples (YMC5v and YMC8v) are normalised to 100%. '-' means 'not detected'.

Sample	Na_2O	MgO	Al_2O_3	SiO_2	P_2O_5	SO_3	Cl	K_2O	CaO	TiO_2	FeO	NiO	CuO	ZnO	BaO	Total	SiO_2/Al_2O_3
YMC3	0.1	0.6	28.7	42.7	0.3	1.3	0.1	2.0	0.9	1.0	2.2	0.1	-	0.1	0.1	80.2	1.5
YMC4	0.2	0.6	23.5	36.4	0.2	0.5	0.1	2.3	1.5	0.4	1.2	-	-	0.1	0.1	67.2	1.5
YMC1	0.1	0.4	10.9	17.9	0.2	1.4	-	1.5	0.3	0.3	0.6	0.1	-	0.1	0.1	33.9	1.6
YMC6	0.2	1.2	18.8	48.0	0.3	0.5	0.1	2.1	0.6	0.7	3.6	0.1	-	0.2	0.1	76.5	2.6
YMC5	0.3	1.2	21.0	58.9	-	0.4	0.1	3.7	0.4	0.7	2.8	-	0.1	0.1	0.1	89.8	2.8
YMC5v	0.7	2.3	18.9	59.9	0.3	0.1	-	5.9	3.4	1.3	7.2	-	-	-	-	100.0	3.2
YMC2	0.3	1.7	16.9	57.7	0.3	0.6	0.1	2.9	0.6	0.8	5.1	-	-	0.1	-	87.1	3.4
YMC8	0.2	0.8	14.6	71.0	0.2	0.4	0.1	2.4	0.2	0.7	2.0	-	-	0.1	0.1	92.8	4.9
YMC8v	0.6	2.9	15.6	64.9	0.1	-	-	3.5	5.5	1.0	5.4	-	-	0.5	-	100.0	4.2
YMC7	0.1	0.8	12.1	61.3	0.1	0.2	0.2	1.8	0.1	0.7	1.8	-	-	0.2	-	79.4	5.1

Figure 4.12 Left: coal *in situ* at Miaobeihou; right: coal samples from Miaobeihou.

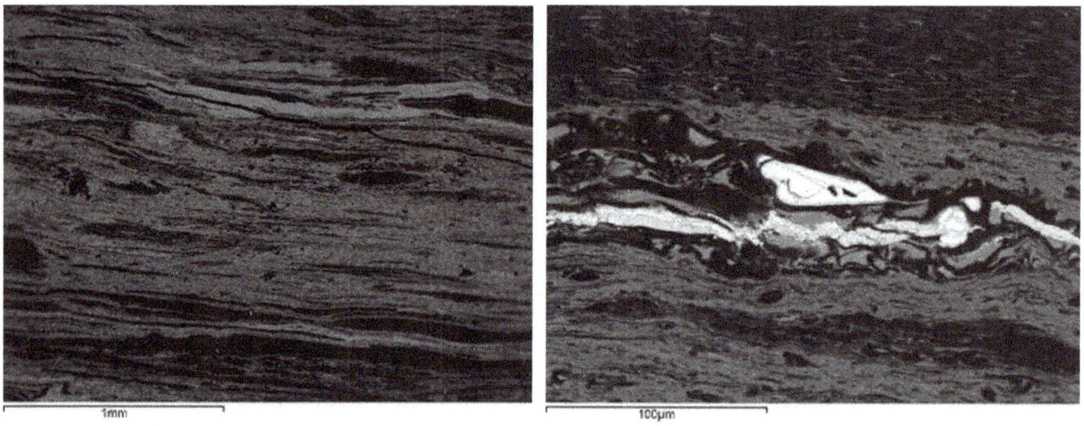

Figure 4.13 BSE images of YMC1, showing the layered structure (left) and pyrite (right).

Figure 4.14 BSE images of YMC3 (left) and YMC4 (right), showing small areas of carbon within the mineral matrices.

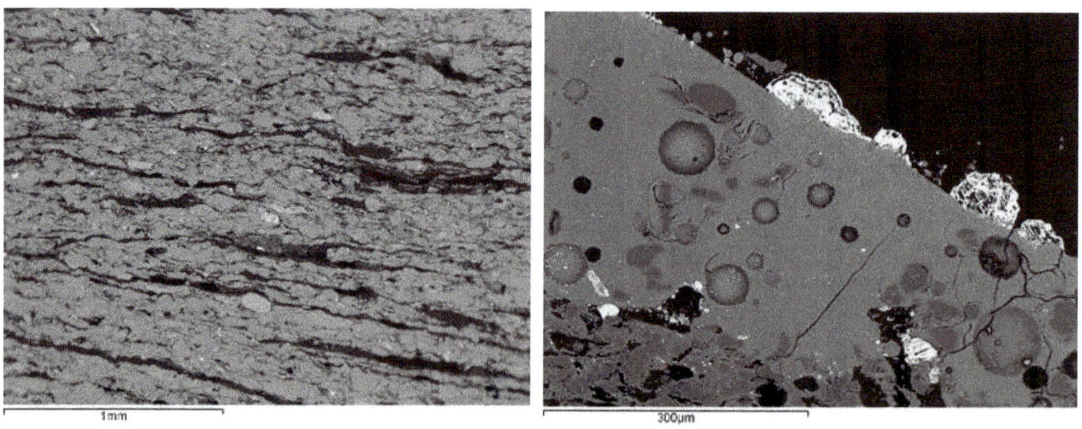

Figure 4.15 BSE images of the fabric of YMC7 (left) and the vitrified surface of YMC8 (right).

In summary, the so-called 'coal samples' studied here are of a very poor grade and thus unlikely to represent the actual quality of the bulk of the fuel/reducing agents employed. Instead they seem more likely to be unused low-quality coal (e.g. YMC1) or coal gangue/ash. Several samples clearly underwent high temperatures, as seen from their vitrified surfaces and partial burning of organic matter. These may have been the residues from the combustion of coal fuel, as they were also found adhering to the external surfaces of some retorts. Whatever the case, these results are useful in that they show the type of gangue/ash components in coal, which helps interpret the analytical results of the retorts and slag.

4.2.2.3 Zinc

In the early 1980s, some local villagers found over 10 silvery pieces of metal when they were digging building foundations at Miaobeihou. These pieces were located exactly at Area II of the site and placed on a stone slab. Three of them were rectangular, 20 by 15 cm and about 5 cm thick; the others were plano-convex, less than 10 cm long and about 6 cm wide. These metals were first taken for silver or lead; however, chemical analysis by atomic absorption spectroscopy (AAS) at Chongqing Iron and Steel Company identified them as metallic zinc of 95-98% purity. The only surviving rectangular ingot (Figure 4.16), now at FCBCH, was analysed by inductively coupled plasma with atomic emission spectrometry (ICP-AES) and found to contain 99.21% zinc and 0.72% lead (Liu *et al.* 2007).

During the recent excavation of Miaobeihou, a few small lumps of zinc metal were discovered. They are irregular in shape, a few centimetres long and bear a white patina and soil on the surfaces. Three of them were sampled (Figure 4.17), one of which was collected from the zinc-rich crust of a condenser fragment (YMZ2).

Figure 4.16 One of the zinc ingots discovered in Area II of Miaobeihou in the 1980s (photo by Liu Haiwang).

Figure 4.17 One of the three lumps of zinc metal from Miaobeihou (YMZ1).

These three metal samples are high purity metallic zinc with a few small lead-rich prills and intermetallic iron-zinc phases. Their EPMA-WDS results show that their main impurities are lead and iron. Due to the inhomogeneous distribution of inclusions of lead prills and iron-zinc intermetallic compounds, the iron and lead contents obtained cannot represent those of the whole sample. However, the data can provide rough estimations of their composition. YMZ1 and YMZ3 contain about 0.4% lead, while YMZ2 has less than 0.1% lead. Iron levels are all below 0.1%. All other elements are below the detection limits (0.03%) (Table 4.5). Laser ablation ICP-AES was also used to quantify the cadmium levels of these samples, but no signals were present for cadmium, indicating very low concentrations (<0.001%).

The lead-rich prills embedded in the zinc matrices are only a few micrometres in diameter. They are primarily composed of lead (60-90%) and zinc (5-30%) with a few percent of oxygen, and less than 0.3% of cadmium, arsenic and silicon. In YMZ1, evenly distributed tiny lead-rich prills up to 3 μm in length and intergranular corrosion (zinc oxide) clearly mark its coarse-grained isometric structure (Figure 4.18). No obvious structures were observed in the other two samples.

A few angular euhedral phases were found, which can be more easily observed in the corrosion. These euhedral phases generally contain around 90% zinc, 6.5% iron, 1-2% oxygen and 0.1-0.5% arsenic in YMZ2 and YMZ3. The phases in YMZ1 contain less iron (4.4-5.9%), but up to 1.8% cadmium. These phases are iron-zinc intermetallic compounds, the zeta (ζ) phase, $FeZn_{13}$, of the iron-zinc system. The ζ phase was identified previously in a Roman zinc tablet from Bern, Switzerland (Rehren 1996). In a zinc melt with only traces of iron, this is the first phase that crystallises at a temperature slightly above 420°C, the melting point of the pure zinc. The lower density ζ phase tends to float on the surface of the melt. When the temperature drops below 420°C, the eta (η) phase, pure zinc, solidifies.

The white patina on the surfaces is primarily composed of basic zinc carbonates. YMZ1 and YMZ3 have a surface layer of corrosion products, composed mainly of zinc carbonates. Platy crystals of zinc sulphates with grain sizes of up to 1 mm intrude deeply into the metal. A thin layer of zinc oxide appears between the metal and zinc carbonates in some parts of YMZ2 and YMZ3 (Figure 4.19). The outer parts exhibit earth minerals (quartz, feldspar, mica, etc.) and coal fragments, which were probably incorporated from soil during burial. YMZ2, however, shows different features. There are calcium carbonates within a zinc-rich crust consisting of zinc oxide and zinc carbonates. The notable feature here is the numerous tiny coal fragments forming a network within this crust (Figure 4.20).

Table 4.5 Average compositions of three zinc samples from Miaobeihou (wt%). Analyses on polished sections at areas of ~80 by ~120 μm by EPMA-WDS. '-' means 'not detected'.

Sample	Zn	Fe	As	Cu	Ag	Pb	In	Bi	Cd	Total
YMZ1	98.08	0.05	0.01	0.01	-	0.41	0.01	0.01	0.01	98.59
YMZ2	98.99	0.02	-	-	0.01	0.06	0.01	-	0.01	99.10
YMZ3	98.41	0.04	-	-	-	0.42	0.01	0.01	0.01	98.90

Figure 4.18 Left: XPL image of the coarse-grained isometric structure of YMZ1 (width of picture 2 mm); right: BSE image of YMZ1, showing intergranular corrosion (dark), iron-zinc intermetallic phases (arrows) and tiny lead-rich prills (white).

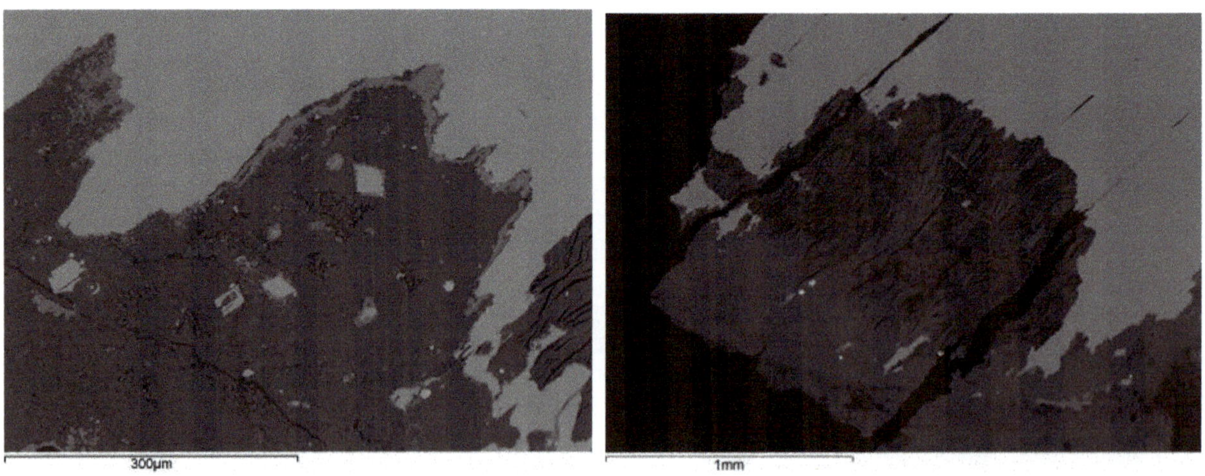

Figure 4.19 BSE images of the corrosion products of YMZ2. Left: a thin layer of zinc oxide between the metal (top) and zinc carbonates (bottom). A few iron-zinc intermetallic phases are embedded in zinc carbonates; right: platy crystals of zinc sulphates.

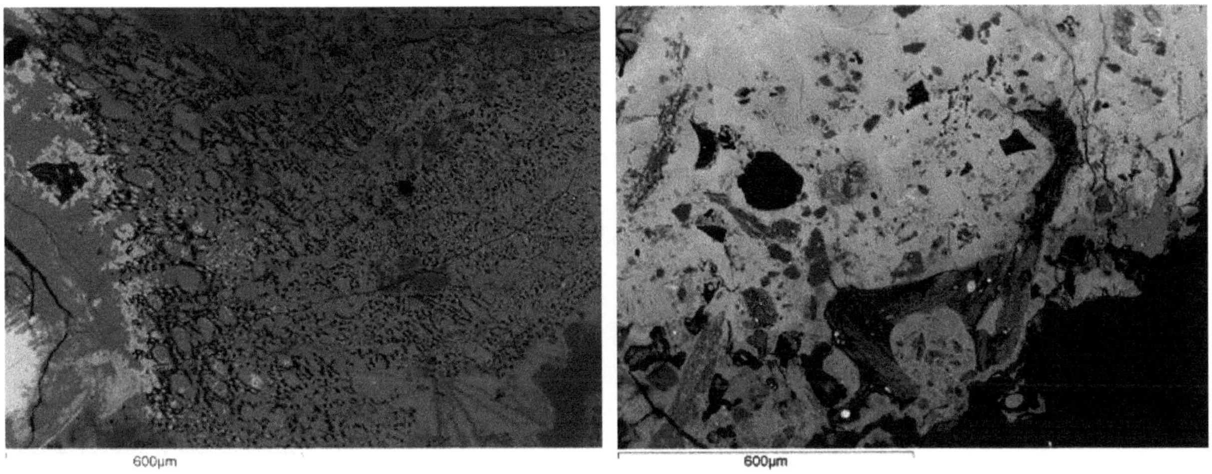

Figure 4.20 BSE images of the outer part of YMZ2. Left: the network of tiny coal fragments; right: large coal fragments within zinc-rich crusts.

4.2.2.4 Retorts

The retorts from Miaobeihou share similar forms and consist of four parts: pots, condensers, pockets and lids (Figure 1.6). The ceramic pots were identified as the main body of the retorts, which would have contained the raw materials. Cylindrical ceramic condensers were built onto the tops of the pots. Inside the condensers there are transversal ceramic partitions, named pockets, with a hole made on one side and a concave upper surface where the zinc would be collected as it condensed. The openings of the condensers were covered by disc-like lids. Analytical results of ceramic pots, condensers and pockets are presented below.

Pots

The pots are flat-bottomed jars, usually 25-29 cm in height. Their flat bases are circular with diameters ranging from 7 to 10 cm; the bodies open gradually to diameters of 11-16 cm and then close progressively; the rims are flared outwards with internal diameters of 6-9 cm. Their capacity is estimated at about 2 L. The pot bodies are about 0.6-1.0 cm thick and gradually become thinner from the bases to the rims.

The pots were formed by wheel throwing, as can be inferred from the surface markings noticeable on unused pots: (a) spiral grooves and ridges across the interior of the bases formed during the opening of the clay lump; (b) spiral grooves and ridges on the external and internal surfaces of the walls resulting from the lifting operation; (c) spiral marking or 'shell pattern' bases, indicating that the pots were cut from wheels when revolving rapidly (Figure 4.21). Throwing also led to the variations in the vertical wall thicknesses of the pots. Thicker walls near the bases were probably required to support the weight of the upper parts of the pots and their charge.

The surface colours of unused pots range from yellow, red and brown to grey. The colours of cross sections of some pot fragments are more complex and variable: yellow surfaces and margins resulting from oxidising atmospheres in the kiln; black cores likely produced by incomplete oxidising atmospheres during firing; black surfaces owing to short-time reducing conditions at the end of the firing. For used pots, the colours were influenced significantly by the high temperatures and strongly reducing conditions required by zinc smelting. They are usually bluish-grey, bluish-black or black (Figure 4.22). Some used pots are covered on the external walls with a luting clay which is now highly vitrified with embedded coal ash fragments, indicating that coal was the fuel employed to fire the retorts.

1. Chemical and mineralogical compositions

Middle sections of three unused and three used pot samples were analysed and found to have similar compositions and characteristics. The ceramic matrices are composed of 66-72% SiO_2, 19-24% Al_2O_3, 3-6% FeO with low levels of alkali and alkaline earth oxides, their sum being about 4% (Table 4.6). Bulk compositions show comparatively higher SiO_2 contents (70-78%) and slightly reduced levels of other oxides, resulting from the presence of abundant quartz inclusions.

Figure 4.21 Left: unused pots collected from Miaobeihou in August 2009; right: a bottom fragment showing 'shell pattern'.

Figure 4.22 Cross sections of unused pots (top left: YMM13; top right: YMM10) and used pots (bottom left: YMM9; bottom right: YMM1). The internal surfaces of these samples are facing up. YMM1 is vitrified on the both surfaces.

Table 4.6 Average matrix (top half) and bulk (bottom half) compositions of six pot fabrics from Miaobeihou (wt%), normalised to 100%. Analyses on polished sections at areas of ~150 by ~200 µm and ~2 by ~2.5 mm respectively by SEM-EDS. YMM1, YMM3 and YMM9 are used pots, while the others are unused. '-' means 'not detected'.

Sample	Na_2O	MgO	Al_2O_3	SiO_2	P_2O_5	SO_3	Cl	K_2O	CaO	TiO_2	MnO	FeO	ZnO
YMM1	0.4	0.6	19.1	72.3	0.1	-	-	2.1	0.3	0.9	-	4.2	-
YMM3	0.3	0.7	20.1	70.7	0.1	-	-	3.1	0.1	1.0	-	3.9	-
YMM9	0.2	0.8	22.1	67.0	0.1	-	-	2.7	0.2	1.1	-	5.8	-
YMM10	0.2	0.7	21.6	68.1	0.2	-	-	2.7	0.2	1.0	-	5.3	-
YMM13	0.3	0.7	23.6	65.9	0.2	-	-	2.8	0.2	1.0	-	5.3	-
YMM14	0.2	0.7	21.9	69.9	0.1	-	0.1	3.0	0.2	0.8	-	3.0	0.1
YMM1	0.3	0.5	14.8	77.7	0.2	-	-	1.7	0.3	0.8	-	3.6	0.1
YMM3	0.2	0.7	17.3	73.8	0.1	-	-	2.8	-	0.9	-	4.1	0.1
YMM9	0.2	0.6	18.7	69.9	0.1	0.1	-	2.4	0.2	1.0	-	6.8	-
YMM10	0.1	0.6	18.2	71.3	0.1	0.1	-	2.4	0.2	0.9	-	6.0	0.1
YMM13	0.2	0.5	18.4	72.1	0.1	-	-	2.2	0.3	0.8	0.1	5.3	-
YMM14	0.1	0.5	16.8	76.5	0.2	0.1	0.1	2.3	0.1	0.7	-	2.5	0.1

The fabrics contain 10-15 vol% of large ill-sorted sand inclusions ranging from 50 μm up to 2 mm in length. They are sub-angular and show low sphericity. Most of the large inclusions are internally cracked quartz grains, partially debonding from or dissolving into the surrounding ceramic (Figure 4.23 and 4.24). The internal cracking of the quartz grains is caused by the large thermal behaviour mismatch between the quartz grains and the surrounding vitrified matrices, indicating firing temperatures of around 1200 °C (Ohya *et al.* 1999; Martinón-Torres *et al.* 2008). In the three unused pots and the used pot YMM9, some of the large inclusions are rock fragments composed of intergrown quartz and alumina-rich phases. The latter phases contain 52-55% SiO_2, 35-39% Al_2O_3, 5-9% K_2O and 1-2% FeO and have at some point been molten (Table 4.7, Figure 4.24 and 4.25). These phases are possibly muscovite [$KAl_3Si_3O_{10}(OH)_2$]. A few feldspar grains, mostly potassium feldspar, were identified; some had been semi-molten, so their K_2O levels were lower than expected (Table 4.8, Figure 4.26). In the other two used pots, no such rock fragments were found, probably because the alumina-rich phases had fully melted and dissolved into the matrices when exposed to higher temperatures during smelting.

In YMM9 and YMM13, a small number of rounded inclusions were identified (Figure 4.25 and 4.27). They include numerous quartz grains which had been semi-molten, the sizes and abundance of which are similar to those in the main body matrices. These inclusions are generally enclosed by a ring of separation between them and the surrounding matrices, albeit partially dissolving into the matrices. They can be interpreted as unhydrated lumps of powdered base clay which were not well mixed with the wet clay.

In addition, ferruginous concretions exist in abundance and in considerable sizes (up to 2 mm large). Some of the ferruginous concretions had been molten, precipitating iron-rich crystals. The matrices are abundant in smaller partially dissolved quartz grains, which are likely to have derived from the clay rather than being intentional temper. A few small inclusions, such as zircon, rutile and monazite [(Ce, La, Pr, Nd, Th, Y)PO_4], were also found in the fabrics.

The mineralogical compositions of two pot samples were analysed by XRD. The unused pot YMM13 contains quartz, mullite ($Al_6Si_2O_{13}$) and hematite, while the used pot YMM2 has quartz, mullite and cristobalite. As seen from their microstructures, quartz is the main type of mineral in both samples. The presence of hematite in YMM13 is consistent with its reddish colour (Figure 4.22 top left), and suggests that it was fired in an oxidising atmosphere. Mullite forms from the decomposition of kaolinite at temperatures ≥1100 °C (Kerr and Wood 2004, 59; Martinón-Torres *et al.* 2006). The presence of mullite in both samples indicates that they were both subjected to temperatures higher than 1100 °C. Cristobalite is a high-temperature polymorph of silica, starting to form from kaolinite decomposition at about 1200 °C (Eramo 2005, 585; Lee *et al.* 1999). The presence of cristobalite in the used pot but its absence in the unused one suggest that the used one experienced a higher temperature (≥1200 °C) than the unused one (<1200°C).

Figure 4.23 BSE image of used pot YMM1, showing internally cracked quartz inclusions.

Figure 4.24 BSE image of used pot YMM14, showing internally cracked quartz inclusions (dark grey), ferruginous concretions (bright grey) and large pores (black). The large inclusion on the top left is a rock fragment composed of quartz and alumina-rich phases.

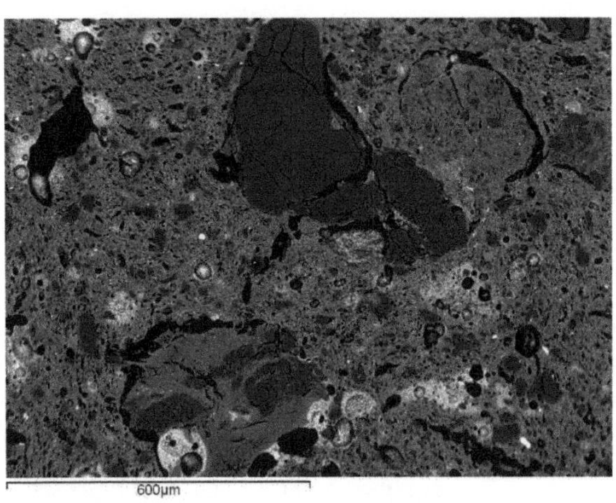

Figure 4.25 BSE image of used pot YMM9, showing some large inclusions. The inclusion on the bottom left is a rock fragment; the one in the middle at the top is mainly quartz; the one on the top right is a clay lump.

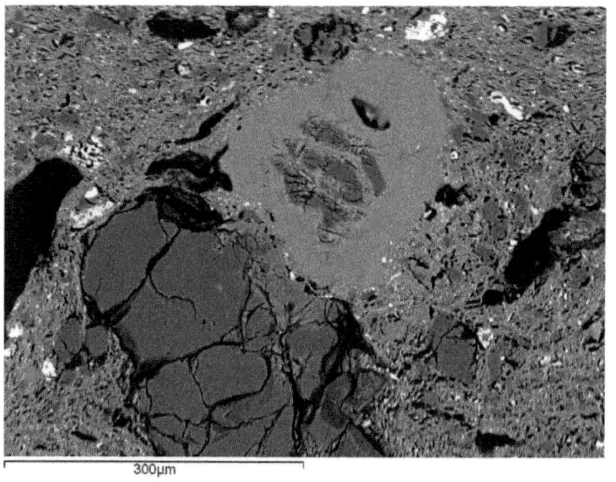

Figure 4.26 BSE image of unused pot YMM10, showing a quartz grain (dark grey) and a potassium feldspar grain (light grey). The dark grey phase within potassium feldspar is albite ($NaAlSi_3O_8$).

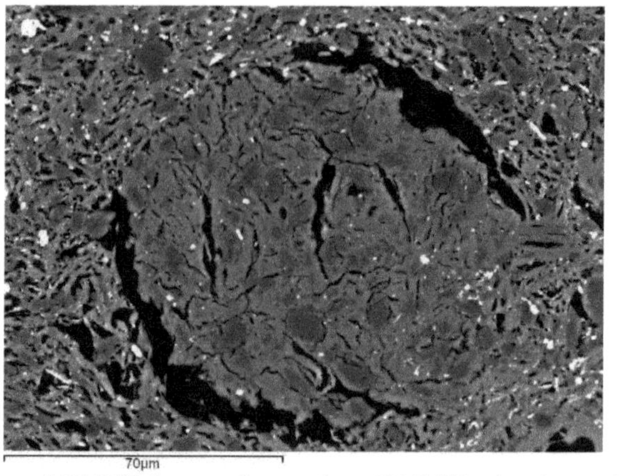

Figure 4.27 BSE image of unused pot YMM13, showing a clay lump.

Table 4.7 Compositions of some alumina-rich phases in some pot samples from Miaobeihou (wt%), normalised to 100%. Analyses on polished sections by SEM-EDS. '-' means 'not detected'.

Samples	Na_2O	MgO	Al_2O_3	SiO_2	K_2O	CaO	TiO_2	FeO
YMM9	0.8	0.5	36.3	52.8	6.9	0.2	0.3	2.3
	0.5	0.7	35.6	54.8	7.0	0.2	-	1.1
YMM10	0.4	0.3	38.6	54.1	5.2	0.1	0.1	1.1
	0.6	0.7	36.0	53.9	7.0	0.2	0.2	1.3
YMM13	0.9	1.0	34.9	52.7	7.6	0.2	0.3	2.2
	0.7	0.5	37.3	53.7	6.6	0.2	-	1.0
YMM14	0.6	0.4	35.1	54.5	8.3	-	-	1.2
	0.9	0.4	37.1	52.9	6.9	0.2	0.3	1.2
	0.8	0.6	36.5	52.6	7.4	0.1	0.3	1.7

Table 4.8 Compositions of some potassium feldspar grains in some pot samples from Miaobeihou (wt%), normalised to 100%. Analyses on polished sections by SEM-EDS. '-' means 'not detected'.

Samples	Na_2O	MgO	Al_2O_3	SiO_2	K_2O	CaO	TiO_2	FeO
YMM9	1.4	0.1	17.4	67.5	12.1	0.3	0.1	1.0
YMM10	1.6	0.3	18.1	68.5	10.5	0.3	-	0.7
	0.8	-	17.8	67.6	12.9	0.5	-	0.5
YMM13	0.7	-	17.7	66.9	14.3	0.2	-	0.3
$KAlSi_3O_8$			18.4	64.7	16.9			

2. Refiring experiments

Two pot samples YMM13 and YMM8 were refired in both oxidising and reducing atmospheres in order to serve as references for estimating the original firing temperatures and operating temperatures of the pot samples. The extents of vitrification seen in the refired YMM13 and YMM8 samples suggest that, at equivalent temperatures, the pot fabrics are more vitrified in a reducing atmosphere than in an oxidising atmosphere and that longer firing enhances the vitrification, resulting in larger bloating pores (Figure 4.28).

To estimate the temperatures at which the pots were originally fired in ceramic kilns prior to smelting, the degree of vitrification of the unused pot samples (YMM10, YMM13 and YMM14) was compared to that of the sample YMM13, refired in an oxidising atmosphere. All the matrices exhibit structures of extensive to continuous vitrification with fine bloating pores, indicating prefiring temperatures of 1100-1200 °C. To estimate the temperatures to which the pots were exposed in zinc smelting furnaces, the degree of vitrification of the used pot samples (YMM1, YMM3 and YMM9) was compared to that of sample YMM8, refired in a reducing atmosphere. YMM9 exhibits the structure of continuous vitrification with medium bloating pores, suggesting that it was exposed to 1200 °C during smelting. YMM1 and YMM3 show continuous vitrification structures with coarse bloating pores, indicating that they were exposed to temperatures 50 °C higher than 1200 °C, or at least 1200 °C but for a longer time. Therefore, it can be inferred that the smelting temperatures were similar to, or slightly higher than, the prefiring temperatures of the pots, estimated at about 1200 °C (Figure 4.29). The temperatures estimated from the refiring experiments are generally consistent with those indicated by the internal cracking of quartz grains and the presence/absence of mullite and cristobalite.

In order to assess whether there were any significant thermal gradients across the height of the pots, the degree of vitrification of several samples from the top and bottom was also investigated. YMT3 from the top and YMB3 from the bottom show similar degree of vitrification to that of the middle parts. However, two samples from the bottom (YMB2 and YMB4) display large bloating pores (10-50 μm in length), suggesting exposure to higher temperatures than 1200 °C. YMB1 exhibits some areas with very large bloating pores up to 1 mm in diameter (Figure 4.30), indicating an even higher temperature.

3. Vitrified surfaces

The used pot YMM1 is vitrified on both surfaces. The internal vitrified surface has a composition similar to the pot fabrics, with higher levels of K_2O, ZnO and BaO (Table 4.9), suggesting that it formed through the reaction of the pot and the charge. Coal or charcoal ash potentially contributed to elevated levels of K_2O, while the ZnO and BaO are likely to have derived from oxidic zinc ores with barite as gangue minerals. Tiny metallic iron prills and semi-molten quartz are scattered within the internal vitrified layer, denoting strongly reducing condition (Figure 4.31 left).

The external vitrified surface contains higher levels of FeO, MgO, CaO, K_2O, ZnO and BaO. CaO, MgO and K_2O were mostly likely introduced by the reaction with coal ash. The elevated level of FeO of this layer could also have originated from coal ash, similar to the vitrified surfaces of the coal samples (see Section 4.2.2.2). Some droplets of iron sulphide within the layer could also have originated from the coal. Enrichment in these oxides has also been found in external vitrified surfaces of glass melting crucibles and cementation brass making crucibles fuelled by coal in the seventeenth-century England (Dungworth 2003; Dungworth and Wilkes 2010). On the outermost surface of the layer, a few iron oxide particles are embedded, which were possibly introduced from contact with iron tools used to handle the retorts. There are also some semi-molten quartz grains and tiny crystals of spinels [(Fe, Mg, Zn)Al_2O_4] in this layer (Figure 4.31 right). The minor amounts of ZnO would have derived from the zinc vapour escaping from the retorts.

Condensers

Besides large quantities of pot sherds, cylindrical accessories were also found in a significant proportion among the zinc smelting assemblage. Although most are fragments, a few are still connected to pot rims. As seen from some intact retorts, the condenser openings have slightly larger diameters than the pots themselves; the condensers are about 5 cm in height above the pot rims. Condenser fragments usually preserve the rim ends, which are thinner than the ends connected to the pots. The condensers show a variety of colours ranging from yellow, brown and grey to black. Whitish or greenish substances adhere to the internal surfaces and to the rims of the condenser fragments, while only a limited amount of the white deposits appear on the external surfaces (Figure 4.32). A few condenser fragments still preserve parts of the pockets attached to their internal surfaces (Figure 4.33).

In contrast with the pots, the condenser fabrics show high levels of ZnO contamination, varying from 6% to 38%. To estimate the original compositions of unused condensers, the ZnO contents were omitted, and the data was renormalised to 100%. Compared to the matrix compositions of the pot fabrics, the condensers show more variable compositions: similar levels of SiO_2 (mostly 64-70%), lower Al_2O_3 (13-17%), but higher FeO (6-9%), K_2O (2-6%), CaO (1-3%) and MgO (1-2%). They have also slightly higher amounts of SO_3, P_2O_5 and Cl (Table 4.10).

In terms of their microstructures, the condensers contain over 30 vol% of fair-sorted inclusions. Most of them are angular quartz grains, ranging from 10 μm up to 500 μm in length (Figure 4.34). There are also a significant number of feldspar grains which show no evidence of thermal distortion. The most remarkable inclusions are coal ash fragments with layered structures, similar to the 'coal' samples analysed (see Section 4.2.2.2). These fragments vary in sizes, up to 2 mm long. A few carbon-rich coal ash fragments were also identified due to their black colour under BSE imaging, low analytical totals and low levels of sulphur (Figure 4.35). In YMT3 and YMT8, there are a few rock fragments composed mainly of silt-sized quartz and feldspar grains (Figure 4.36 left), possibly from Type C 'coal' (see Section 4.2.2.2). Three samples (YMT1, YMT3 and YMT17) contain higher proportions of large coal ash fragments (Figure 4.34 right), resulting in their relatively higher contents of CaO. A prill of iron sulphide was found in YMT17, which likely derived from the coal. There are only a few coal ash fragments in the other samples (Figure 4.34 left).

In some parts of the condensers, zinc-rich phases, mostly zinc oxide and zinc carbonates, were deposited within the matrices and filling cracks (Figure 4.36 right). Even some mineral inclusions are heavily contaminated with zinc oxide. The zinc-rich condenser fabrics contrast with the pot fabrics which are almost free of zinc.

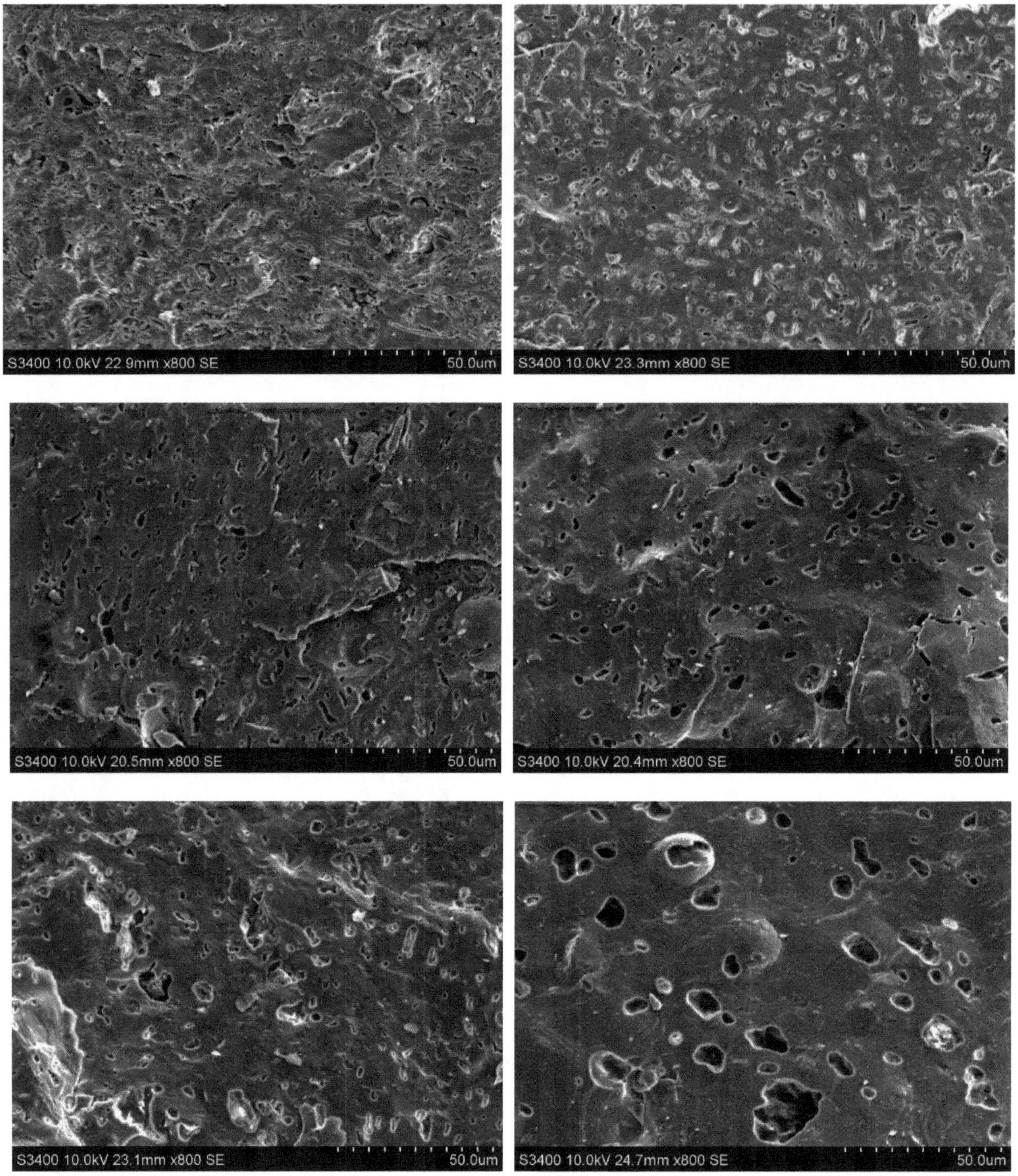

Figure 4.28 SE images of the fresh fractured surfaces of refired pots YMM13 and YMM8 taken by a Hitachi S-3400N SEM. The left column shows the degree of vitrification of pot YMM13, refired in an oxidising atmosphere. Top left: extensive vitrification (V), refired at 1100 °C for 1 h; middle left: continuous vitrification with fine bloating pores [CV(FB)], refired at 1200 °C for 1 h; bottom left: continuous vitrification with medium bloating pores [CV(MB)], refired at 1200 °C for 4 h. The right column shows the degree of vitrification of pot YMM8, refired in a reducing atmosphere. Top right: continuous vitrification with fine bloating pores [CV(FB)], refired at 1100 °C for 1 h; middle right: continuous vitrification with medium bloating pores [CV(MB)], refired at 1200 °C for 1 h; bottom right: continuous vitrification with medium to coarse bloating pores [CV(MB/CB)], refired at 1200 °C for 4 h.

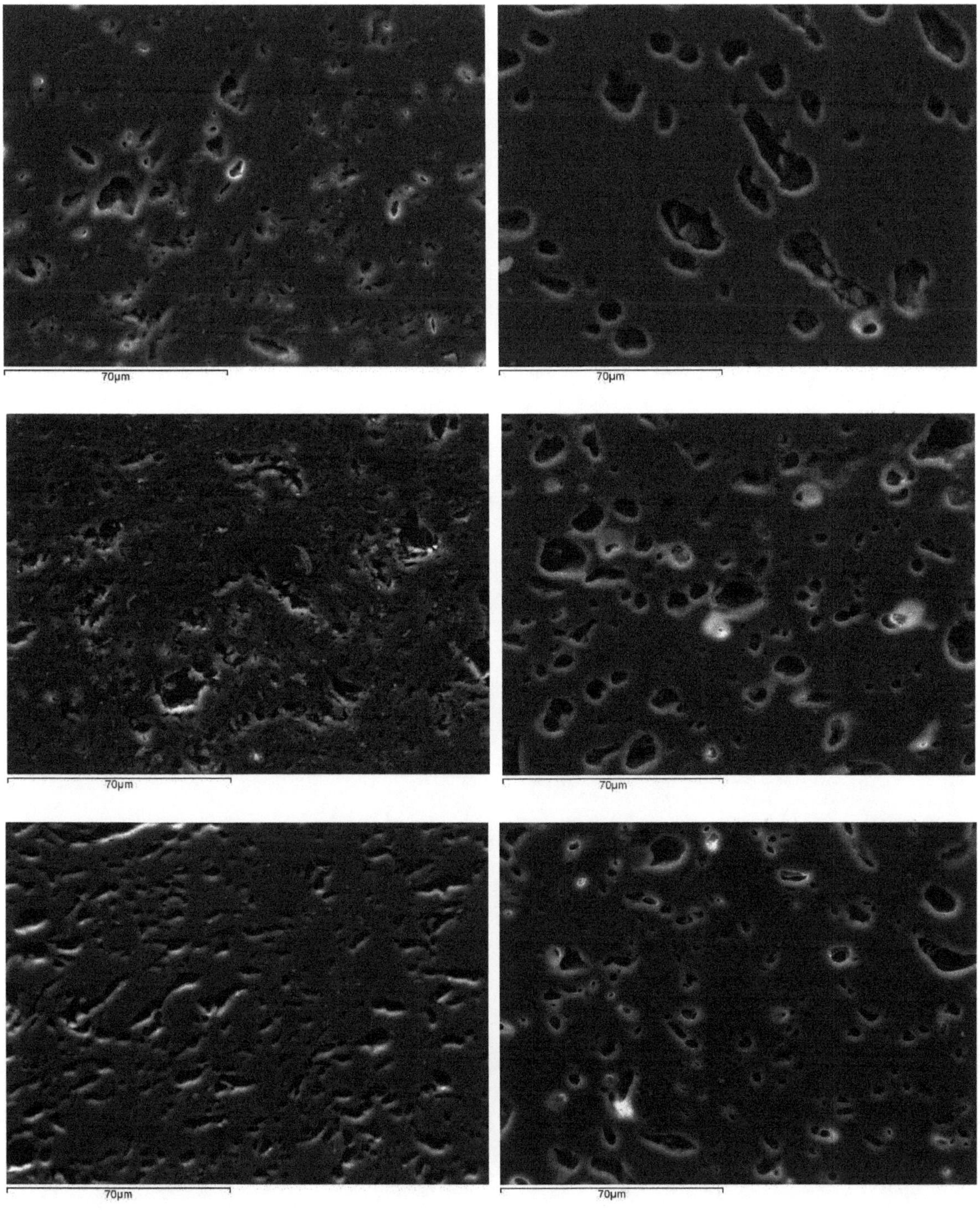

Figure 4.29 SE images of cross sections of six pot samples from Miaobeihou. The left column shows the degree of vitrification of unused pot samples. Top left: continuous vitrification with fine bloating pores [CV(FB)] in YMM10; middle left: extensive vitrification (V) in YMM13; bottom left: continuous vitrification with fine bloating pores [CV(FB)] in YMM14. The right column shows the degree of vitrification of used pot samples. Top right: continuous vitrification with coarse bloating pores [CV(CB)] in YMM1; middle right: continuous vitrification with coarse bloating pores [CV(CB)] in YMM3; bottom right: continuous vitrification with medium bloating pores [CV(MB)] in YMM9.

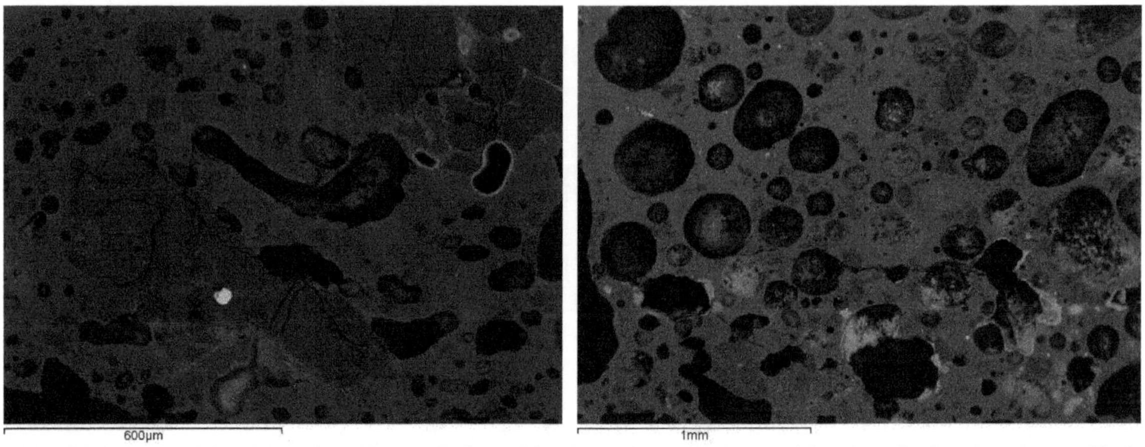

Figure 4.30 BSE images of YMB4 and YMB1. Left: large bloating pores with iron oxides deposited and an iron prill (bright) in YMB4; right: larger bloating pores with iron oxides deposited in YMB1.

Most of the condenser samples taken from the uppermost rim are not vitrified, except YMT20, which partly exhibits small bloating pores. This suggests that the top parts of condensers were generally exposed to temperatures lower than 800 °C. However, the lower parts are highly vitrified with large bloating pores, the sizes of which increase from the pot rims downwards (Figure 4.37). It is clear that there are distinct temperature gradients above and below the rims of the pots, which is consistent with the functioning of the retorts as discussed later: the condensers would have to be kept relatively cool, thereby facilitating the condensation of zinc vapour.

Condenser YMT8 has a glassy layer on the external surface (Figure 4.37 left and 4.38 left), which contains higher levels of CaO, MgO and K_2O than the condenser fabric. This layer is likely to be the lute applied to seal the condenser to the pot itself. The equivalent material between the condenser and the pot in YMT3 has a similar composition to that of the condenser, although some large coal ash fragments were found in this heavily vitrified material (Figure 4.37 right and 4.38 right). This coal could have derived from the fuel used at Miaobeihou, as large coal ash fragments were frequently found embedded in the vitrified outer layers around the pots.

Table 4.9 Average compositions of vitrified surfaces of YMM1 (wt%), normalised to 100%. Analyses on polished sections for areas of ~400 by ~600 μm by SEM-EDS. i=internal surface; e=external surface; m=matrix. '-' means 'not detected'.

Sample	MgO	Al_2O_3	SiO_2	P_2O_5	SO_3	K_2O	CaO	TiO_2	MnO	FeO	ZnO	BaO
YMM1i	0.7	19.1	72.0	0.1	-	3.8	0.5	1.0	0.1	2.2	0.2	0.3
YMM1m	0.6	19.1	72.3	0.1	-	2.1	0.3	0.9	-	4.2	-	-
YMM1e	1.8	18.0	63.1	0.2	0.2	3.0	1.3	1.0	0.1	10.4	0.7	0.2

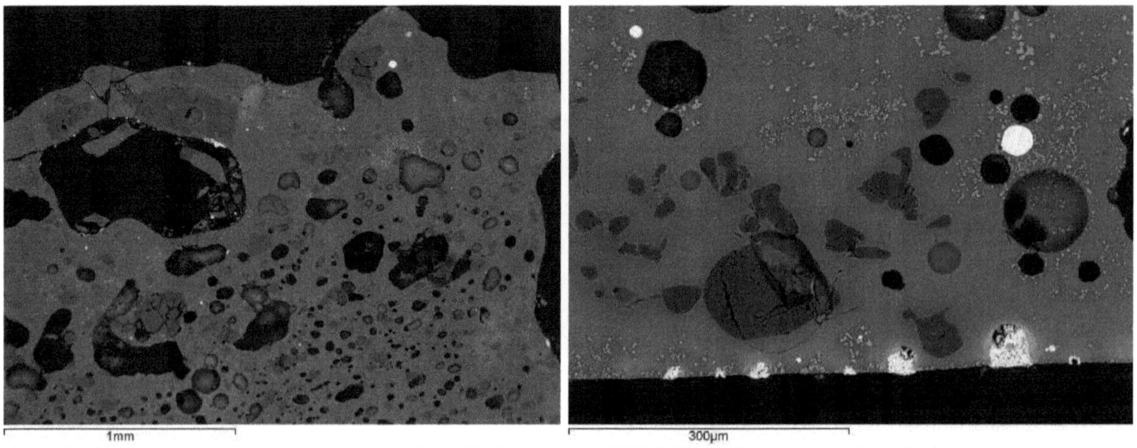

Figure 4.31 BSE images of the internal and external vitrified surfaces of YMM1. Left: quartz grains (dark grey) and tiny metallic iron prills (white); right: quartz grains (dark grey) and iron sulphide droplets (white), and several iron oxide particles (light grey) on the outermost surface.

Figure 4.32 Internal (top left) and external (top right) surfaces of condenser fragments from Miaobeihou. The scale bar is 5 cm. Vertical cross sections of three condenser fragments with their internal surfaces facing left (bottom).

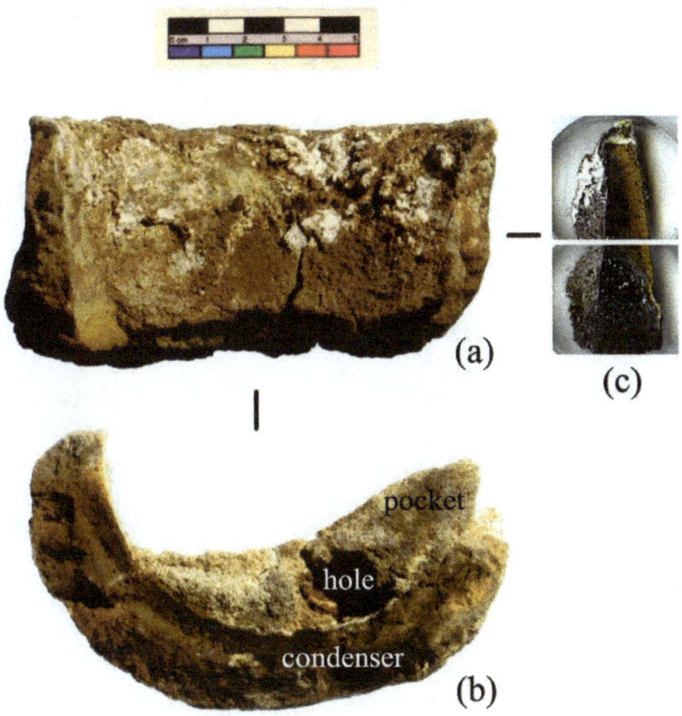

Figure 4.33 A condenser fragment YMT1 with part of the pocket and white deposits. (a) internal surface of the condenser; (b) bottom view; (c) vertical cross section.

Table 4.10 Average bulk compositions of eight condenser fabrics from Miaobeihou (wt%), normalised to 100%. Analyses on polished sections for areas of ~2 by ~2.5 mm by SEM-EDS. The rows at the bottom show the same results after omitting ZnO and renormalising to 100%. '-' means 'not detected'.

Sample	MgO	Al_2O_3	SiO_2	P_2O_5	SO_3	Cl	K_2O	CaO	TiO_2	MnO	FeO	ZnO
YMT1	1.3	12.3	46.1	0.2	0.3	0.1	2.9	2.0	0.5	0.1	5.8	28.4
YMT3	1.4	14.4	57.7	0.2	0.1	0.3	2.3	2.2	0.8	0.1	7.9	12.6
YMT8	0.8	9.1	50.8	-	0.1	-	1.6	0.8	0.7	-	4.1	32.0
YMT12	1.2	13.7	55.8	0.2	0.3	0.1	2.9	0.9	0.8	-	5.7	18.4
YMT17	1.3	16.5	63.8	0.1	0.1	0.1	2.5	1.9	0.9	0.1	7.0	5.7
YMT19	1.0	10.2	43.1	0.2	0.6	0.2	1.2	0.9	0.5	0.1	3.7	38.3
YMT20	0.9	11.5	56.0	0.3	1.1	0.2	4.7	0.9	0.7	0.1	5.0	18.6
YMT22	1.1	14.5	59.8	0.3	0.2	-	2.1	1.3	0.9	0.1	7.0	12.7
YMT1	1.8	17.2	64.5	0.2	0.4	0.2	4.0	2.7	0.7	0.2	8.1	
YMT3	1.6	16.5	65.9	0.2	0.1	0.4	2.7	2.5	0.9	0.2	9.0	
YMT8	1.2	13.4	74.5	-	0.1	0.1	2.4	1.2	1.0	0.1	6.0	
YMT12	1.5	16.8	68.3	0.2	0.4	0.1	3.6	1.1	1.0	-	7.0	
YMT17	1.4	17.5	67.6	0.1	0.1	0.1	2.7	2.0	1.0	0.1	7.4	
YMT19	1.6	16.5	69.9	0.3	1.0	0.3	1.9	1.5	0.8	0.2	6.0	
YMT20	1.1	14.1	68.8	0.4	1.4	0.2	5.8	1.1	0.9	0.1	6.1	
YMT22	1.3	16.6	68.6	0.3	0.2	-	2.4	1.5	1.0	0.1	8.0	

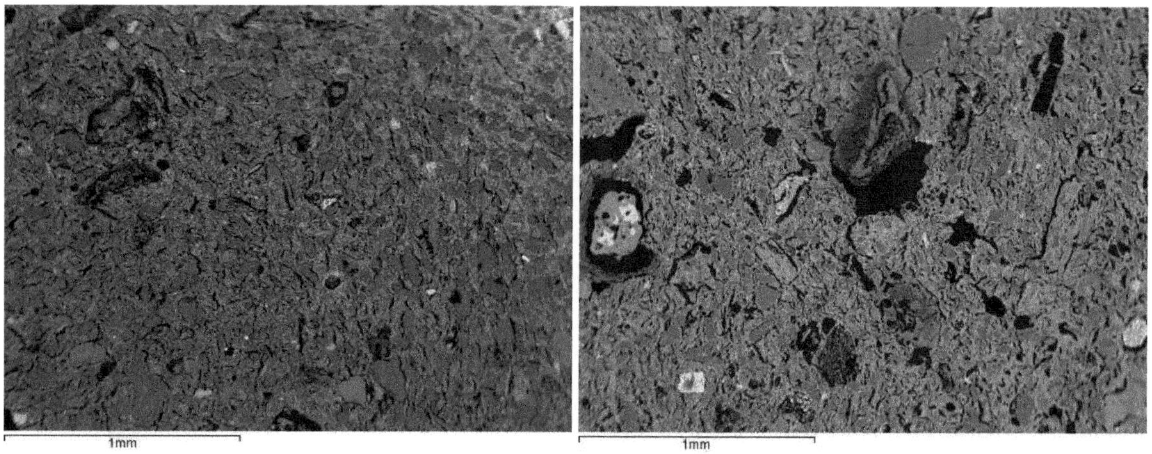

Figure 4.34 BSE images of condenser YMT22 (left) and YMT3 (right).

Figure 4.35 BSE images of different coal ash fragments in condenser YMT17, showing a burnt coal ash fragment (left) and a mineral-rich coal fragment with a small area rich in carbon (right).

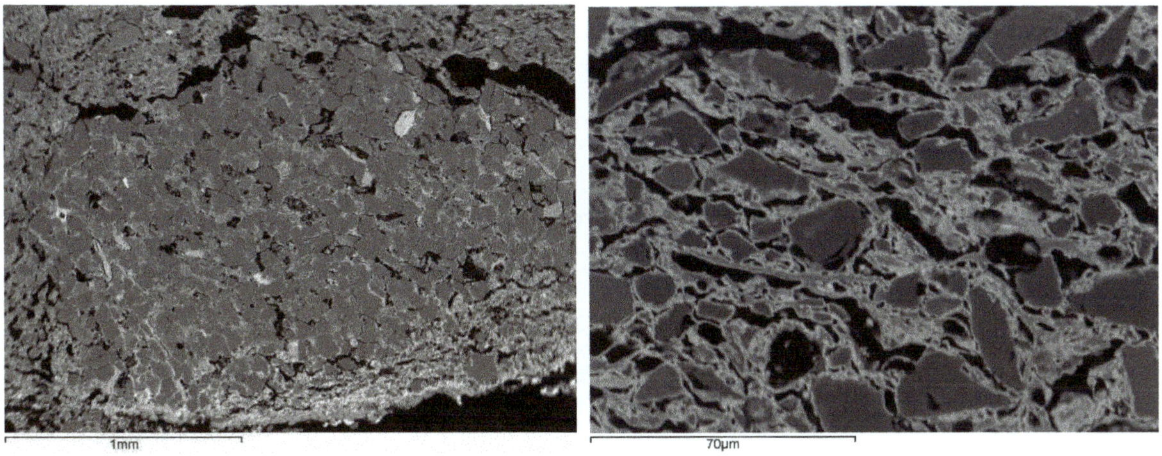

Figure 4.36 BSE images of condenser YMT8, showing a rock fragment mainly composed of silt-sized quartz (left) and the quartz grains in the zinc-rich matrix (right).

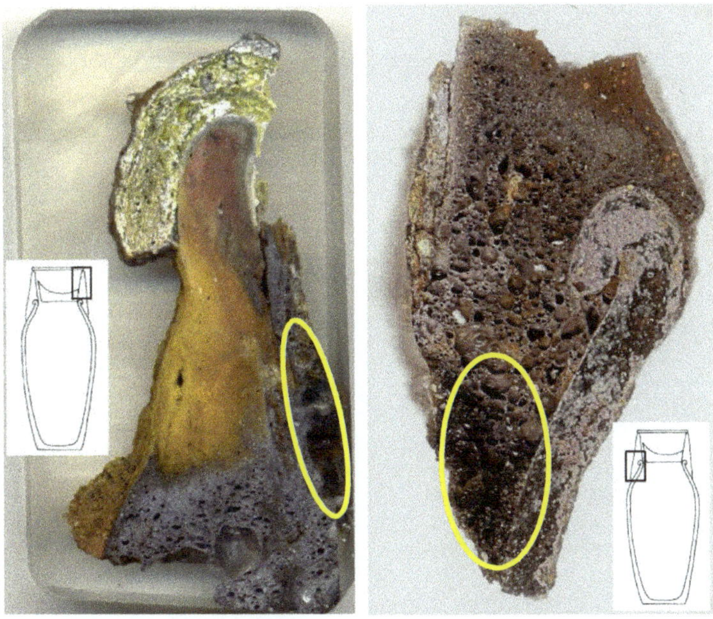

Figure 4.37 Left: cross section of YMT8, showing that the top yellowish and reddish part of the condenser is not vitrified but the lower bluish part is heavily vitrified, due to different temperatures and redox conditions. A thick zinc-rich crust adheres to the inner side of the rim, and part of pocket fabric is attached to the internal surface of the condenser. The glassy outer layer is marked; right: cross section of YMT3, showing bloating pores in the condenser fabric developing from the area slightly above the pot rim and increasing in size downwards. The black material or lute is marked.

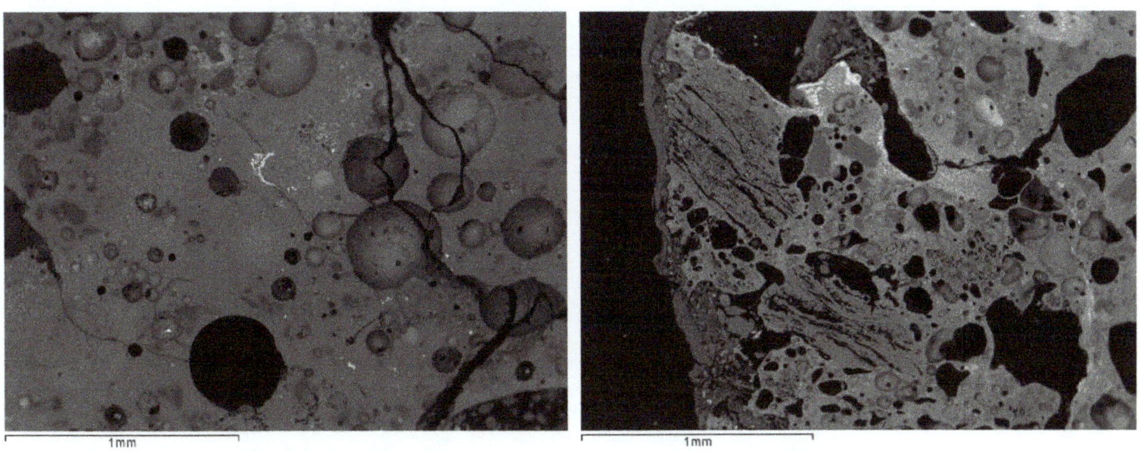

Figure 4.38 BSE images of the glassy layer with quartz (grey) and zinc sulphide (white) in YMT8 (left) and the black material with coal ash fragments in YMT3 (right).

Pockets

Pockets are the ceramic dishes with concave upper surfaces, placed horizontally within the condensers. The metallic zinc collects in pockets. A rectangular hole, generally 2 cm long and 1 cm wide, can be found on one side of the pocket, where the thickness of the pocket is larger than that of the opposite side. The hole connects the reaction zone with the condensation zone, allowing zinc vapour to rise. The uneven bottom surfaces of the pockets display imprints of the charge below, suggesting that fresh clay was applied to form pockets after filling the charge to the rims of the pots. The bottom surfaces are generally more vitrified than the upper parts, so they are better preserved. In most cases, the upper parts of pockets are partially missing, which was perhaps caused by craftspeople scraping off zinc-rich materials for resmelting after zinc ingots were collected at the end of the process (Figure 4.39 and 4.40).

The pocket fabrics of YMT2 and YMT8 are vitrified, porous with small round bloating pores and larger irregular voids (Figure 4.41 left). It is possible that some of the larger voids may have been caused by the plucking out of mineral inclusions during sample preparation. Their bulk compositions are similar to those of the condenser fabrics, except for their higher Al_2O_3 and lower SiO_2 content (Table 4.11). A few large inclusions with layered structures are likely to be coal (ash) fragments. The main inclusions are quartz grains: a few large ones and a high number of smaller ones which had been semi-molten. A few tiny prills of iron and zinc sulphide were detected in YMT2, showing that the pocket experienced a strongly reducing atmosphere.

The pocket fabrics of YMT1 and YMT12 are heavily contaminated with zinc oxide (Figure 4.41 right). In these samples, large inclusions up to 1 mm in diameter consist of quartz, potassium feldspar and coal (ash) fragments, while small inclusions are mainly quartz grains.

A thin layer of fine-grained fabric was detected on the internal surface of the condenser YMT17 (Figure 4.42). It is completely different from the other pocket fabrics studied and seems to be unfired clay, perhaps from the post-deposition.

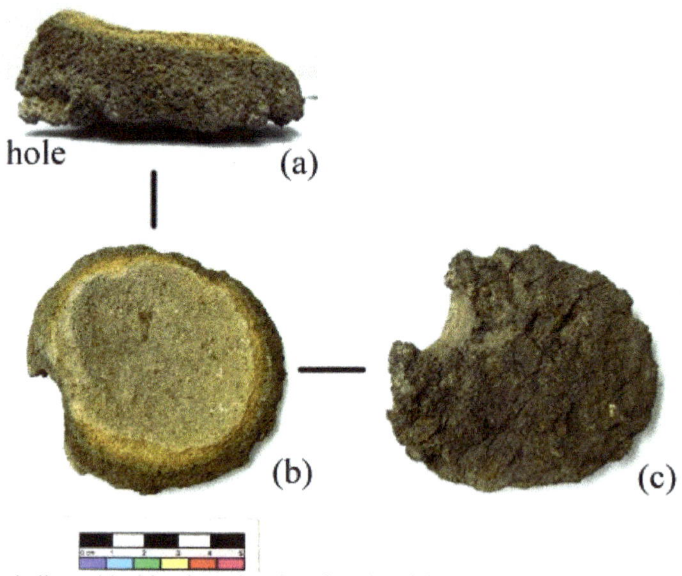

Figure 4.39 A pocket from Miaobeihou. (a) side view showing that the side with the hole is thicker than the opposite side; (b) the upper surface; (c) the bottom surface showing clear imprints of the charge.

Figure 4.40 A pocket still preserved inside the retort YMT2. Left: top view; right: detail of the pocket adhering to the condenser and pot.

Table 4.11 Average bulk compositions of five pocket fabrics from Miaobeihou (wt%), normalised to 100%. Analyses on polished sections for areas of ~2 by ~2.5 mm by SEM-EDS. The rows at the bottom show the same results after omitting ZnO and renormalising to 100%.

Sample	MgO	Al_2O_3	SiO_2	P_2O_5	SO_3	Cl	K_2O	CaO	TiO_2	FeO	ZnO
YMT2	1.5	18.8	63.1	0.2	0.3	0.3	2.9	2.4	1.0	7.6	1.9
YMT8	1.6	18.0	60.2	0.2	0.3	0.6	4.6	1.9	1.0	5.0	6.6
YMT1	1.0	12.0	34.2	0.3	0.3	0.1	0.9	1.6	0.5	3.4	45.7
YMT12	0.7	7.2	25.3	0.3	1.1	0.2	0.6	0.4	0.2	1.7	62.3
YMT17	2.8	15.2	51.8	0.2	0.1	0.2	2.2	1.9	0.7	6.0	18.9
YMT2	1.5	19.2	64.4	0.2	0.3	0.3	3.0	2.4	1.0	7.7	
YMT8	1.8	19.3	64.2	0.2	0.3	0.7	4.9	2.1	1.1	5.4	
YMT1	1.8	22.1	63.0	0.5	0.6	0.2	1.7	3.0	0.9	6.2	
YMT12	1.8	19.2	67.0	0.7	3.0	0.6	1.5	1.2	0.6	4.4	
YMT17	3.5	18.8	63.9	0.2	0.2	0.2	2.7	2.4	0.8	7.3	

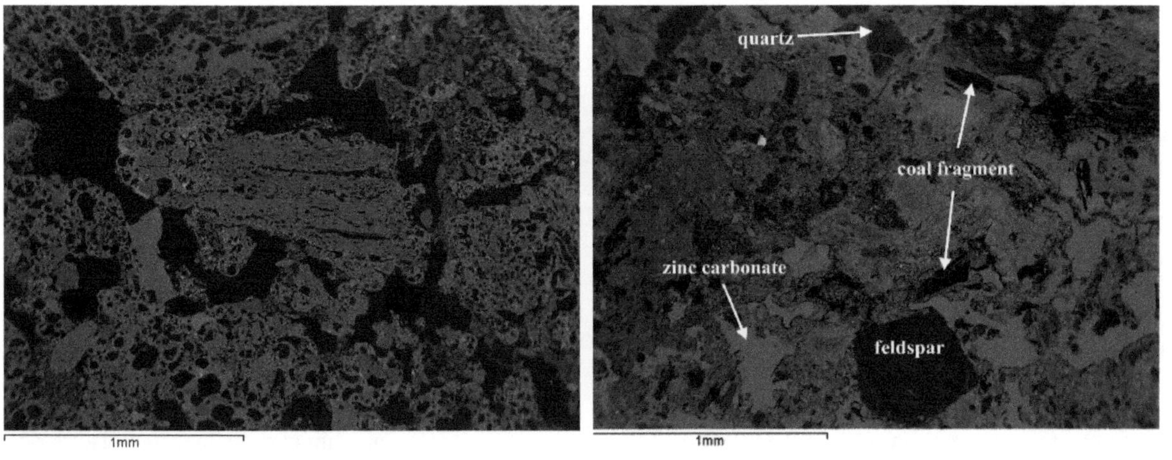

Figure 4.41 BSE images of the vitrified fabric of YMT2 (left) and the zinc-rich fabric of YMT1 (right) with quartz, feldspar, coal fragments and zinc carbonates.

Figure 4.42 BSE images of the fine-grained layer in YMT17. Left: the layer on the top is adhering to the internal surface of the condenser on the bottom; right: an enlarged picture of the layer.

4.2.2.5 Slag

Slag here refers to the metallurgical residues left within the pots after the smelting process. This slag is rusty, porous and quite vitrified. It is found generally adhering to the internal surfaces of the lower parts of the pots, and in some cases small fragments of coal and charcoal can still be seen trapped inside it (Figure 4.43). Small discrete lumps of porous and fragile slag were also found together with the retorts, indicating that more residues could have formed within the pots. The loose slag was probably removed by the craftspeople in order to reuse the pots. Five pieces of slag still adhering to the pots were selected: two samples were prepared and analysed for both YMB1 and YMB2; single samples for YMB3, YBM4 and YBM5.

These slag samples are mainly composed of a silica-rich glass, with the SiO_2 levels ranging from 29% to 54% and Al_2O_3 from 6% to 13%. They are also rich in FeO, ranging from 10% to the extreme value of 46% in YMB5. These samples contain variable levels of ZnO (3-10%), CaO (3-10%), MgO (2-4%), BaO (0.2-6%) and SO_3 (0.5-2%) (Table 4.12). The relative chemical heterogeneity is also manifested in their microstructures. The most important phases in the slag are described below.

The internal surfaces of the pots are partly dissolving into the slag. The matrices had been molten, exhibiting numerous partly dissolved quartz grains. Lumps of ceramic from the pots were apparent in the slag melt (Figure 4.44).

Within the glassy phases of slag, a variety of primary phases were crystallised from the silicate melt, including anorthite ($CaAl_2Si_2O_8$), celsian ($BaAl_2Si_2O_8$), spinel [$(Zn, Fe)Al_2O_4$], olivine [$(Fe, Mg, Zn)_2SiO_4$], pyroxene [$(Ca, Fe, Mg, Zn)_2Si_2O_6$] and melilite [$Ca_2(Al, Mg, Fe)Si_2O_7$]. Although mineralogical terminology is still used to describe these phases, it should be noted that these anthropogenic phases are not natural minerals. All are generally richer in zinc than the corresponding natural minerals.

Clusters of zinc sulphide and metallic iron are frequent (Figure 4.45). Zinc sulphides appear as round globules containing about 6% iron, and as tiny flower-like crystals which have certainly crystallised from the melt. The metallic iron contains minor amounts of arsenic, zinc, copper and phosphorous. In some areas, iron prills have corroded away, leaving voids where iron oxides were sometimes redeposited.

These slag samples are weathered to some extent and contain secondary phases, mostly iron oxides, as superficial coatings, along fractures and within pores. These oxides are probably derived from the corrosion of metallic iron formed within the slag. The presence of secondary iron oxides in cracks, pores and surfaces denotes that metallic iron was more abundant than is now apparent.

In addition, residues of coal fragments were identified within the rust, as well as some relicts of charcoal fragments in YMB1, YMB2 and YMB3, which can be clearly recognised through their cell structures (Figure 4.46). This indicates that both mineral and vegetal carbon were part of the charge.

There is a significant degree of variability among the samples, not only with regards to the relative abundance of the phases described but also in terms of the presence/absence of particular phases. YMB3 contains a large number of crystals of barite ($BaSO_4$), consistent with its higher BaO level. YMB4 is almost free of metallic iron and rust, while YMB5 used to contain abundant metallic iron prills, which have all corroded to rust. In YMB5, calcium carbonates have also deposited in the pores, resulting in its high CaO level.

Figure 4.43 Left: the bottom of a pot and the slag (side view), YMB1; right: the bottom of a pot and the slag (top view), YMB2.

Table 4.12 Bulk compositions (top half) and glassy phases (bottom half) of slag samples from Miaobeihou (wt%), normalised to 100%. Analyses on polished sections for areas of ~2 by ~2.5 mm and ~400 by ~600 μm respectively by SEM-EDS. '-' means 'not detected'.

Sample	MgO	Al_2O_3	SiO_2	P_2O_5	SO_3	Cl	K_2O	CaO	TiO_2	MnO	FeO	CuO	ZnO	BaO	PbO
YMB1-1	4.6	10.0	40.2	0.3	2.7	0.2	1.1	6.0	0.4	0.1	25.6	0.2	7.6	1.0	-
YMB1-2	2.5	10.8	48.7	0.3	1.9	0.2	1.8	4.6	0.6	0.2	15.9	0.1	10.0	2.6	-
YMB2-1	2.1	12.4	53.7	0.2	1.5	0.2	2.3	4.4	0.7	0.2	10.0	0.1	9.1	3.1	-
YMB2-2	2.4	12.8	53.5	0.3	0.7	0.2	2.3	4.5	0.7	0.2	12.7	0.1	7.7	1.7	0.1
YMB3	1.8	8.3	31.7	0.2	2.1	0.3	1.0	3.2	0.4	0.3	42.3	0.2	2.6	5.5	0.1
YMB4	2.3	10.8	53.2	0.1	1.6	0.3	1.8	5.2	0.7	0.1	11.9	0.1	9.6	2.0	0.1
YMB5	4.0	5.6	28.7	0.4	0.5	0.2	0.3	9.7	0.2	0.2	45.8	-	4.1	0.2	-
YMB1-1	3.0	13.0	52.9	0.1	0.7	0.1	1.8	6.0	0.6	0.2	13.7	-	6.6	1.2	-
YMB1-2	2.9	11.1	50.9	0.1	0.8	0.1	1.7	5.3	0.8	0.2	13.8	0.1	9.9	2.2	-
YMB2-1	2.4	12.4	54.6	-	0.5	-	2.3	4.6	0.7	0.2	9.9	-	8.6	3.5	-
YMB2-2	2.7	13.2	56.7	0.1	0.5	-	2.4	5.3	0.9	0.2	8.6	-	7.9	1.4	
YMB3	2.5	12.5	57.1	0.2	0.3	0.1	2.4	6.6	0.7	0.4	8.2	-	5.5	3.4	-
YMB4	3.0	10.5	54.0	0.1	0.9	0.2	2.0	5.7	0.6	0.2	10.8	-	9.8	2.1	-
YMB5	4.8	8.8	49.3	0.4	1.0	0.3	0.8	12.8	0.4	0.4	14.0	-	6.6	0.4	-

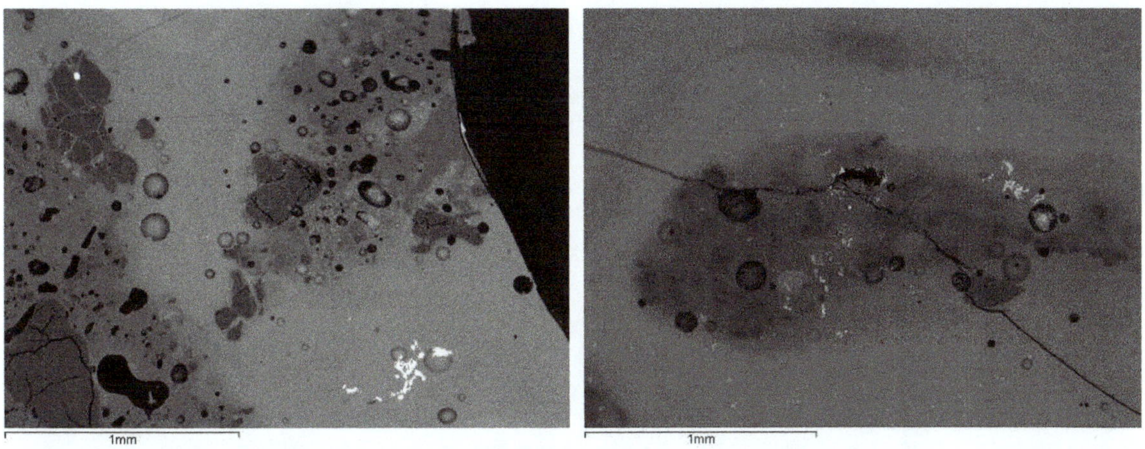

Figure 4.44 BSE images of the interaction between the pots and the slag. Left: the interface between the ceramic (left) and the slag (right) in YMB2-2; right: a lump ceramic from the pot in YMB1-2 which had been semi-molten.

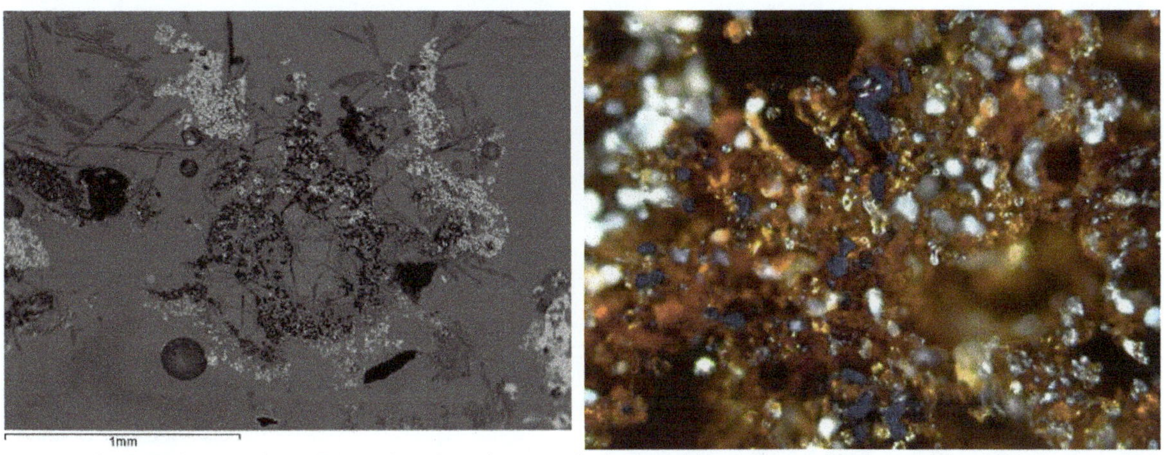

Figure 4.45 Left: BSE image of YMB1-1, showing clusters of zinc sulphide and metallic iron; right: XPL image of YMB2-1, showing zinc sulphide (yellow), metallic iron (dark) and voids (white). Width of picture is 0.5 mm.

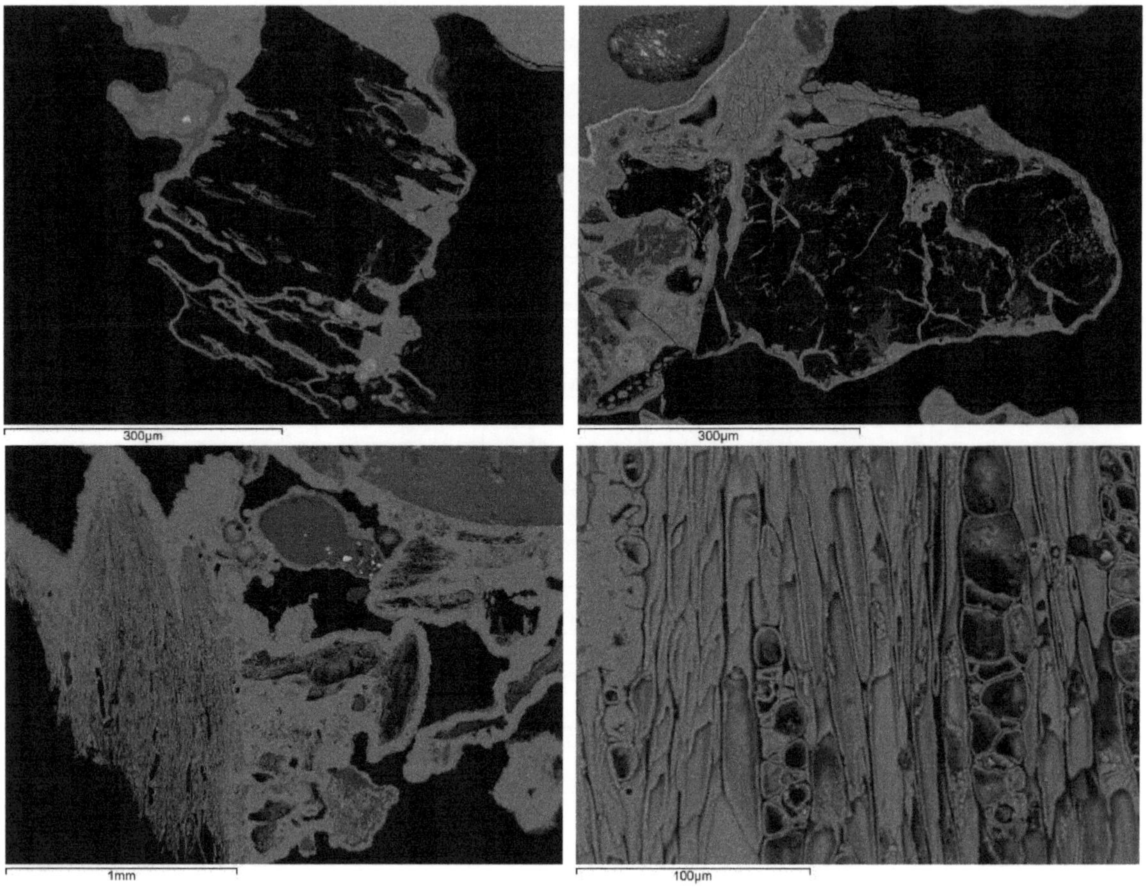

Figure 4.46 BSE images of coal and charcoal fragments in the slag samples from Miaobeihou. Top left: coal in YMB1-1; top right: coal in YMB2-1; bottom left: charcoal in YMB1-1; bottom right: the cell structure of charcoal in YMB1-1. Note that these fragments are embedded within the rust.

4.2.2.6 Zinc-rich crusts

The greenish and whitish crusts adhering to the condensers are mostly zinc-rich minerals (Figure 4.47). The greenish substances, zinc oxide, are the primary components of these crusts, which were formed by the oxidation of zinc vapour during the distillation process. Whitish materials, mostly zinc carbonate or zinc carbonate hydroxide, were often found around or within voids of the zinc oxide crusts. Zinc sulphate and chloride minerals were also detected in cavities. These white crusts are unlikely to have formed during distillation, but are rather the corrosion products of metallic zinc and zinc oxide.

Although metallic zinc is susceptible to corrosion, a few zinc prills are still trapped within the crusts as spherical droplets of sizes ranging from 50 to 500 μm. Two droplets of metallic zinc in YMT20 are different: one has tiny lead-rich prills, while the other has a number of small inclusions rich in cadmium (about 75%) and contains on average 4.1% cadmium, 0.4% lead and 0.3% arsenic (Figure 4.48). The metallic zinc and surrounding zinc oxide are called 'blue powder' in some traditional processes (see Section 2.3) and also in modern zinc metallurgy (Ingalls 1903, 521).

In addition, a few siliceous minerals and coal ash fragments are embedded within the crusts. In some cases, these might have derived from the condenser fabrics, especially those near the condensers. Siliceous materials and coal ash fragments far from the condensers are more likely to have been introduced from the charge together with zinc vapour. In addition, layers of tiny zinc sulphide particles were found around coal fragments (YMT1), along with zinc-rich crusts (YMT20).

4.2.3 Discussion

The analytical characterisation of the various ceramic parts involved in zinc distillation and the associated metallurgical residues allow a detailed understanding of the manufacture and performance of the retorts and the charge as well as an insight into the conditions of reduction and condensation of zinc distillation.

4.2.3.1 Retort design

The retorts from Miaobeihou are composed of four parts: pots, condensers, pockets and lids. The different parts of the retorts all exhibit the appropriate formal and material properties suitable for their specific performance characteristics.

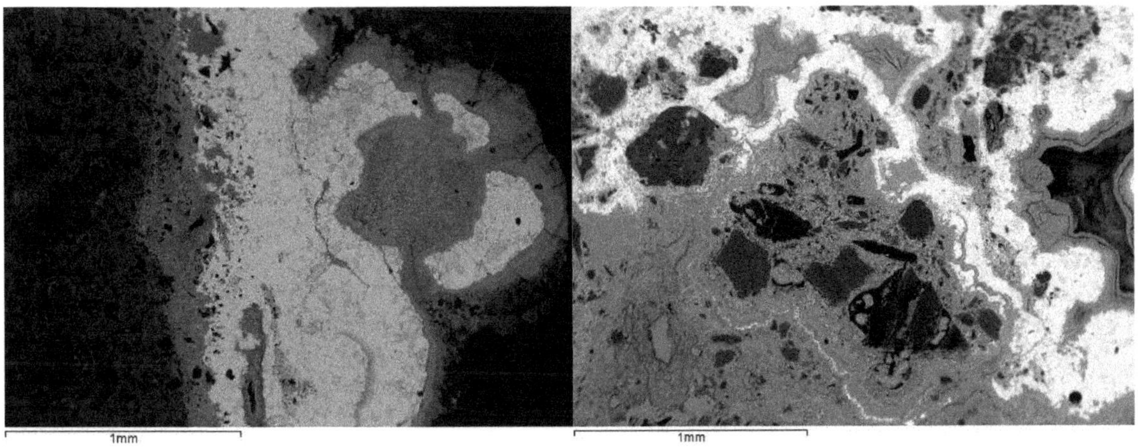

Figure 4.47 BSE images of zinc-rich crusts on the surfaces of the condensers from Miaobeihou. Left: zinc oxide (white) and zinc carbonates (grey) adhering to the internal surface of the condenser (dark grey) in YMT17; right: minerals and coal fragments in YMT1.

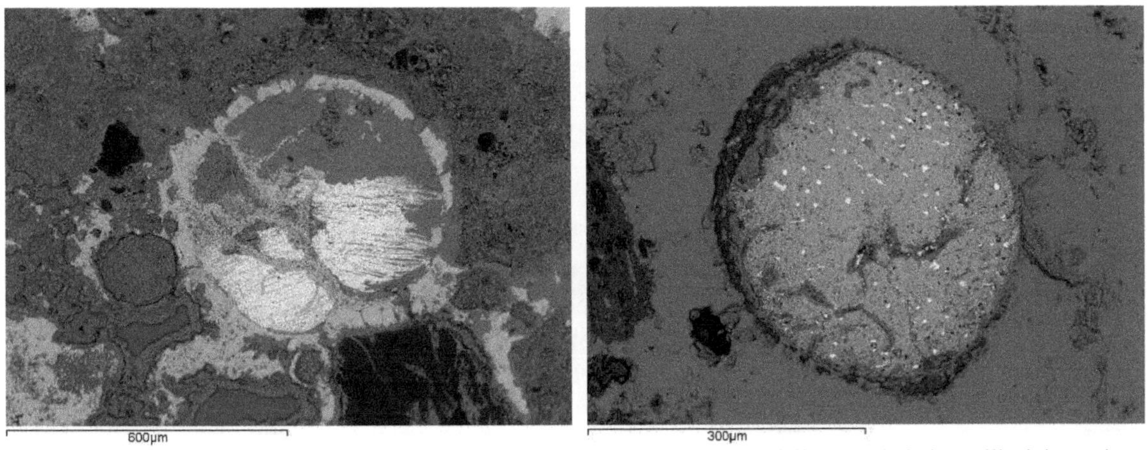

Figure 4.48 BES images of zinc prills trapped within the crusts in YMT20. Left: a partially corroded zinc prill; right: a zinc prill rich in cadmium.

Pots

The pots have the shape of ordinary storage jars, with a large capacity (~2 L) to hold the charge. They are stable on their flat bottoms. The whole or part of the bases of used pots show few firing traces, indicating that, in the case of rectangular furnaces, the pots were sitting on the furnace bars. The increasing wall thickness of the jars from the top to the bottom would not only maintain structural stability, but also provide enough strength to hold the charge, as well as improving resistance to chemical attack from the reacting charge at the bottom.

Considering materials, the clay used for the pots was quite refractory, with about 21% Al_2O_3. However, the relatively high levels of FeO (about 5%) and K_2O (about 3%) suggest that they were not made of the most refractory fireclay. It is not clear whether this clay was deliberately chosen for making the pots for the retorts, or whether a common domestic pottery clay was used. This problem requires further studies on local domestic pots. What is certain is that the addition of quartz sand improved the physical and thermal properties of the pots, as has been demonstrated by mechanical tests on replicates tempered with sand (Kilikoglou *et al.* 1998; Tite *et al.* 2001). The cracks and pores developed in the clay-quartz system can arrest and stop crack propagation under mechanical and thermal stresses, thereby increasing the toughness and thermal shock resistance of these pots. In addition, the pots were fired at high temperatures of about 1100-1200 °C before being used for smelting. This prefiring produced extensively vitrified fabrics with closed porosity which were more resistant to the corrosive attack of the charge, in addition to limiting the penetration of zinc vapour formed during smelting. It is noteworthy that the ZnO levels in three used pot fabrics were below the detection limit (0.3%) in spite of the high vapour pressure of zinc that must have developed during the reaction and the high chemical reactivity of zinc oxide and silica at high temperatures (Kearns *et al.* 2010). As a comparison, a similar clay tempered with quartz was used for making crucibles for brass cementation in Zwickau, Germany, in the late fifteenth century. These crucible fabrics had 7-8% ZnO, probably due to lower prefiring temperatures (Martinón-Torres and Rehren 2002). The Miaobeihou pot fabrics developed some mullite crystals, which could also improve the thermal shock resistance, thermal and chemical refractoriness, high-temperature strength and toughness (Martinón-Torres *et al.* 2008). The pots may have been reused after removing pockets, condensers and slag.

Condensers

The pots alone could not be used as retorts for zinc distillation, as it would be difficult to separate the low temperature condensation zone from the high temperature reaction zone. The solution was to attach a separate chamber for condensation by adding cylindrical extensions onto the mouths of the pots. The relatively thicker walls in the lower parts of the condensers would not only contribute to structural stability, but also act – together with the pockets – as thick insulating barriers to contain the strong heat below the pot rims. The mouths of the condensers are slightly bigger than those of the pots, facilitating the loading of the charge and the collection of products.

Compared to the pots, the condensers were less refractory, with higher levels of alkali and alkaline earth oxides and FeO (Table 4.13 and Figure 4.49). It was not necessary for the condensers to be refractory as they were exposed to much lower temperatures (<800 °C) than the pots (~1200 °C) during the process. They were probably not prefired prior to smelting, as evidenced by their high levels of ZnO.

The choice of coal ash fragments as temper for condensers might be due to two reasons which are not mutually exclusive. On the one hand, coal ash fragments could have contributed to the mechanical properties of the condensers like other platy or fibrous inclusions (such as mica and shell), which are effective at stopping crack propagation, thus increasing both toughness and thermal shock resistance (Tite *et al.* 2001). On the other hand, it should be noted that coal ash was produced in abundance in zinc smelting furnaces. If it was not recycled, it would have to be dumped as waste. Thus the use of coal ash as temper in the condensers was probably also a practical consideration. The use of coal ash in the manufacture of condensers suggests that the condensers were made at the zinc smelting site.

Table 4.13 Average bulk compositions of different parts of retorts from Miaobeihou (wt%), normalised to 100%. Analyses on polished sections for areas of ~2 by ~2.5 mm by SEM-EDS. '-' means 'not detected'.

Sample	MgO	Al_2O_3	SiO_2	P_2O_5	SO_3	Cl	K_2O	CaO	TiO_2	MnO	FeO
Pot	0.6	17.4	73.5	0.1	0.1	0.1	2.3	0.2	0.9	0.1	4.7
Condenser	1.4	16.1	68.5	0.2	0.5	0.2	3.2	1.7	0.9	0.1	7.2
Pocket	2.1	19.7	64.5	0.4	0.9	0.4	2.8	2.2	0.9	-	6.2

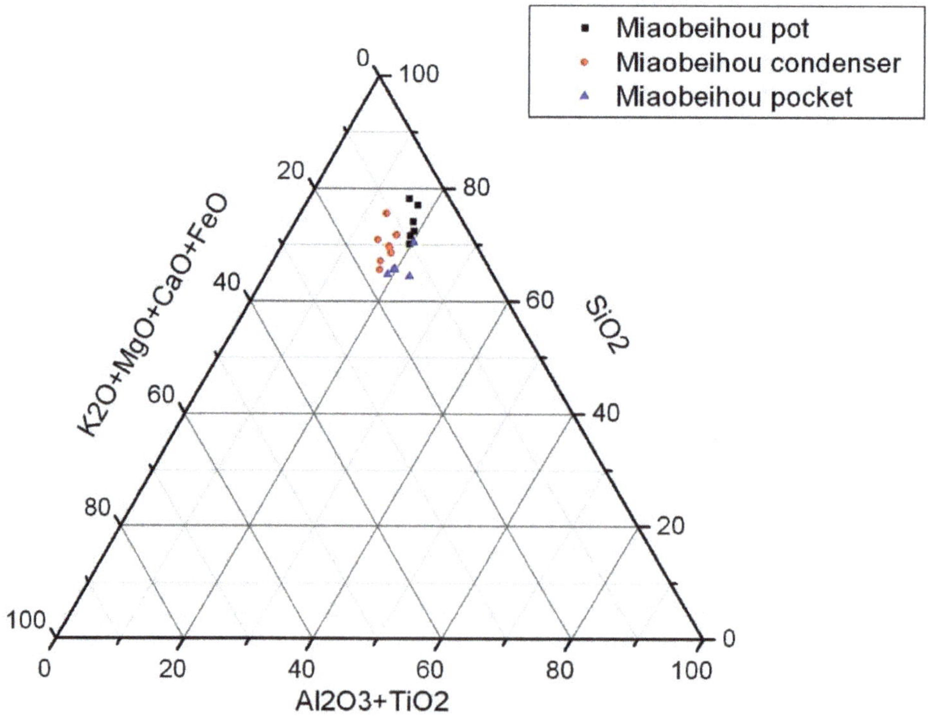

Figure 4.49 Ternary diagram of SiO_2-(Al_2O_3+TiO_2)-(K_2O+CaO+MgO+FeO) showing the bulk compositions of the pot, condenser and pocket fabrics from Miaobeihou. The condensers and pockets contain higher levels of alkali and alkaline earth oxides and FeO than the pots, but lower levels of SiO_2.

Pockets

The pockets were added after the pots were charged. They were pressed onto the top of the charge and supported by it. A hole was made on one side of each pocket, allowing zinc vapour to enter the condensers. Once zinc vapour entered the condensation zone, it would liquify when in proximity with the lower and cooler surfaces of the lids, dripping onto the upper surfaces of the pockets. The pockets with concave upper surfaces, served as containers for collecting metallic zinc and as insulating barriers from the strong heat in the reaction zone. Pockets are thicker on the side of the hole than on the opposite side, thus preventing the liquid zinc from flowing back to the pots via the hole.

The pocket fabrics were also less refractory than the pots but similar to the condensers (Table 4.13 and Figure 4.49). However, the mineral and coal ash inclusions in the pocket fabrics are more coarse-grained. The presence of coal ash suggests that the pockets were made at the smelting site. The fabrics were not compressed as tightly as the pots and the condensers, as shown by the porous vitrified lower parts and the upper parts which are heavily contaminated with zinc oxide. The pockets, similar to the condensers, would not be exposed to high temperatures, so less refractory non-prefired fabrics would be sufficient.

Lids

The top openings of the condensers must have been closed by lids. A small hole would probably have been present in the lids, opposite the hole of the pockets, in order for gases generated by the process to escape. At Miaobeihou, no lids have been found, although an iron fragment, 0.4 mm thick, could indicate a lid. A heavily corroded iron lid, 11.6 cm in diameter and 0.8 cm thick, was excavated at Zhangjiahe near Miaobeihou (HPCHAI *et al.* 2007). More iron lids have also been discovered at Dafengmen in Shizhu, one of which was analysed (see Section 5.3.2.1).

As iron has good thermal conductivity, lids made of iron could transfer heat fast into the air and keep the condensation zone at appropriate temperatures for condensation. In addition, iron lids could be reused many times, which would also explain their general scarcity in the archaeological records. It is possible that clay lids were used as in some traditional zinc smelting processes (Craddock and Zhou 2003), but these could be recycled as part of the charge and are therefore not likely to have survived.

4.2.3.2 The charge

The slag in the pots is the most informative material to reconstruct the makeup of the original charge. The analytical results of the slag indicate that a mixture of zinc ores, coal/charcoal and recycled zinc-impregnated ceramics may have constituted the charge.

The strong interaction interface between the slag and the pots suggests that the pot fabric is one of the source materials for the slag. Compared to the pot fabric, the slag is much richer in FeO, ZnO, CaO, MgO, BaO and SO_3, but poorer in SiO_2 and Al_2O_3 (Figure 4.50). The SiO_2 and Al_2O_3 concentrations of the pots and the slag show a positive correlation and relative proportions similar to those in the ceramics (Figure 4.51), indicating that SiO_2 and Al_2O_3 in the slag could have originated from the thermal degradation of the pots. However, the pots in use could not have contributed much SiO_2 and Al_2O_3; therefore, other source materials with similar SiO_2 to Al_2O_3 ratios must have been introduced. A plausible explanation is that zinc-rich condensers, pockets and clay lids, if any, were added to the charge for recycling, as documented in traditional zinc smelting, such as the Hezhang process in the 1950s (MIP 1958). A small amount of SiO_2 and Al_2O_3 could also have derived from zinc ores and coal.

The enrichment in FeO, ZnO, CaO, MgO, BaO and SO_3 in the slag clearly came from the charge, comprised of zinc ores and reducing agents. The high amounts of FeO and ZnO most likely derived from the ores, which are iron-rich oxidic zinc ores according to the analytical results of the archaeological mineral samples (see Section 4.2.2.1). The elevated levels of CaO, MgO, BaO and SO_3 very likely originated from the gangue of the zinc ores, i.e. dolomite, calcite and barite, which are the typical gangue minerals of the zinc ores in the Laochangping lead-zinc deposits (see Section 5.1.1).

The reducing agents were coal and charcoal, as their residues were identified in the slag. Charcoal ash typically introduces CaO, MgO, K_2O, Na_2O, FeO, MnO and P_2O_5, but different types of wood produce ash with different compositions (Misra *et al.* 1993; Jackson and Smedley 2004). Some of the CaO, MgO and P_2O_5 in the slag would have come from the charcoal ash.

The use of coal as a reducing agent was anticipated more than charcoal, because most traditional zinc smelting processes used coal (see Section 2.3). The coal used as a reducing agent was generally restricted to high quality anthracite and bituminous coal with little ash or sulphur. The residual coal fragments in the slag are rich in carbon with only a small proportion of sulphur (less than 1%). Some of the CaO, MgO and FeO in the slag could also have originated from the coal ash.

In the slag, most of the sulphur exists as globules and tiny crystals of zinc sulphide. At first sight, one might think that these rounded phases are the residues of sphalerite minerals from the zinc ores. However, sphalerite cannot be oxidised under the strongly reducing conditions of the zinc smelting process and would therefore have remained if there were some sulphidic minerals in the ores. The tiny crystals were obviously crystallised from the slag melt. Moreover, a few globules of zinc sulphide were also detected in some vitrified condenser and pocket fabrics (YMT3 and YMT8). Thus, it seems more likely that all phases of zinc sulphide were crystallised from the melt which was rich in zinc and sulphur. Sulphur has two possible sources: coal and zinc ore. In addition to barite, which has been found in YMB3, it is possible that the oxidic zinc ores contain some sulphidic minerals, although no sphalerite was found in the mineral samples studied. Irrespective of its source, sulphur readily reacted with zinc to form zinc sulphide, which directly reduced the yield.

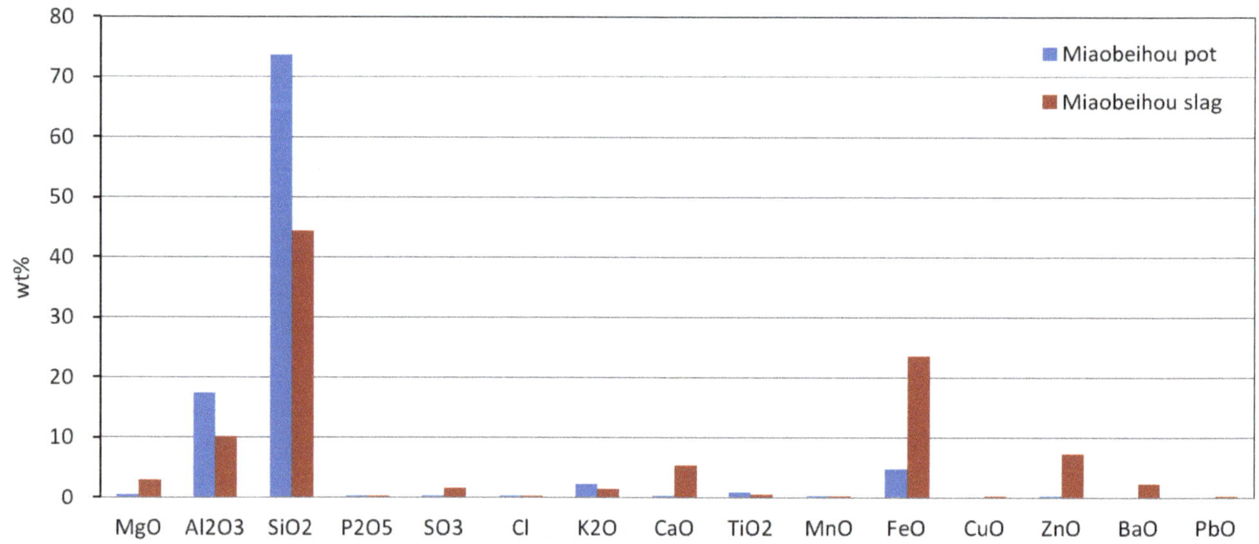

Figure 4.50 Bar chart comparing average bulk compositions of the pots and slag from Miaobeihou.

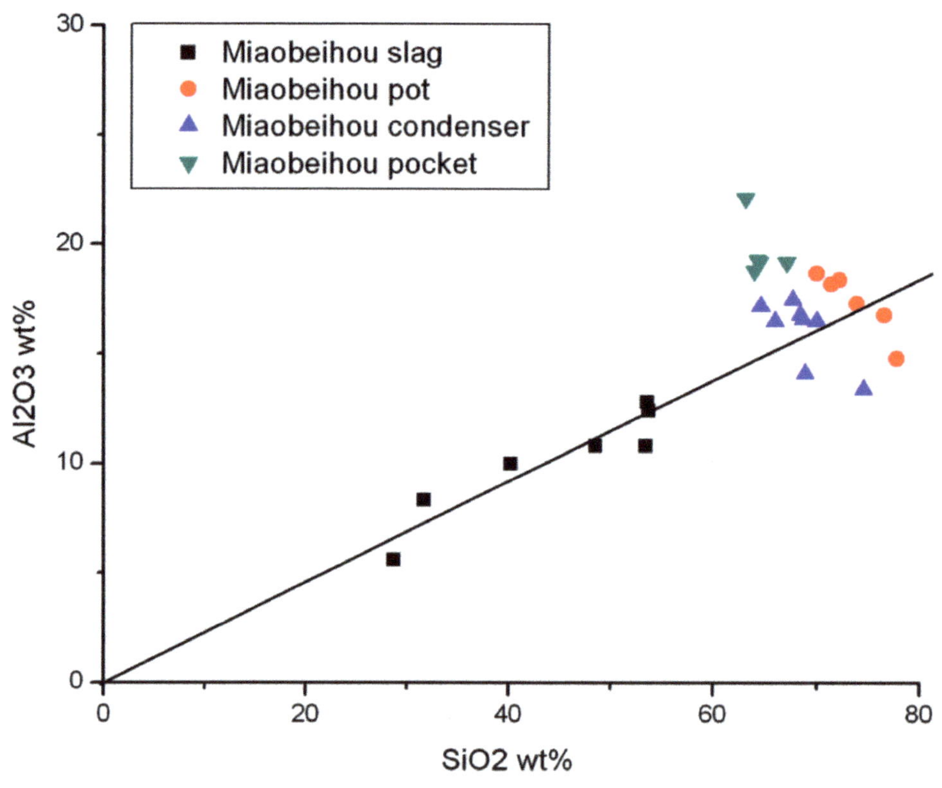

Figure 4.51 The SiO_2 and Al_2O_3 contents of the slag, pot, condenser and pocket samples from Miaobeihou. Note the linear correlation of SiO_2 and Al_2O_3 of the slag samples and the similar relative proportions to those of the ceramics.

4.2.3.3 Conditions of reduction and condensation

The metallurgical process taking place within the retorts involved two steps: the reduction of zinc ores in the lower reaction zone and the condensation of zinc vapour in the upper condensation zone (Figure 4.52).

There is no evidence of a separate ore roasting or calcining stage before smelting. It can therefore be hypothesised that in the reaction zone, zinc ores, mainly smithsonite, would be first decomposed to zinc oxide and carbon dioxide at a low temperature of about 300 °C.

$$\text{(a)}\ ZnCO_3(s) = ZnO(s) + CO_2(g)$$

The carbon dioxide from zinc carbonates can react with carbon from coal and charcoal to produce carbon monoxide; in addition, initial oxygen present in the pots can also generate carbon monoxide through reaction with the solid carbon provided by charcoal and coal.

$$\text{(b)}\ C(s) + CO_2(g) = 2CO(g)$$

$$\text{(c)}\ 2C(s) + O_2(g) = 2CO(g)$$

At temperatures over 1000 °C, zinc oxide was reduced to the gaseous metallic zinc through its reaction with carbon monoxide. The carbon dioxide released produced more carbon monoxide, by reacting with coal and charcoal, thus maintaining the reversible reaction to reduce more zinc oxide to zinc vapour.

$$\text{(d)}\ ZnO(s) + CO(g) \rightleftharpoons Zn(g) + CO_2(g)$$

Generally an excess of carbon equal to 2-3 times the theoretical value is required in the charge to reduce zinc oxide (Hopkins 1954, 187). The presence of this carbon excess is confirmed by the unreacted coal residues identified in most of the slag.

The reduction of zinc oxide requires highly reducing conditions, as seen from the Ellingham diagram: the metals further down in the line are more difficult to smelt, requiring more reducing conditions (Morgan 1985, 57). Zinc is more difficult to smelt than copper, tin, lead and iron. Thus iron oxide in the charge can be more readily reduced to metal than zinc oxide. The presence of metallic iron in the slag is therefore expected since iron-rich oxidic zinc ores were likely used at Miaobeihou. The reduction of iron oxide also needs considerable heat and carbon, consuming more fuel and reducing agents.

The temperature of operation was estimated by the degree of vitrification of the pot walls and bottoms. It seems that temperatures of about 1200 °C were generally maintained, while even higher temperatures were reached at the bottom parts of pots. Under the strong heat and highly reducing conditions, the charge sintered and partially vitrified, especially near the bottoms of the pots. The glassy phases of the slag formed a skeletal adhesive which stabilised the charge, while the open porosity facilitated the reaction because it would allow the zinc vapour to escape to the condensation zone.

The zinc produced, in the form of vapour, ascended via the hole of the pockets to the condensation zone. Zinc vapour liquified first below its boiling point of 907 °C when it reached the cooler lids and the condenser walls, dripping into the pockets. The zinc droplets trapped in the zinc-rich crusts adhering to the condensers indicate that zinc was in a liquid state before solidification below its melting point of 420 °C.

$$\text{(e)}\ Zn(g) \longrightarrow Zn(l) \longrightarrow Zn(s)$$

Three gases entered the condensation zone: zinc vapour, carbon monoxide and carbon dioxide. Carbon monoxide would probably have been predominant over carbon dioxide and oxygen, otherwise the metallic zinc would have been readily reoxidised to zinc oxide. However, it is likely that some zinc was oxidised, forming zinc-rich crusts and permeating into condensers and pockets – which could be recycled – as well as being lost in the fumes.

$$\text{(f)}\ 2Zn(g) + O_2(g) = 2ZnO(s)$$

$$\text{(g)}\ Zn(g) + CO_2(g) = ZnO(s) + CO(g)$$

The key to zinc distillation is to control the temperature gradient between the reduction and condensation zones. As seen from the degree of vitrification of condensers and pockets, there must have been a sudden temperature drop slightly above the mouths of pots. It has been suggested that a narrow temperature range of 420-550 °C was required in the condensation zone, which needed to be maintained in order for the zinc to condense (Craddock *et al.* 1998, 41). To control the temperature sufficiently, no heat was supplied around the condensers, i.e. the external surfaces of the condensers are not vitrified but the surfaces below the pot rims are heavily vitrified, with some coal ash fragments clearly visible. In addition, the thick lower parts of the condensers and the pockets would have helped to contain strong heat below the condensation zone.

At Miaobeihou, the metallic zinc produced was of high purity, with involatile materials left in the slag. Lead was the main impurity of the metal. Although only traces of PbO were detected in the mineral and slag samples, zinc ores richer in lead are expected to have been used to produce these zinc samples with about half a percent of lead. The zinc ores used were rich in iron, but only traces of iron entered the products with most of it ending up in the slag.

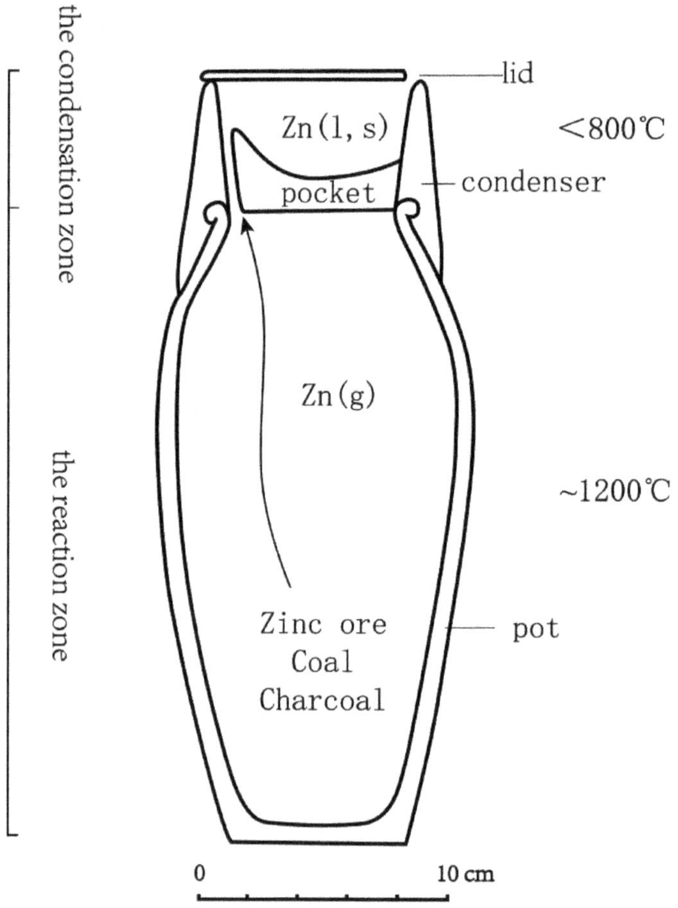

Figure 4.52 Reconstruction of a retort from Miaobeihou. The retort, consisting of pot, condenser, pocket and lid, is divided into two zones: the lower reaction zone and the upper condensation zone. In the reaction zone, zinc ores are reduced by coal/charcoal at a temperature of about 1200 °C; the zinc vapour formed rises to the condensation zone (<800 °C) where it first liquifies and then solidifies (g-gas, l-liquid, s-solid).

4.3 Puzihe (YP) and Muxiexi (CM)

4.3.1 Site description

Puzihe is also located at Yangliusi village, just next to Miaobeihou (SPAI and CMBCH 2007). As with Miaobeihou, the Puzihe site is divided by five ditches into six areas, named Area I to VI from the west to the east.

The retorts excavated at Puzihe are of similar shapes and sizes to those from Miaobeihou. Two rectangular furnace foundations (Y1 and Y2) were excavated in Area III. Y2 was described and illustrated in the archaeological report of the site (SPAI and CMBCH 2007). It is 15.2 m in length and 1.3 m in width. The furnace bed is 0.1 m above the contemporary ground level. There are two surviving walls built with bricks and red clay at the short ends of the furnace and dozens of rows of Han bricks, acting as furnace bars, on the base. The distance between each pair of rows of Han bricks is 12-14 cm.

Some porcelain bowls were found with the smelting remains. On the bottoms of four blue and white bowls are inscriptions showing the periods when they were manufactured. Two bowls were made in the Xuande period (AD 1398-1435) and the other two were made in the Chenghua period (AD 1447-1487). Accordingly, the smelting remains could broadly date to the Ming Dynasty (SPAI and CMBCH 2007).

Muxiexi is located at Changjiang village, northeast of Puzihe. Smelting remains, concentrated in two areas separated by one ditch, include a large deposit of smelting remains and one rectangular furnace foundation. The retorts found at Muxiexi show similar shapes and sizes to those from Miaobeihou and Puzihe, indicating that the site was probably contemporary to Miaobeihou and Puzihe.

Surface materials were collected from Puzihe and Muxiexi during fieldwork in August 2009 (Figure 4.53). Only fragments of pots and condensers were found and analysed, but neither pockets nor lids were found.

Figure 4.53 The landscape of Puzihe (left) and Muxiexi (right) in August 2009.

4.3.2 Results

4.3.2.1 Retorts

Pots

The pots from Puzihe and Muxiexi are flat-bottomed jars of similar shapes and sizes to those from Miaobeihou. In the archaeological report of Puzihe there are illustrations of three intact pots (SPAI and CMBCH 2007). One was probably unused, showing the spiral grooves and ridges on the external surface; the other two are heavily vitrified on the external surfaces with parts of the condensers still attached.

The pot fragments collected show that these pots were wheel thrown: the wall fragments bear grooves and ridges on the surfaces, while the base fragments display spiral markings. The thicknesses of the fragments increase from the top to the bottom, with an average of about 1 cm. The unused pots, such as YMM1 and YMM4, underwent similar firing conditions: the inner half having been oxidised and the outer half being reduced. The used pots are bluish-grey, with grey vitrified layers on the internal surfaces and black vitrified layers on the external surfaces (Figure 4.54).

Four pot samples from Puzihe and four from Muxiexi were analysed. They show similar compositions and characteristics to those from Miaobeihou. Their ceramic matrices are composed of 67-72% SiO_2, 17-22% Al_2O_3 3-5% FeO and low levels of alkali and alkaline earth oxides (Table 4.14). Their bulk compositions display slightly higher levels of SiO_2 (70-78%), owing to the presence of quartz inclusions, but lower levels of other oxides. Some of them contain over 1% ZnO, such as YPB5 (1.7%), CMT3 (2.0%) and CMB4 (2.4%).

Figure 4.54 Cross sections of unused pots (top left: YPM1; top middle: YPM4) and used pots (top right: YPM3; bottom left: CMT3; bottom middle: CMM3; bottom right: CMB5). The internal surfaces of these samples are facing up. The used ones are vitrified on both surfaces.

Table 4.14 Average matrix (top half) and bulk (bottom half) compositions of four pot fabrics from Puzihe and four from Muxiexi (wt%), normalised to 100%. Analyses on polished sections for areas of ~150 by ~200 μm and ~2 by ~2.5 mm respectively by SEM-EDS. YPM1 and YPM4 are unused pots, while the others are used. '-' means 'not detected'.

Sample	Na_2O	MgO	Al_2O_3	SiO_2	P_2O_5	SO_3	Cl	K_2O	CaO	TiO_2	MnO	FeO	ZnO
YPM1	0.2	0.7	21.9	67.3	0.2	-	-	3.0	0.3	1.2	-	5.1	0.1
YPM3	0.5	0.6	21.7	67.4	0.2	0.1	-	3.0	0.1	1.1	0.1	5.1	0.1
YPM4	0.2	0.6	20.0	70.7	0.1	0.1	-	2.6	0.1	1.1	-	4.4	0.1
YPB5	0.7	0.5	19.8	70.0	0.1	0.1	-	2.1	0.3	1.1	-	4.1	1.2
CMT3	0.7	0.6	21.0	67.2	0.1	-	-	3.1	0.2	1.0	-	5.2	0.9
CMM3	0.3	0.5	20.2	71.9	0.1	-	0.1	2.2	0.2	1.0	-	3.3	0.2
CMB4	0.5	0.5	16.6	72.0	0.1	-	-	2.3	0.2	0.8	-	4.4	2.6
CMB5	0.4	0.5	19.7	71.7	0.1	-	-	2.4	0.3	1.0	-	3.3	0.6
YPM1	0.3	0.7	19.0	70.3	0.1	-	-	2.8	0.3	1.0	-	5.4	0.1
YPM3	0.5	0.4	17.1	73.2	0.1	0.1	-	2.3	0.1	1.0	-	4.8	0.4
YPM4	0.1	0.5	16.1	75.6	0.1	-	-	2.3	0.1	0.9	0.1	4.1	0.1
YPB5	0.7	0.5	15.9	73.0	0.0	0.1	-	1.8	0.3	0.9	-	5.1	1.7
CMT3	0.8	0.5	15.4	73.9	0.2	-	-	2.0	0.1	0.8	0.1	4.2	2.0
CMM3	0.4	0.3	15.1	78.2	0.1	-	0.1	1.5	0.2	0.9	-	2.8	0.4
CMB4	0.5	0.4	13.1	75.6	0.1	0.1	-	1.8	0.2	0.8	-	5.0	2.4
CMB5	0.3	0.5	15.9	76.6	0.1	-	-	1.7	0.2	0.8	-	3.1	0.8

Like the pots from Miaobeihou, the pot fabrics from Puzihe and Muxiexi contain ill-sorted inclusions, up to 2 mm in size. They are sub-angular and show low sphericity. Most of the large inclusions are internally cracked quartz grains with a few ferruginous concretions present (Figure 4.55). Some of the large inclusions in YPM4 are rock fragments composed of intergrown quartz and alumina-rich phases (Figure 4.56 left). In addition, there are abundant small quartz grains partially dissolved into the matrices and a few small inclusions, such as zircon, rutile and monazite. In YPM1 and CMM3, some unhydrated clay lumps were identified, ranging in sizes from 200 μm to 1 mm (Figure 4.56 right).

The unused pots (YMM1 and YPM4) show extensive vitrification structures, while the four used ones generally exhibit continuous vitrification structures with medium (2-10 μm diameter) and coarse (10-50 μm diameter) bloating pores (Figure 4.57). Therefore, it can be estimated that the pots were prefired at 1100-1200 °C and used at around 1200 °C.

The four used samples are vitrified on both surfaces (Table 4.15 and Figure 4.58). The internal surfaces of YPM3 and CMT3 are richer in MgO, SO_3, K_2O, CaO, ZnO and BaO than the pot fabrics, an enrichment clearly derived from the charge. There are quartz grains which had been semi-molten and tiny metallic iron prills. In CMM3 and CMB5, the internal vitrified layers contain significant concentrations of CaO (~18%) and MgO (4-7%) but lower levels of FeO (0.5-1.0%). There are abundant crystals of diopside ($MgCaSi_2O_6$), tiny prills of metallic iron in both samples and some crystals of anorthite ($CaAl_2Si_2O_8$) in CMB5. The extremely high levels of CaO were not only from the coal or charcoal ash, but more likely derived from the ores rich in calcite and dolomite. Alternatively, it is also possible that the internal surfaces of the pots were glazed before use. Intentional glazing was practiced in some traditional processes; for example, retorts were soaked in a muddy limewater to form lime-rich glazes in Bijie in 1994-95 (Craddock and Zhou 2003). However, no unused pots with glazes have been found.

The external vitrified surfaces contain higher levels of MgO, CaO and FeO, again indicating the use of coal as fuel. The FeO levels of YPM3 and CMM3 are so high (~14%) that it is suspected that they may also have been introduced from other sources. They have some quartz which had been semi-molten and abundant tiny crystals of spinel. The external vitrified surface of CMB5 is distinct, with higher levels of CaO and MgO but only occasional quartz grains present in the layer. The local enrichment in these oxides was a consequence of the coal ash or perhaps through intentional glazing.

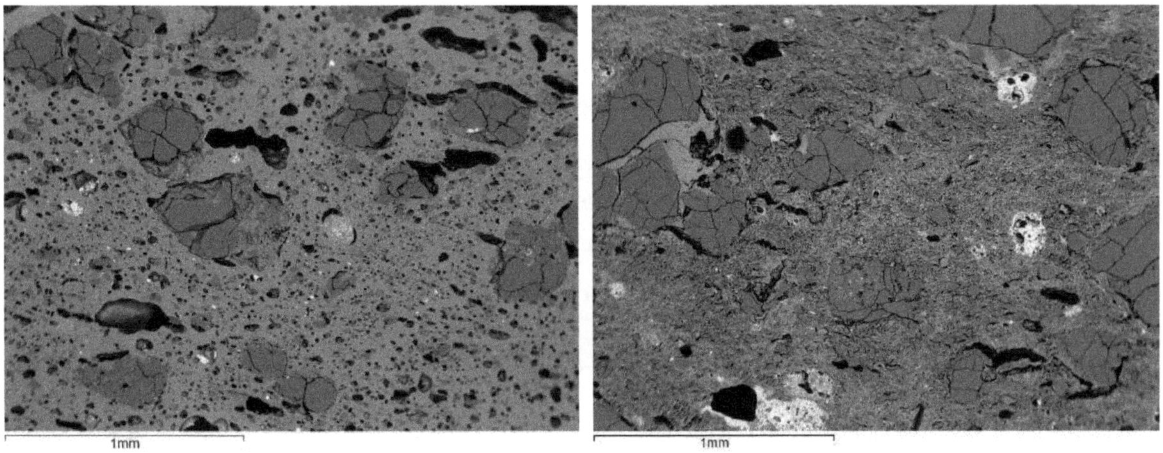

Figure 4.55 BSE images of used pot YPM3 (left) and unused pot YPM4 (right).

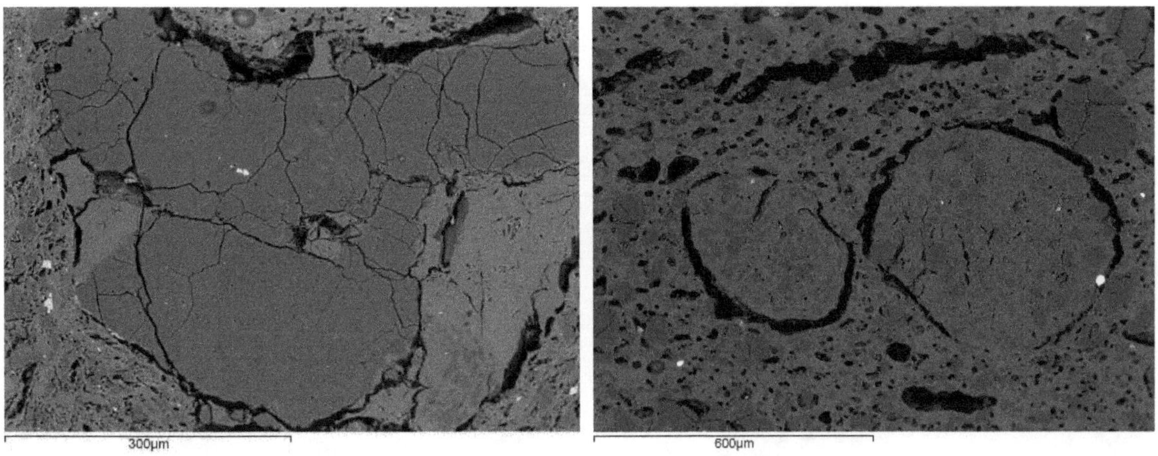

Figure 4.56 BSE images of a rock fragment in YPM4 (left) and two clay lumps in CMM3 (right).

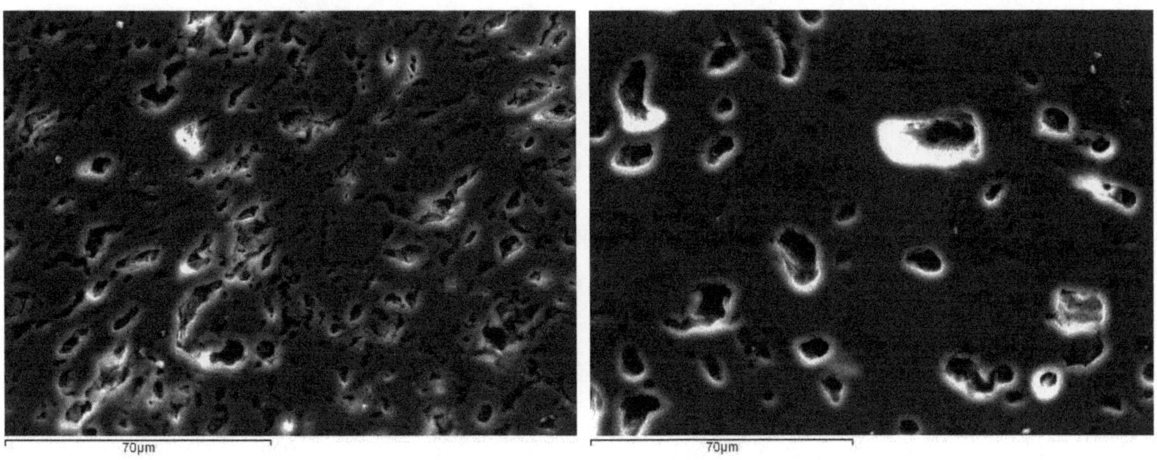

Figure 4.57 SE images of unused pot YPM4 (left), showing the extensive vitrification structure (V), and used pot CMM3 (right), showing a continuous vitrification structure with coarse bloating pores [CV(CB)].

Table 4.15 Average compositions of vitrified surfaces of the pots from Puzihe and Muxiexi (wt%), normalised to 100%. Analyses on polished sections for areas of ~400 by ~600 µm by SEM-EDS. i=internal surface; e=external surface. Matrix=average matrix composition of the pots from Puzihe and Muxiexi. '-' means 'not detected'.

Sample	MgO	Al_2O_3	SiO_2	P_2O_5	SO_3	Cl	K_2O	CaO	TiO_2	MnO	FeO	ZnO	BaO
YPM3i	1.8	16.4	67.6	-	0.1	-	3.8	3.0	0.9	0.1	4.3	1.8	0.1
CMT3i	1.4	18.3	69.1	-	-	0.1	4.1	1.5	0.9	0.1	2.8	1.0	0.5
CMM3i	6.7	14.2	56.4	-	0.2	0.4	1.6	18.1	0.7	0.2	1.0	0.6	-
CMB5i	3.8	13.9	60.0	-	0.3	0.6	2.1	17.9	0.6	0.1	0.5	0.2	0.1
Matrix	0.6	20.1	69.8	0.1	-	-	2.6	0.2	1.0	-	4.4	0.7	-
YPM3e	0.8	15.2	64.5	0.1	0.1	-	3.5	0.9	0.9	0.1	13.6	0.4	0.1
CMT3e	1.5	14.0	70.9	0.2	0.1	-	3.2	1.3	1.0	0.1	5.8	1.5	0.3
CMM3e	1.0	12.4	64.6	0.2	-	-	2.8	2.1	0.8	0.1	14.7	1.1	-
CMB5e	3.2	13.0	61.4	0.3	0.1	-	2.1	8.7	0.9	0.1	7.9	2.3	0.1

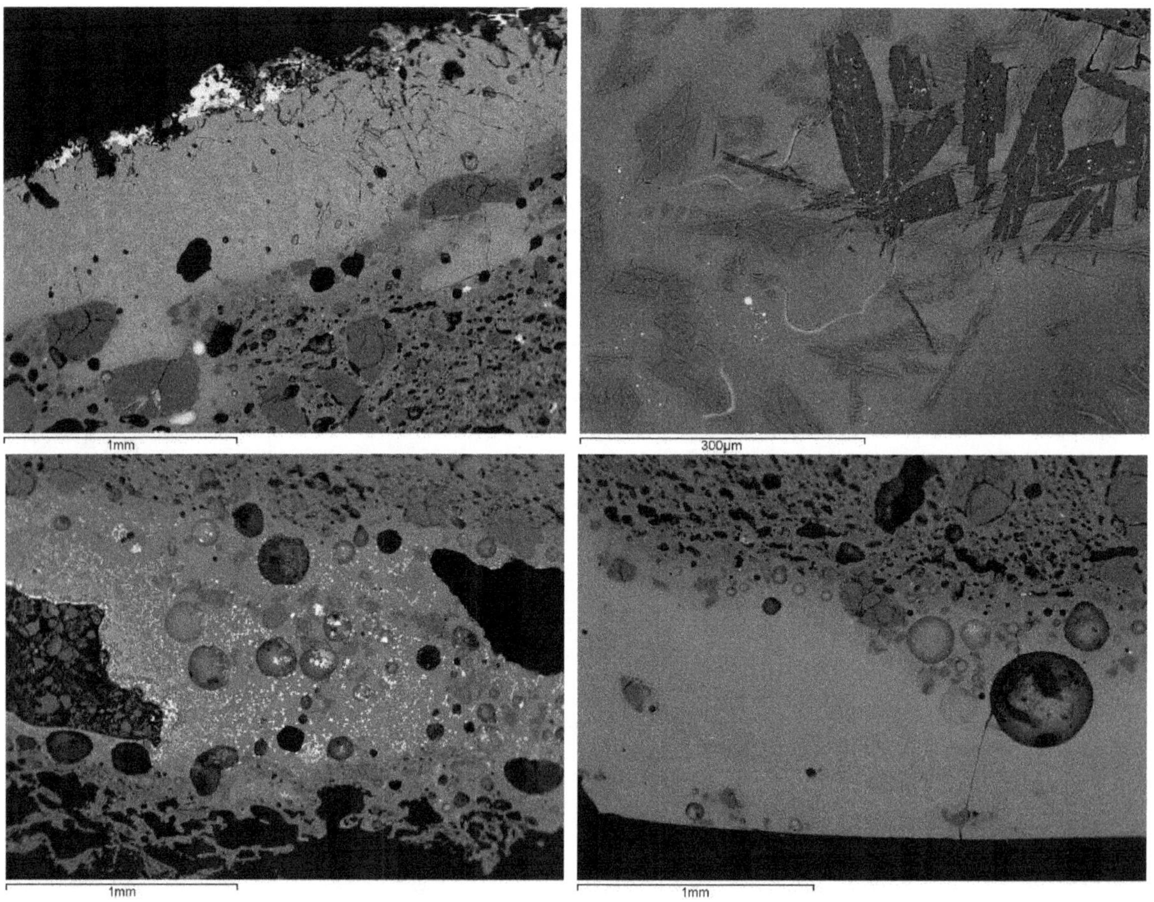

Figure 4.58 BSE images of vitrified surfaces of CMM3 and CMB5. Top left: the internal surface of CMM3, showing numerous diopside crystals within the layer, and rust on the innermost surface (bright grey) and the pot fabric (bottom); top right: the internal surface of CMB5, showing diopside crystals (mid grey) and anorthite crystals (dark grey); bottom left: the external surface of CMM3, showing quartz which had been semi-molten and tiny spinels (bright) within the layer, the outermost bloating lute layer (bottom) and the pot fabric (top); bottom right: the external surface of CMB5, showing occasional quartz grains within the layer and the pot fabric (top).

Condensers

The condensers collected are either connected with the pots (Figure 4.59) or detached from them. They are about 5 cm in height, cylindrical and thinner at the upper parts. Their colours range from yellow and brown to grey (Figure 4.60). Thin layers of whitish substances adhere to their internal surfaces.

Three condenser samples from Puzihe and three from Muxiexi were analysed. Their fabrics are contaminated with variable levels of ZnO, ranging from 5% up to 39%. To estimate the original compositions, the ZnO contents were omitted, and the data renormalised to 100%. This resulted in compositions of 68-73% SiO_2, 13-18% Al_2O_3, 6-8% FeO, 2-4% K_2O, 1-2% CaO and 1-2% MgO (Table 4.16).

They contain over 30 vol% fair-sorted inclusions, most of which are angular quartz grains and feldspar grains with sizes ranging from 10 to 500 μm in length. Two samples (YPT1 and YPT6) contain higher proportions of large coal ash fragments (Figure 4.61 top right), while the others have fewer large coal ash fragments (Figure 4.61 top left). In YPT1 and YPT6, there are a few large inclusions of sizes and abundance similar to those in the surrounding matrices (Figure 4.61 bottom left). They might be unhydrated lumps of powdered base clay which were not mixed well with the wet clay. A few fragments of vitrified materials were identified within the fabrics of YPT1, CMT1 and CMT4 (Figure 4.61 bottom right). The clear interfaces between the vitrified lumps and the surrounding unvitrified matrices indicate that the vitrified lumps were introduced to the condensers after they sintered. Their compositions (Table 4.17) are similar to the vitrified surfaces of the coal samples (Table 4.4) and the external vitrified surfaces of the used pots (Table 4.15). The vitrified fragment in YPT1 is particularly rich in iron, with two types of iron silicates and quartz grains. The presence of the coal ash and these vitrified lumps both indicate that the condensers were made at the site.

Most of the condensers are not vitrified, suggesting that they were exposed to temperatures below 800 °C. CMT2 and YPT2 from the lower part of the condensers are, in contrast, heavily vitrified with large bloating pores, indicating that they were exposed to higher temperatures, certainly over 800 °C. In YPT2 and CMT2, black vitrified layers with higher levels of CaO, MgO and K_2O than the condenser fabrics were found on the outermost parts; they are remains of the lute applied around the junction of the pots and the condensers.

Figure 4.59 The condenser attached to the pot, YPT1.

Figure 4.60 Vertical cross sections of some analysed condenser fragments from Puzihe and Muxiexi, with their internal surfaces facing left. Top left: YPT2; top right: YPT6; bottom left: CMT1; bottom right: CMT2.

Table 4.16 Average bulk compositions of three condenser fabrics from Puzihe and three from Muxiexi (wt%), normalised to 100%. Analyses on polished sections for areas of ~2 by ~2.5 mm by SEM-EDS. The rows at the bottom show the same results after omitting ZnO and renormalising to 100%. '-' means 'not detected'.

Sample	MgO	Al_2O_3	SiO_2	P_2O_5	SO_3	Cl	K_2O	CaO	TiO_2	MnO	FeO	ZnO
YPT1	1.2	17.0	65.1	0.2	0.2	0.2	1.9	1.1	1.0	0.1	6.9	5.1
YPT2	0.7	9.1	44.5	0.2	-	0.1	1.3	0.6	0.4	-	3.7	39.4
YPT6	1.1	15.3	64.3	0.3	0.2	0.2	2.4	0.7	0.9	0.1	6.9	7.6
CMT1	1.3	12.4	60.6	0.5	0.3	0.1	2.5	1.8	0.6	0.2	5.1	14.6
CMT2	1.4	13.0	64.7	0.3	0.1	-	2.9	1.1	0.8	0.1	5.2	10.4
CMT4	1.5	11.2	60.8	0.2	-	0.2	2.4	0.7	0.7	0.2	5.3	16.8
YPT1	1.3	17.9	68.5	0.2	0.2	0.2	2.0	1.2	1.1	0.1	7.3	
YPT2	1.2	14.9	73.3	0.4	0.1	0.1	2.1	0.9	0.7	0.1	6.2	
YPT6	1.2	16.6	69.6	0.3	0.2	0.2	2.6	0.8	1.0	0.1	7.4	
CMT1	1.5	14.6	70.9	0.6	0.4	0.1	2.9	2.1	0.7	0.2	6.0	
CMT2	1.6	14.5	72.1	0.4	0.1	-	3.3	1.2	0.9	0.1	5.8	
CMT4	1.8	13.5	73.2	0.2	-	0.2	2.8	0.8	0.9	0.2	6.4	

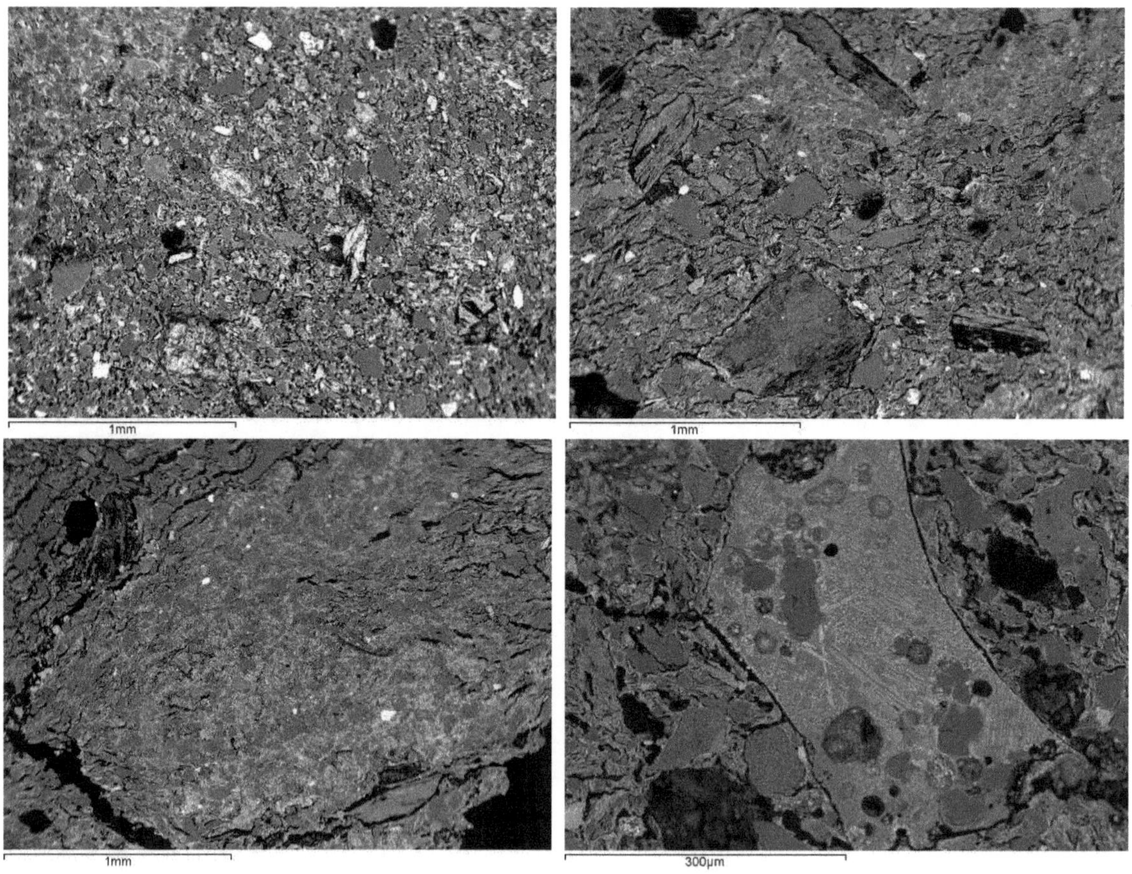

Figure 4.61 BSE images of the condensers from Puzihe and Muxiexi. Top left: the fine fabric in CMT4; top right: the coarse fabric in YPT6; bottom left: a large clay lump in YPT6; bottom right: a vitrified fragment in YPT1, showing crystals of iron silicates and quartz grains.

Table 4.17 Compositions of fragments of vitrified ceramics (wt%), normalised to 100%. Analyses on polished sections by SEM-EDS. '-' means 'not detected'.

Sample	MgO	Al_2O_3	SiO_2	P_2O_5	SO_3	K_2O	CaO	TiO_2	MnO	FeO	ZnO	BaO
YPT1	1.1	14.7	53.2	0.1	0.7	2.7	1.3	1.1	0.1	24.5	0.4	0.1
CMT1	2.4	13.3	63.3	0.9	0.1	3.7	7.0	2.3	0.3	5.2	1.4	0.1
CMT4	2.2	13.9	65.4	0.6	0.1	4.1	3.1	1.8	0.1	7.8	0.9	-

4.3.2.2 Slag

Two slag samples from Puzihe and three from Muxiexi were analysed. YPB1 is a large block of porous metallurgical residues still remaining inside the pot (Figure 4.62 left). CMB1 is thick vitrified dark slag attached to the bottom part; the slag clearly attacked part of the pot base, with a large pot ceramic fragment embedded within the slag (Figure 4.62 middle). The other slag samples (YPB5, CMM1 and CMB4) are small lumps attached to the bottoms or walls of the pots (Figure 4.62 right).

The five slag samples from Puzihe and Muxiexi generally contain variable levels of major and minor oxides: SiO_2 (33-58%), Al_2O_3 (7-15%), FeO (10-33%), ZnO (3-15%), CaO (2-8%), MgO (1-4%), BaO (0.3-2%) and SO_3 (0.4-1.4%) (Table 4.18). Each sample displays a different microstructure, as shown in the various phases identified. YPB1 contain s a significant amount of residues from the coal and charcoal (Figure 4.63), tiny crystals of olivine [$(Zn, Fe)_2SiO_4$], a few ceramic lumps and rust within pores.

YPB5 contains abundant crystals of pyroxene [$(Ca, Fe, Mg, Zn)_2Si_2O_6$] in the glassy matrix and rust, with a small number of metallic prills and some zinc sulphides.

In CMM1, olivine [$(Fe, Mg, Zn)_2SiO_4$], iron prills, zinc sulphides, rust and coal residues were identified. In addition, a few residual inclusions of zinc silicates (2-3 mm in length) were found. They occasionally appear to have recrystallised internally as angular, cored crystals with lower FeO levels in the cores and higher FeO levels (up to 10%) towards their surfaces. However, these inclusions still retain the shapes of the parent minerals but with cracks throughout (Figure 4.64 top left and top right). It is likely that zinc silicate minerals from the ores partially melted into the slag and later recrystallised internally to these iron-bearing zinc silicates. Their presence here suggests that zinc silicate minerals, such as those identified among the archaeological minerals from Miaobeihou, could not be reduced to metal as easily as zinc oxide or zinc carbonates, which probably explains the higher ZnO content in this sample.

In CMB1, abundant metallic iron prills and post-depositional iron oxides are clustered. The appearance of zinc silicate inclusions made of recrystallised cored crystals again proves the difficulty of reducing zinc silicates. Within a cluster of iron-bearing zinc silicate crystals (~12% FeO), are embedded metallic iron prills, zinc oxide with ~15% FeO and zinc-rich spinel ($ZnAl_2O_4$) (Figure 4.64 bottom left). Furthermore, a lump of pure zinc oxide (~500 µm in diameter) was identified with zinc silicate crystals growing between it and the slag melt (Figure 4.64 bottom right). The presence of zinc oxide and the absence of coal/charcoal suggest that perhaps in this case there were insufficient reducing agents for the complete reduction of the zinc oxide.

CMB4 is mainly composed of a glassy matrix, with rust deposited in the fractures and holes.

Figure 4.62 Slag samples YPB1 (left), CMB1 (middle) and CMM1 (right).

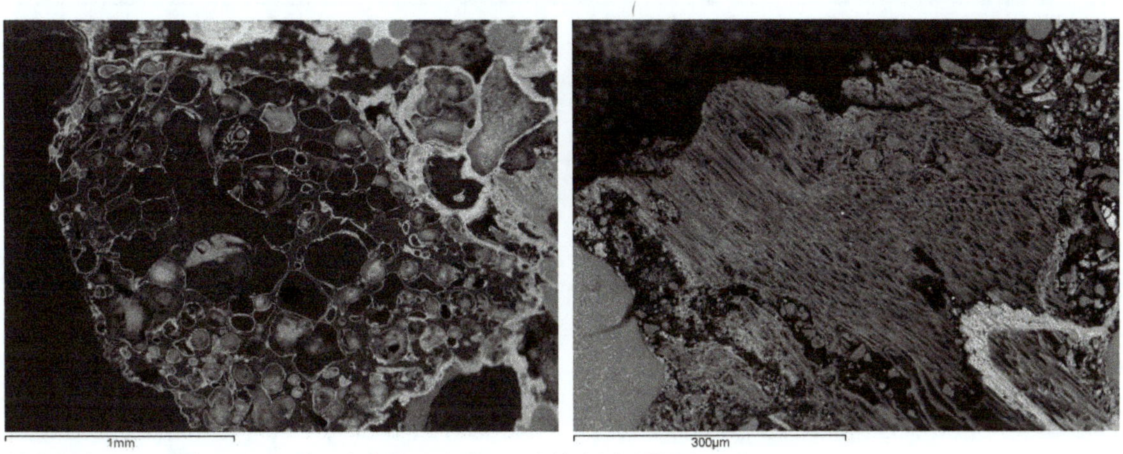

Figure 4.63 BSE images of fragments of coal (left) and charcoal (right) in YPB1.

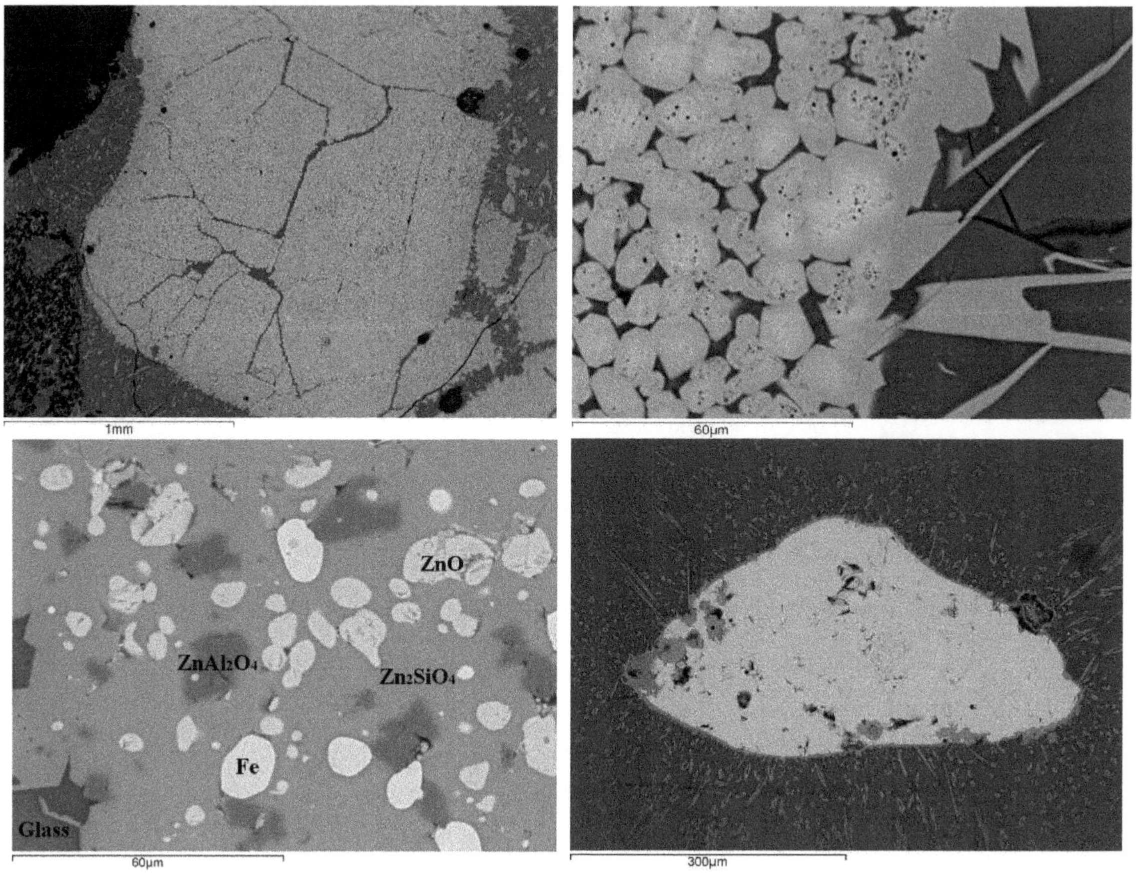

Figure 4.64 BSE images of slag CMM1 and CMB1.Top left: an inclusion of zinc silicate in CMM1; top right: enlarged detail of the zinc silicate inclusion in CMM1; bottom left: metallic iron prills (Fe), zinc oxide (ZnO) with ~15% of FeO and zinc-rich spinel ($ZnAl_2O_4$), in iron-bearing zinc silicates (Zn_2SiO_4) with ~12% FeO in CMB1; bottom right: a lump of zinc oxide in CMB1.

Table 4.18 Bulk compositions (top half) and glassy phases (bottom half) of slag samples (wt%), normalised to 100%. Analyses on polished sections for areas of ~2 by ~2.5 mm and ~400 by ~600 µm respectively by SEM-EDS. '-' means 'not detected'.

Sample	MgO	Al_2O_3	SiO_2	P_2O_5	SO_3	Cl	K_2O	CaO	TiO_2	MnO	FeO	CuO	ZnO	BaO	PbO
YPB1	1.3	15.3	58.2	0.3	0.4	0.3	3.4	2.1	0.8	0.1	9.7	0.1	6.8	1.2	-
YPB5	3.3	13.1	41.3	0.2	0.6	0.1	1.0	7.5	0.6	0.1	28.6	0.2	3.1	0.3	-
CMM1	3.8	7.3	33.1	0.2	0.8	0.2	0.5	5.1	0.3	0.2	32.8	0.1	14.3	1.3	-
CMB1	3.0	9.1	45.8	0.3	1.4	0.1	1.4	5.3	0.3	0.2	15.6	0.1	15.2	2.1	0.1
CMB4	2.3	12.1	57.9	0.2	0.6	0.1	1.8	3.5	0.4	0.2	15.6	0.1	3.4	1.8	-
YPB1	1.7	16.3	59.8	0.2	0.3	0.2	3.9	2.6	1.0	0.2	5.1	-	7.8	0.8	0.1
YPB5	3.7	16.6	58.1	0.2	0.5	0.1	2.2	10.0	0.9	0.2	4.0	0.1	3.1	0.3	-
CMM1	5.8	9.2	42.8	0.2	1.1	0.3	0.9	7.0	0.4	0.2	15.0	-	15.4	1.6	0.1
CMB1	3.0	9.7	51.3	0.3	0.8	0.1	1.7	5.7	0.5	0.2	10.0	-	14.2	2.5	-
CMB4	2.7	13.2	58.9	0.1	0.5	0.1	1.9	3.9	0.6	0.3	12.3	-	3.6	1.9	-

4.3.3 Discussion

Given their similar makeup, the formal and material properties of the retorts from Puzihe and Muxiexi are generally commensurate with those from Miaobeihou. The pots are wheel-thrown flat-bottomed jars made of a quite refractory clay and quartz temper. The pots were prefired at 1100-1200 °C in specialised ceramic kilns and exposed to 1200 °C or higher temperatures during smelting. They have good strength, toughness and resistance to the chemical attack of the charge. But the used ones show slightly higher levels of ZnO than those from Miaobeihou. The condensers were built upon the pot rims with less refractory ceramic materials (clay and coal ash) than the pot fabrics (Figure 4.65). They were probably not prefired and experienced lower temperatures (<800 °C) during smelting. Although no pockets from Puzihe and Muxiexi have been found, similar pockets to those from Miaobeihou are likely to have been used. Iron lids might also have been employed at these sites, but no lids have been found so far.

Based on the analytical results of the slag samples from Puzihe and Muxiexi, the zinc ores used were mainly zinc carbonates with variable levels of FeO. Gangue minerals might have also included calcite, dolomite and barite. The reducing agents used are likely to have combined coal and charcoal, as residues of both were identified in YPB1. The slag samples from Puzihe and Muxiexi have more variable microstructures than those from Miaobeihou. However, only a few samples were available for analysis for each site, so generalising about a higher variability in operating parameters at these sites would be premature.

It is worth noting three important aspects concerning the raw materials and the metallurgical processes which can be inferred from the slag samples. First, the presence of residual zinc silicates in CMM1 and CMB1 indicates that hemimorphite cannot be as easily reduced as zinc carbonates. Hemimorphite would need to be first dehydrated and then converted to willimite (Zn_2SiO_4). Decomposition would not occur until 1430 °C because of the strong bond between zinc oxide and silica. From the perspective of modern zinc metallurgy, zinc silicate is more difficult to reduce than zinc oxide. It can only be reduced by solid carbon and the reduction only takes place when the mineral is intimately mixed with carbon (Ingalls 1903, 206). In CMM1 and CMB1, the zinc silicate ores were probably not in direct contact with the reducing agents, but surrounded by the ferruginous slag. Thus the zinc silicates partially melted and recrystallised to iron-bearing zinc silicates by taking in some iron oxides from, and releasing some zinc oxide into, the slag melt. Nevertheless, the presence of these unsuccessfully reacted hemimorphite particles suggests that some might have been reduced by carbon. Hemimorphite is a kind of zinc ore reported to have been utilised in some of the traditional zinc smelting practices (see Section 2.3), indicating that its reduction is not impossible.

Secondly, generally an excess of reducing agents was required to make the reduction of zinc oxide efficient, as confirmed by the coal and charcoal residues identified in YPB1; while insufficient carbon would leave unreduced zinc oxide, as identified in samples CMB1 and CMM1. If not reduced, the zinc oxide would readily react with silica from the slag melt to form zinc silicate crystals containing a few percent of FeO.

Thirdly, as already shown for Miaobeihou, the SiO_2 and Al_2O_3 levels in the slag samples show a positive correlation with those of the pot and condenser samples (Figure 4.66). The SiO_2 and Al_2O_3 in the slag partly derived from the thermal degradation of the pot fabrics. However, considering the high levels of these oxides in the slag, it is likely that the bulk was introduced by the recycled zinc-rich condensers and pockets added to the charge, which is most evident in the ceramic lumps within YPB1. Some SiO_2 and Al_2O_3 could also have been introduced by the zinc ores and coal.

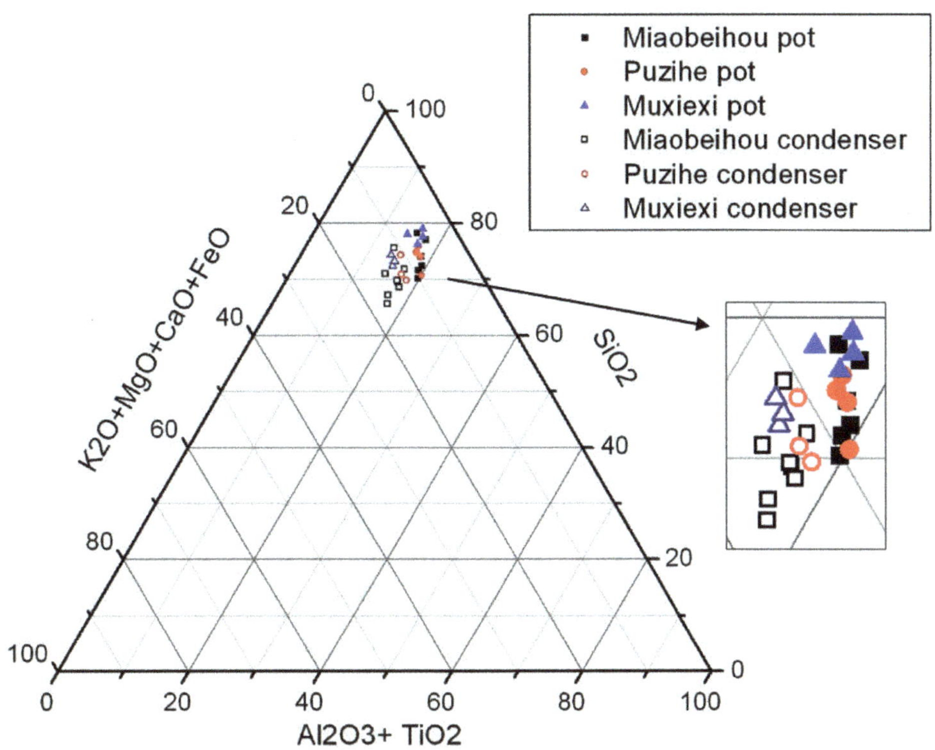

Figure 4.65 Ternary diagram of SiO_2-(Al_2O_3+TiO_2)-(K_2O+CaO+MgO+FeO) showing the bulk compositions of the pot and condenser fabrics from three Fengdu sites. The condensers contain higher levels of alkali and alkaline earth oxides and FeO than the pots, but lower levels of SiO_2.

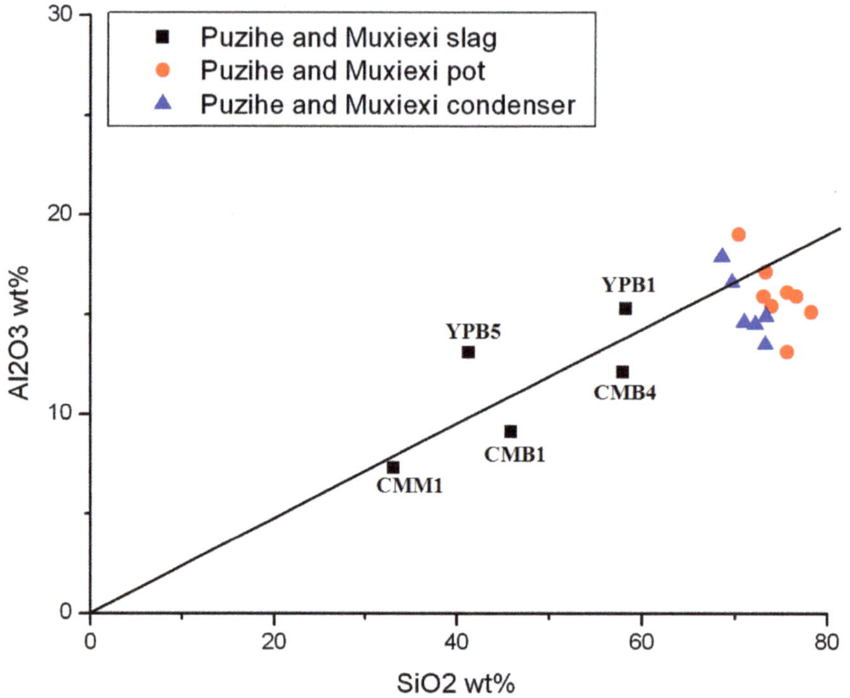

Figure 4.66 The SiO_2 and Al_2O_3 levels in the slag, pot and condenser samples from Puzihe and Muxiexi. Note the rough linear correlation of SiO_2 and Al_2O_3 of the slag samples with relative proportions similar to those in the ceramics.

4.4 Zinc production in Fengdu

From the archaeological finds and analytical results of production remains from three sites in Fengdu, it has been possible to reconstruct the technology of zinc distillation employed at a much higher resolution. This section first summarises the *chaîne opératoire* and technological choices of zinc production in Fengdu. The technological reconstruction enables a reinterpretation of the description and illustration in *Tian Gong Kai Wu*. This section then attempts a more dynamic picture of the organisation of production, followed by an estimation of the scale of production. An even broader perspective will be presented in chapter six, facilitated by the comparison with the *chaîne opératoire* at Dafengmen, Shizhu.

4.4.1 Zinc distillation technology in Fengdu and Tian Gong Kai Wu

The main stages of zinc production, i.e. the *chaîne opératoire*, included retort production, mining and ore preparation, reducing agent and fuel preparation, furnace construction and smelting.

Retorts used at the Fengdu sites were made up of three ceramic parts, so several steps were involved in their production. The first step was the manufacture of jar-shaped pots. The similarities among the pots from the three sites in Fengdu in terms of shapes, sizes, materials (clay and temper) and manufacturing methods suggest that they were not made by the zinc smelters, but rather manufactured in specialised ceramic workshops. To produce large numbers of wheel-thrown high-fired pots, the ceramic workshops must have been equipped with large ceramic kilns to facilitate large-scale production. In the traditional process in Bijie in 1994-95, kilns with several linked chambers were used and a few hundred pots were fired simultaneously (Craddock and Zhou 2003). There have been ceramic kilns in the areas where these zinc workshops were situated since the Neolithic period (see Section 4.1.1). Although no kilns for firing the smelting pots have been found, they may have been located in the area. The condensers and pockets made of clay and coal ash would have been added to the pots at each zinc workshop, as reflected in their more variable makeup and the presence of coal ash fragments in their fabrics.

Zinc ores used were predominantly iron-rich oxidic zinc ores, mainly smithsonite. They might have been transported from the nearest zinc mines located in the Laochangping region, Shizhu (see Section 5.1). The ores were beneficiated and crushed to small fragments before being mixed with reducing agents, coal and charcoal. Coal was used as fuel, but whether it was in the form of raw coal or coal briquettes remains uncertain. Charcoal could have been produced locally, but the coal must have been transported from nearby coal mines. Furnaces were rectangular-shaped and built with Han bricks, with dozens of furnace bars arranged at regular intervals on the bases.

Before smelting, the pots, connected with the condensers, were charged with mixed zinc ores and coal/charcoal. Clay lumps tempered with coal ash were pressed onto the top of the charge to form the pockets, leaving a hole on one side. The retorts were then covered with lids and luted. They were fired within the rectangular furnaces fuelled by coal. In the reaction zone, high temperatures of about 1200 °C and strongly reducing conditions were achieved. The zinc vapour formed passed through the pocket hole and condensed in the pockets of the cooler condensation zone. Metallic zinc of high purity was produced and probably remelted to cast into larger zinc ingots.

The processes employed at the Fengdu sites show some similarities to that depicted in *Tian Gong Kai Wu* (see Section 1.2): the shapes of the earthen jars illustrated are very similar to the Fengdu pots; the ores charged into the jars were also oxidic zinc ores; the fuel to heat the jars was coal. Thus we may infer that other characteristics might have been similar. It has been proposed that the apparent heap of stacked pots shown in this source does not represent an actual pile, but may instead be just an effect of the perspective drawing of several rows of pots sitting at the same level (Craddock 2009a). However, there is evidence at other installations of the piling of items in a heap for firing, such as brick firing and lime burning, which are very similar to the illustrations in *Tian Gong Kai Wu* (Figure 4.67). As such, it is possible that the illustration does show an actual heap of zinc smelting retorts – which would not be functional given the necessary thermal gradients that have to be maintained between pots and condensers. This potentially inaccurate representation may indicate that Song Yingxing, the author of *Tian Gong Kai Wu*, did not fully understand the process.

The apparently inaccurate representation may be explained by considering the authorship and aims of the book itself. Song Yingxing (AD 1587-1666) was born into a family of bureaucratic landlords in Jiangxi province in 1587 and spent his early years studying for the imperial examinations in pursuit of a political career, but failed five times to pass the highest level exam in the capital. In the 1630s, he ended up working as a local education official and turned his interest to writing about science, technology, philosophy, politics and society (Pan 1990). Between 1636 and 1638, Song published his broad thematic writings using woodblock printing, including *Tian Gong Kai Wu*. However, Song's aim was not mainly to produce a technical handbook, but to enclose the all-encompassing scope of the traditional Chinese literati (Schäfer 2005, 53). The documentation of technical details in *Tian Gong Kai Wu* was just a medium to uphold his idea that the congruence of knowledge and action shaped mankind's world (Schäfer 2005, 55). This could explain why there are obvious mistakes and misapprehensions in the texts and illustrations of *Tian Gong Kai Wu*, such as zinc smelting and also iron and steel production technology (Wagner 2007; Wagner 2008, 339-346). It is likely that the author did not witness these processes but learnt them from others or from written sources. This point further stresses the need to critically assess a range of documentary sources and, where possible, contrast their information with archaeological data.

4.4.2 Organisation of production

It is difficult to discuss the organisation of production because the archaeological reports of many of the zinc sites are not available for consultation. With only limited and fragmentary archaeological information, we can obtain a preliminary picture of how zinc production was organised in Fengdu, which can be developed and refined in the future.

Figure 4.67 The illustrations of zinc smelting (left), brick firing (middle) and limestone burning (right) in *Tian Gong Kai Wu* (Song 1637). These processes were all stacked with alternate layers of coal briquettes and jars/bricks/limestone.

The approximately 20 zinc smelting sites were all situated on the first terraces along the Yangtze River. Although only the Miaobeihou site has been dated to the fifteenth to seventeenth centuries by radiocarbon dating, the similarities of the retorts, furnace structures and other relevant debris suggest that all these workshops were in use in similar periods. Considering the historical background of brass and zinc in China (see Section 2.1), it is more likely that these zinc workshops were in operation during the late Ming Dynasty. However, at the very end of the Ming period the region suffered the turmoil of war, so production would have ceased. Thus zinc production was probably at its full potential for a short period from the late sixteenth to early seventeenth centuries.

Zinc production in Fengdu was a nucleated industry with dozens of workshops, which were centred around the furnaces. The layout of zinc workshops in Fengdu can be reconstructed based on the excavation of a relatively complete workshop at the Jiudaoguai site (Figure 4.1, no.11). The furnace was surrounded by a working shed, a working platform and several pits (Figure 4.68 and 4.69). The working shed was a two-room house built of rammed earth. It was 7.2 m in length and 4.6 m in width. It had a door opening towards the furnace and one of the rooms had a hearth. The platform outside the shed was paved with Han bricks and stones. The pits around the furnaces were areas for storing or preparing raw materials and discarding spent retorts and metallurgical residues. A similar workshop layout was also recorded at Miaobeihou (see Section 4.2.1). The layout of these sites resembles that of the traditional zinc smelters (Figure 4.70), suggesting that the majority of the smelting activities, such as preparing ores and coal, mixing the charge, charging pots and dumping waste, were taking place in the immediate vicinity of the furnaces. Activities such as building condensers, adding pockets and luting retorts could also have been carried out next to the furnaces, as indicated by the illustration in *Tian Gong Kai Wu*, where two craftspeople appear to be adding some clay to the pots (Figure 1.4).

At another zinc workshop at Jiudaoguai, six furnace foundations were excavated (Figure 4.71). More than one furnace could have been in operation at the same time at this workshop. One of the furnace foundations overlapped another one, suggesting that furnaces might also have been rebuilt. At Jiudaoguai, three areas with concentrated smelting remains were located at intervals of a dozen metres. Whether they were individual workshops or belonged to a large production unit remains uncertain. It is also unknown how the approximately 20 zinc smelting sites were related to one another. However, since all these workshops nucleated along the Yangtze River within Fengdu county, it is very likely that they were organised in the same manner and shared the same sources of zinc ores, coal and pots.

It is difficult to comment from the archaeological evidence on how many craftspeople were involved or the division of labour in operating one furnace and the various activities by the furnace. Ethnographic evidence of the traditional processes suggests that each furnace was most likely attended by one highly experienced craftsperson, called the furnace master, who was responsible for crucial operations, and several less skilled workers who assisted the master and carried out some simple tasks (see Section 2.3).

Figure 4.68 One furnace foundation at Jiudaoguai (photo by Liu Haiwang). The rectangular furnace was surrounded by a working shed, a working platform and several pits.

Figure 4.69 A zinc workshop model reconstructed based on the archaeological finds at Jiudaoguai, displayed at the China Three Gorges Museum. A shelter should have been built for the furnace to protect from the rain, as practiced in the traditional processes (Xu 1986; Craddock and Zhou 2003).

Figure 4.70 A traditional zinc smelting workshop at Ya Kou near Zhehai, Yunnan province (photo by Paul T. Craddock, May 1995). The two sheltered saddle furnaces at the front were surrounded by piles of retorts, coal and possible ores/mixed charge. The machine in the middle was used for mixing ores and coal.

Figure 4.71 Six rectangular furnace foundations at Jiudaoguai (photo by Liu Haiwang).

4.4.3 Scale of production

The capacity of an ordinary retort from Miaobeihou is estimated at about 2 L, roughly calculated from a cylinder with an internal diameter of 10 cm and a height of 26 cm. Assuming the charge consisted of zinc ores with a grade of 35% ZnO as seen in the archaeological minerals (see Section 4.2.2.1) and an excess of coal, three times the theoretical value (see Section 4.2.3.3), the ratio of ore to coal in the volume was about 2:1 (using specific gravities of 2.5 for ores and 0.8 for coal). Allowing for a 20% loss of zinc as mentioned in *Tian Gong Kai Wu* (Sun and Sun 1966, 247), about 0.7 kg of zinc metal would have been produced per retort. If the reducing agent was charcoal alone (specific gravity 0.5), 0.6 kg zinc could have been produced per retort. Therefore, an ordinary retort could have produced 0.6-0.7 kg metallic zinc per firing, roughly corresponding to the weight of the planoconvex ingots described above (see Section 4.2.2.3). If a furnace bar held three retorts as with traditional trough furnaces, then an ordinary furnace with 30 furnace bars could have accommodated 90 retorts and produced 54-63 kg zinc. Assuming each firing lasted one day, as for the traditional process, one furnace could have produced 11-13 tonnes of zinc for 200 working days a year.

At present, it is difficult to estimate the scale of production in Fengdu any further due to the lack of detailed archaeological information about site dimensions and relative chronologies. Although it is possible that retorts would be reused when the slag could be detached from the inside, the uncertainty about this point makes it even harder to gauge the yield of the site even by counting the number of pots preserved. There is no doubt, in any case, that the scale of production during the late Ming Dynasty must have been impressive, since 20 sites have hitherto been excavated by the river and thousands of retorts litter the area.

It is therefore surprising that no historical documents have been found recording the mass production of zinc in this region in the Ming period. It is known that, in the late Ming Dynasty, zinc was mostly used to make brass coins in official mints (see Section 2.2.1). The large-scale production of zinc in Fengdu is most likely to have been set up to supply zinc for minting coins. The location of these sites by the river, rather than in closer proximity to the mines, may have been driven by the government's desire to control production. The reasons for the choice of location as well as the potential destination of the zinc produced and its broader implication will be further discussed in Section 6.1.

Chapter 5

Analyses of Zinc Smelting Remains in Shizhu

Following on from the discussion on the reconstruction of technology at the river sites in Fengdu, this chapter turns to the Laochangping region, Shizhu. The geology, historical records and archaeological finds of this area are introduced. Analytical results of geological zinc minerals from Yushi Cave, an old mine working, and retorts and slag from the up-land Dafengmen zinc smelting site are presented, followed by a discussion on retort design and the charge.

5.1 Zinc mines and smelting sites in Shizhu

5.1.1 Geology

Zinc ores usually coexist with lead ores. Lead-zinc ore deposits are widespread in most provinces in China, forming twelve primary metallogenetic zones (IMRCAGS 2007). The nearest lead-zinc deposits to the zinc smelting sites at Fengdu discussed in the previous chapter were found about 50 km away from the smelting sites in the Laochangping region, Shizhu county, southeast Chongqing.

The southeast Chongqing area is located in the northeast fold belts within the upper Yangtze Syneclise of the Yangtze Platform, which includes three structural units: the Qiyao Mountain, the Qian River and the Xiu Mountain. Lead-zinc deposits generally occur in the Paleozoic carbonate rock, mainly related to the strata of the Lower and Middle Cambrian period (Yang *et al.* 2008). Lead-zinc deposits in southeast Chongqing have been found in the counties of Shizhu, Youyang, Qianjiang and Xiushan. The largest deposits lie in the Laochangping region, located at the northern end of Laochangping Anticline, i.e. Qiyao Mountain.

A geological survey in the Laochangping region in the 1980s revealed two mining areas: Xijiaoxi and Laochangping. The Xijiaoxi area is composed of 15 small lead-zinc deposits, all located along the Xijiaoxi 洗脚溪 (a small river for washing feet) - local folklore says that the miners used to rest and wash their feet by the river. Liuhuang Cave is an old mine working which was exploited by ancient miners. In the Laochangping area, 31 old mine workings have been found, including Yushi Cave and Chuan Cave. The lead-zinc deposits of the two areas usually include primary sulphidic ore minerals, such as sphalerite (ZnS), galena (PbS), pyrite (FeS_2), chalcopyrite ($CuFeS_2$) and chalcocite (Cu_2S); secondary oxidic ore minerals, such as limonite [$FeO(OH)\cdot nH_2O$], smithsonite ($ZnCO_3$), hemimorphite [$Zn_4Si_2O_7(OH)_2\cdot H_2O$], hydrozincite [$Zn_5(OH)_6(CO_3)_2$], cerussite ($PbCO_3$), anglesite ($PbSO_4$), malachite [$Cu_2CO_3(OH)_2$] and otavite ($CdCO_3$); and gangue minerals, such as barite ($BaSO_4$), fluorite (CaF_2), calcite ($CaCO_3$), dolomite [$CaMg(CO_3)_2$], quartz (SiO_2) and clay. Lead and zinc ores are usually associated with Fe, Ag, Hg, Cd, Ge and Cu (Fan and Liu 1982).

5.1.2 Historical records

Many ethnic minority groups lived in the mountainous areas of southwest China. In these areas, the *Tusi* 土司 (native chieftain) system, a special local administrative system was in operation during the Yuan, Ming and Qing Dynasties. Native chieftains were tribal leaders recognised as imperial officials by the central government to exercise indigenous administrative power. They were obliged to pay tax and tribute to the court, provide troops for the court and suppress local rebellions. They also had the right to reign over their areas as tribal leaders. The post was confirmed as hereditary, but succession, promotion, and demotion were all controlled by the central Chinese administration (Gong 1992).

Shizhu is mountainous and deeply cut through by small streams. In the Ming and early Qing Dynasties, Shizhu was under the *Tusi* system. The Ma family *Tusi* governed the local indigenous *Tujia* 土家 people through hereditary succession. Shizhu was a relatively isolated region which had little contact with the outside world. The *Tujia* people mostly depended on farming, gathering, hunting and fishing. They did not have a particularly good reputation for crafts, but some industries also developed on the basis of family workshops, such as wine making, papermaking, textiles, mining and metalworking. As the transport conditions were harsh in such a mountainous region, commercial activities were carried out only in some market towns (Peng and Li 2009; CCLGSC 1994; Li 2008a).

Shizhu was rich in mineral resources, such as iron, copper, lead, silver, zinc and coal. There was a long history of mining and smelting in the Laochangping region, as indicated by the name Laochangping 老厂坪, literally 'a flat area with old factories'. As early as the Tang and Song Dynasties, lead and silver were exploited in the region (Huang and Lang 1992, 595). During the *Tusi* period, the *Tusi* family and the ingenious people had already participated in the exploitation of these resources. The earliest direct evidence is recorded in the official historical work *Ming Shi* 明史 (History of the Ming Dynasty), compiled by a number of officials commissioned by the Qing court (Zhang 1974):

> In 1480, Sichuan provincial governor Sun Ren reported to the Emperor: 'In March, 300 interior robbers came to Shizhu, killed *Tusi* Ma Cheng and over 20 yamen runners, set fire, plundered and then left. This is because Shizhu and neighbouring

> Fengdu fought for areas producing silver, but officials did not punish either side, resulting in their hostility and vendetta between them'. The Emperor then commanded officials to arrest the robbers. Sun also reported: 'Shizhu annually collected a lead tax of 5130 catties [c. 3 tonnes], which ceased after the Zhengtong period (AD 1436-1449). Neighbouring soldiers and civilians arrogated to themselves the right of taxation and seized the opportunity to steal the profits, leading to disasters. Please exempt the lead tax, close the mines and remove the patrol and inspection official of Linjiang, Zhongzhou to Jiangchi, Nanbinli, Fengdu to facilitate defence.' The Emperor followed Sun's request. (see Appendix 4.1)

The 'lead' in the text most likely referred to black lead, i.e. lead, as *Comprehensive Gazetteer of Sichuan Province* of the Jiajing period (AD 1522-1566) clearly documents that Shizhu produced 'black lead' (Liu and Yang 1996, 284). It can be inferred from these two accounts that silver and lead had been mined and produced at least since the fifteenth century. However, local conflicts arose over the proceeds of the mining and smelting activities in the Laochangping region, which at that time were situated between Shizhu and Fengdu. The issues were so serious that they were reported to the Emperor for judgments. In order to control the situation, the mines were closed and an office was set up in proximity to Shizhu, at Fengdu for defence purposes.

In the Wanli period (AD 1573-1620), the Emperor sent many officials all over the country to inspect mining and smelting activities and to collect tax. It is recorded that in 1613, *Tusi* Ma Qiansheng offended Qiu Shengyun, the official sent to Sichuan province, and was put into prison because of mining affairs (RGALHS 2009). What types of ore were mined is not mentioned in the text. No clear evidence of zinc mining and smelting activities at Shizhu has been found in Ming documents.

During the Yongzheng period (AD 1723-1735) of the Qing Dynasty, the *Tusi* system in southwest China was mostly replaced by the *Liuguan* 流官 (officials with limited tenure) system. In Shizhu, the *Tusi* system was finally abolished in 1761 because of severe conflicts inside the *Tujia* family. The central government directly controlled the area by appointing officials regularly (Peng and Li 2009). With the change of administrative system, numerous Han people, mostly from the neighbouring provinces of Jiangxi, Hunan, Hubei and Guizhou, migrated into this region, bringing Han production technology and culture (Tan 2009). Agriculture, animal husbandry, craft and commerce activities flourished. The mining and smelting industries developed very fast, because the Qing central government encouraged merchants and commoners to invest in exploiting ore resources (Li 2010).

According to *Local Gazetteer of Shizhu*, compiled in 1775 (RGALHS 2009), after the change of the administrative system, copper, lead and zinc were produced for minting on a large scale in Shizhu. The mining and smelting of minting materials were directly administered by *Tongzhi* 同知, the governor of Shizhu. The factories were managed by government officers from nearby areas. At the Baishaling factory, white lead, i.e. zinc, was first smelted in 1768 and mass-produced from 1770. The production of white lead consumed coal; the coal mines were 130 *li* (c. 73 km) away from ore mines and it took three days for a round trip on foot. Part of the white lead produced had to be paid as tax to the government; another part had to be sold to the government at the reduced price of 1.85 taels silver per 100 catties (c. 60 kg); the rest could be sold to merchants at higher prices for profit. Over 300,000 catties (c. 179 tonnes), including the parts for tax and government purchase, were sent to the Sichuan provincial mint in 1770. Initially, white lead commanded good prices on the local market, but later it became cheaper in order to compete with that imported from neighbouring Guizhou province. However, the costs of production were so high compared to those of Guizhou zinc that the Baishaling factory stopped production for the market and simply yielded the amount required for tax and government purchase. In 1775, smelting 100 catties of white lead resulted in a loss of 0.8 to 1 tael silver, which upset both the government and the smelters. The Baishaling factory also occasionally produced black lead, taking advantage of the fact that white lead ores sometimes contained black lead minerals (see Appendix 4.2).

A revised edition of *Local Gazetteer of Shizhu*, complied in 1843 (RGALHS 2009), records that old mine workings were later destroyed by flooding, and that the factory was moved quite frequently to many localities: Leigutai, Lengshuixi, Heitianchi, Shenxigou, Dafuping, Luziping, Dahuotian and Zhongxigou[4]. It is also stated that the production of white lead was constrained not only by supply of ores but also by that of coal, i.e. the frequent relocation of factories was not only related to floods but also the need to move the furnaces closer to the coal mines. In 1783, the tax was cut and 95,600 catties (c. 57 tonnes) of zinc were sent to the provincial mint. In 1838, the tax increased gradually again to 127,500 catties (c. 76 tonnes) before decreasing to 60,000 catties (c. 36 tonnes) in 1839. Finally, the production of zinc declined due to the exhaustion of zinc ores and the unprofitable production (see Appendix 4.3).

In addition to the official archives recording the management and evolution of zinc production, a local historical document provides more information (Ran and Ran 1991). According to this source, the government set up offices of Sanfutou and Fengshui Yamen at Dafengmen and Fengshuisi, respectively, to administer mining and smelting affairs. It is also recorded that at Dafengmen there was a Waguanchang 瓦罐厂 (earthenware jar factory) which could have provided

[4] These localities should be not far away from Baishaling, in the Laochangping region. Further investigations should be made to find out their specific locations.

crucibles for hundreds of furnaces. The elders said that there used to be a market town with over 100 houses in the Laochangping region, where dozens of pigs were killed every day to serve the miners. Some places in this region may also have been named because of the large-scale mining and smelting activities, such as Jieshangbazi 街上坝子 (a market for trading), Dianziping 店子坪 and Fanjiadianzi 范家店子 (hostels for accommodating merchants), Zhanfang 栈房 (hostels for serving the porters) and Zafangwan 醡房湾 (workshops for making local alcohol).

According to the two Shizhu gazetteers mentioned above, mining and smelting in Shizhu from the Qianlong period (AD 1736-1795) were supervised by the local governor, *Tongzhi*, who was responsible for collecting tax and solving civil disputes. Private merchants invested in and ran factories, some with raised capital (Huang and Lang 1992, 595). One of the old mine workings in Laochangping, Liujiadadong 刘家大洞 (a large cave of the Liu family), was probably run by a merchant named Liu. In 1796, Zeng Zhaoqian, a merchant from Jiangxi province, came to Laochangping to exploit lead ores, which, according to his gravestone, made him a fortune within a few years (Figure 5.1).

Figure 5.1 Left: Zeng Zhaoqian's grave; right: the gravestone with inscriptions saying that he came to Laochangping to exploit lead ores in 1796.

5.1.3 Field surveys

While searching for the possible zinc mines which supplied the zinc smelting sites at Fengdu in late 2004, the first field trip to Longtan township[5], Shizhu, led to the discovery of three smelting sites: Laochangping, Guanziyao and Dafengmen. During the following fieldwork season in 2005, eleven rectangular furnace foundations were found at the sites of Longdongwan and Laoyaodong. Some of these were excavated. In 2009, the national archaeological survey in Longtan revealed heaps of slag and retorts and old mine workings at seven more sites. In total, eleven smelting sites have so far been discovered in Longtan (Table 5.1).

Table 5.1 Smelting sites in Longtan, Shizhu and relevant finds.

Village	Site	Finds
Duhui 都会	Dafengmen 大风门	Retorts
	Guanziyao 罐子窑	Retorts, iron lids
	Longdongwan 龙洞湾	Retorts, zinc smelting furnaces
	Laoyaodong 老窑洞	Retorts, zinc smelting furnaces
Longtan 龙潭	Laochangping 老厂坪	Tap slag, old mine workings
	Baolongtang 宝龙塘	Unknown
Muping 木坪	Tianxianqiao 天仙桥	Old mine workings, slag
Wanbao 万宝	Baishanling 白山岭	Tap slag
	Luzhashi 炉渣石	Tap slag
	Luohedong 罗河洞	Unknown
	Guanhuwan 官湖湾	Unknown

[5] Township is a rural administrative division under county, but above village.

With limited access to the reports of these field surveys, the kinds of smelting activities involved at each site still remain unclear to the author. However, the abundance of retorts and remains of rectangular furnaces indicates that a given site was smelting zinc, while the presence of tap slag heaps suggests a different metallurgical activity, possibly lead smelting. More fieldwork and scientific examination of smelting remains are necessary to reveal the activities at each site and reconstruct the whole mining and smelting history of this area.

In August 2010, the author visited four of these sites: Laochangping, Baishaling, Luzhshi and Dafengmen. At the first three sites, all located in the Laochangping area, there are many large heaps of tap slag with clear flow structures (Figure 5.2 top left and top right). A lead fragment was recovered at the Laochangping site (Figure 5.2 bottom right). However, no evidence of furnaces, crucibles or tuyères was found. Charcoal was found trapped inside some tap slag (Figure 5.2 bottom left). Radiocarbon dating of charcoal samples from the three sites yields dates which mostly fall in the Ming and Qing Dynasties; only one charcoal fragment from Baishanling dates the site to the Song Dynasty (AD 960-1276) (Table 5.2). Five tap slag samples collected from the Laochangping site in 2004 were analysed: three samples were identified as slag from lead smelting, while the other two samples suggest arsenical copper smelting (Xie and Rehren 2009). The interpretation was that two physically separate operations were taking place, probably at different locations.

The fourth site, Dafengmen, located about 7-8 km away to the west of the Laochangping region, is littered with zinc smelting debris (Figure 5.3). Radiocarbon dating of charcoal samples from the site indicates that the production mainly dates to the Qing Dynasty (Table 5.2). One charcoal sample was dated to the Song Dynasty, but it is highly unlikely that this date corresponds with the zinc smelting activity, i.e. the charcoal was perhaps made from an old tree or simply derived from earlier activities in the area.

Although three more zinc smelting sites, Guanziyao, Longdongwan and Laoyaodong, have been found near Dafengmen (Table 5.1), only zinc production remains from Dafengmen were collected during fieldwork in June 2010. Analyses from this site allows an understanding of zinc distillation technology practiced in this region during the Qing Dynasty, which can then be compared with the technology used in Fengdu in the Ming Dynasty. We also visited Yushi Cave, one of the largest old zinc mine workings in the Laochangping region, and collected a few mineral samples, which were also analysed in order to help the interpretation of the potential ores exploited. The results of the mineral samples from Yushi Cave are first presented, followed by those of different parts of retorts and slag from Dafengmen.

Figure 5.2 Top left: slag heaps from the Laochangping site; top right: tap slag with clear flowing structures from the Laochangping site; bottom left: charcoal trapped inside the slag from the Luzhashi site; bottom right: a lead fragment from the Laochangping site.

Table 5.2 AMS radiocarbon dates of charcoal samples from the sites of Laochangping, Baishanling, Luzhashi and Dafengmen conducted at Peking University AMS lab, calibrated by OxCal 4.1 and IntCal 09.

Site	Lab code	^{14}C date (BP, 1σ)	Calibrated date (AD)		Period
			68.2% probability	95.4% probability	
Laochangping	BA07556	385±35	1448 (54.6%) 1515, 1599 (13.6%) 1617	1441 (61.5%) 1527, 1555 (33.9%) 1633	Ming
	BA07557	215±30	1649 (28.3%) 1675, 1778 (28.0%) 1799, 1941 (11.9%) 1953	1644 (33.2%) 1684, 1735 (46.7%) 1806, 1931 (15.5%) 1955	Qing-modern
	BA101503	350±30	1481 (29.2%) 1523, 1572 (39.0%) 1630	1457 (95.4%) 1635	Ming
Baishanling	BA101485	895±30	1049 (27.7%) 1085, 1123 (9.5%) 1138, 1199 (27.0%) 1188, 1199 (3.9%) 1206	1040 (39.1%) 1110, 1116 (56.3%) 1215	Song
	BA101486	335±30	1492 (21.8%) 1528, 1552 (32.7%) 1603, 1611 (13.7%) 1633	1474 (95.4%) 1641	Ming
	BA101487	245±30	1642 (47.5%) 1668 1782 (20.7%) 1797	1523 (9.3%) 1570, 1631 (53.7%) 1681, 1739 (0.3%) 1742, 1763 (25.8%) 1802, 1938 (6.4%) 1955	Ming-modern
	BA101488	305±25	1522 (48.0%) 1575, 1584 (4.1%) 1590, 1625 (16.2%) 1644	1492 (72.0%) 1603, 1615 (23.4%) 1649	Ming
	BA101489	280±30	1523 (38.3%) 1571, 1630 (29.9%) 1660	1499 (0.4%) 1502, 1513 (54.2%) 1601, 1616 (38.4%) 1666, 1784 (2.3%) 1796	Ming-Qing
	BA101490	265±25	1529 (13.1%) 1544 1634 (55.1%) 1664	1521 (27.5%) 1575, 1584 (0.8%) 1590, 1626 (59.1%) 1669, 1781 (7.9%) 1799	Ming-Qing
	BA101492	130±30	1683(12.1%) 1706, 1720 (8.6%) 1737, 1758 (1.1%) 1761, 1804 (8.0%) 1820, 1833 (27.0%) 1883, 1914 (11.4%) 1936	1675 (37.9%) 1778, 1799 (42.5%) 1894, 1905 (15.0%) 1942	Qing-modern
Luzhashi	BA101504	370±30	1454 (49.1%) 1519, 1594 (19.1%) 1619	1447 (55.1%) 1528, 1553 (40.3%) 1634	Ming
	BA101505	385±25	1450 (55.0%) 1495, 1602 (13.2%) 1615	1444 (71.3%) 1523, 1574 (24.1%) 1626	Ming

Table 5.2 (cont.)

Site	Lab code	^{14}C date (BP, 1σ)	Calibrated date (AD)		Period
			68.2% probability	95.4% probability	
Dafengmen	BA07558	810±30	1215 (68.2%) 1261	1175 (95.4%) 1271	Song
	BA101496	155±30	1669 (13.3%) 1694, 1728 (32.1%) 1781, 1798 (8.0%) 1812, 1919 (14.7%) 1945	1666 (16.6%) 1709, 1718 (33.7%) 1785, 1795 (27.3%) 1890, 1910 (17.8%) 1953	Qing-modern
	BA101497	235±30	1645 (42.0%) 1668, 1782 (26.2%) 1798	1528 (2.3%) 1544, 1633 (48.3%) 1683, 1737 (2.6%) 1756, 1761 (32.6%) 1804, 1936 (9.6%) 1955	Qing-modern
	BA101499	280±30	1523 (38.3%) 1571, 1630 (29.9%) 1660	1499 (0.4%) 1502, 1513 (54.2%) 1601, 1616 (38.4%) 1666, 1784 (2.3%) 1796	Ming-Qing
	BA101500	160±25	1669 (12.4%) 1689, 1730 (36.1%) 1780, 1798 (7.4%) 1809, 1926 (12.3%) 1944	1665 (16.5%) 1700, 1721 (39.5%) 1785, 1793 (10.8%) 1819, 1832 (9.9%) 1880, 1915 (18.8%) 1954	Qing-modern
	BA101501	160±25	1669 (12.4%) 1689, 1730 (36.1%) 1780, 1798 (7.4%) 1809, 1926 (12.3%) 1944	1665 (16.5%) 1700, 1721 (39.5%) 1785, 1793 (10.8%) 1819, 1832 (9.9%) 1880, 1915 (18.8%) 1954	Qing-modern
	BA101502	110±35	1691 (19.0%) 1729, 1811 (41.2%) 1892, 1907 (8.0%) 1924	1680 (31.3%) 1764, 1800 (64.1%) 1939	Qing-modern

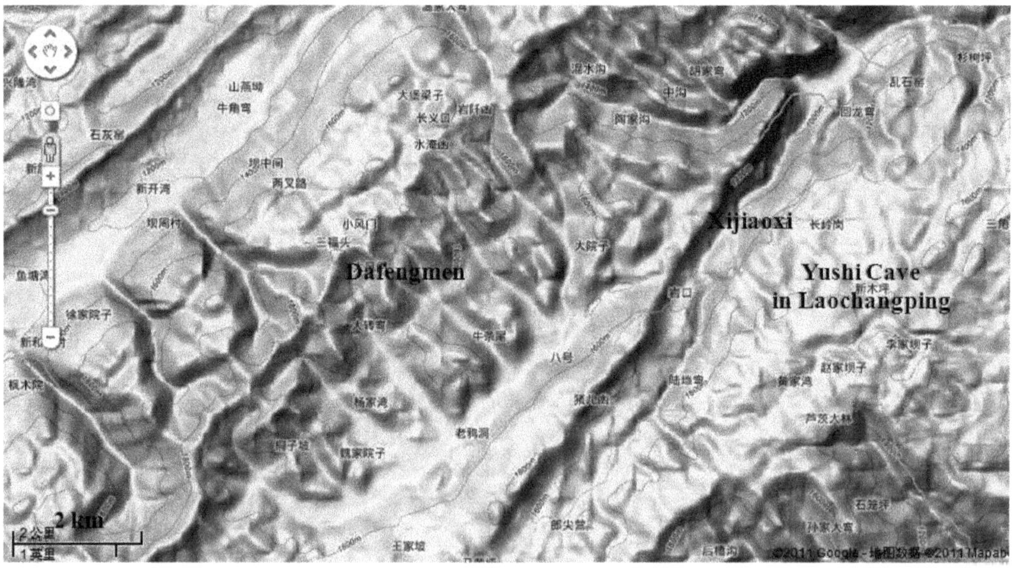

Figure 5.3 The location of Dafengmen and Yushi Cave. From Google Maps.

5.2 Yushi Cave (Y)

5.2.1 Site description

The exploitation of Yushi Cave can be traced back to at least the late Ming Dynasty. It is said locally that the cave was so large that it took seven days and nights to walk through. The geologist Wang Wanzhi visited Yushi Cave in 1980. He reported that the opening of the cave was large enough for three persons to walk through, shoulder to shoulder. There was a spacious area which was called Wanrendalitang 万人大礼堂, literally 'a great hall for ten thousand people'. At the northern end of the 'great hall', Wang found a stela with inscriptions indicating that there was an opening ceremony for exploiting the cave in the fourth year of the Wanli period (AD 1576). But the stela was missing when he visited the cave again in 1985. Near the stela, a bridge had been built over a deep underground stream. Wang encountered over 80 branch tunnels during his visit (Wang 1991).

The mine was still being worked until recently, as reported by the local guide. During our visit in 2010, abandoned mining facilities could be seen, such as timber props for supporting the roof, timber ladders and bridges and timber frames to support tunnels. Mining tools included wickerwork scoops for shovelling ores and wickerwork baskets for carrying ores out of the cave. Exposed light bulbs were hanging from cables for lighting. Modern equipment was also found left inside the cave, including air compressors and scales. On the roof of the mine, traces of chiseling with pickaxes by the ancient miners could be observed. Generally the methods employed here were the traditional zinc mining technologies, resembling those practiced in Huize in the 1990s (Craddock 1997). The ores exploited in the cave were reddish oxidic zinc ores, visually similar to the mineral fragments discovered at Miaobeihou (see Section 4.2.2.1). A few small fragments of greenish and bluish copper minerals were found together with zinc minerals on the floor (Figure 5.4).

5.2.2 Results: minerals

Four mineral samples from Yushi Cave were analysed (Figure 5.5). YO1 and YO2 are fragments collected from the ground inside Yushi Cave. They are weathered brick-red minerals with white or shiny minerals filling cavities, visually similar to mineral samples from Miaobeihou. YO3 and YO4 were collected from zinc ore fragments piled nearby the cave, which had been mined recently but not used. They are yellowish, whitish and greyish, some with layered structures. Local villagers informed us that zinc ores with layered structures are of a high grade.

YO1 and YO2 contain significant levels of ZnO (25-32%) and FeO (26-27%), a few percent of SiO_2 (2-8%) and minor amounts of CaO, MgO, Al_2O_3, CdO and PbO, with analytical totals of 62-63% (Table 5.3). XRD results show that they are dominated by smithsonite and hemimorphite. YO1 also contains hydrozincite, with hematite, calcite, quartz, goethite as gangue minerals; YO2 has calcite, quartz, dolomite and goethite (Table 5.4).

Matrices of both YO1 and YO2 show finely intergrown oolithic structures of iron oxides, hematite or goethite, and zinc carbonates such as smithsonite. YO1 has small amounts of zinc silicates, such as hemimorphite, filling the cavities and gangue minerals, including quartz and calcium carbonates, i.e. calcite (some contain 6-8% CdO). Some of the quartz grains appeared to have been partly replaced by calcium carbonates. Furthermore, two hydrozincite minerals with layered radial aggregations were seen on the surface and a kind of needle-like zinc carbonate mineral with 5.9% SiO2 had accumulated in one cavity. In contrast, YO2 has considerable amounts of zinc silicates, quartz and calcium carbonates (some contain 3-4% CdO), resulting in its higher levels of SiO2 (8.3%) and CaO (2.4%); some of the quartz grains were partly replaced by zinc silicates and calcium carbonates (Figure 5.6). These two geological mineral samples are generally similar in composition and microstructure to the archaeological mineral samples (YMO1 and YMO2) from Miaobeihou (see Section 4.2.2.1).

The sample with a clear layered structure, YO3, contains the highest level of zinc (ZnO 56.9%) with 5.3% SiO_2 and less than 1% other oxides. Analytical totals were 64% (Table 5.3). Both XRD and SEM results show that the predominant minerals are zinc carbonates, i.e. smithsonite, with only small amounts of zinc silicates, hemimorphite, hydrozincite and quartz grains (Table 5.4, Figure 5.7 left). The levels of zinc are consistent with the view of local people that this layer-structured ore is of a higher grade than YO1 and YO2.

YO4 is not only rich in zinc (ZnO 60.5%), but also in SiO_2 (32%), with an analytical total of around 100%. It consists primarily of zinc silicates with minor amounts of calcium carbonates, calcite, calcium/magnesium carbonates, dolomite, and lead carbonates (Table 5.4, Figure 5.7 right).

The analytical results thus indicate that the geological mineral samples are oxidic zinc ores, with zinc carbonates and silicates being the main zinc minerals. YO1 and YO2 are iron-rich zinc carbonates, indistinguishable from the archaeological mineral samples YMO1 and YMO2. YO4 is composed mainly of zinc silicates, similar to the archaeological mineral sample YMO3 from Miaobeihou, which is dominated by zinc silicates and quartz. Lastly, YO3 is a relatively pure zinc carbonate mineral, low in iron. No parallel has been found in archaeological contexts (Figure 5.8). Overall, the similarities between the geological and archaeological mineral samples are consistent with the possibility that the Miaobeihou zinc smelter utilised the oxidic zinc ores exploited from the Laochangping region.

Figure 5.4 The finds in Yushi Cave. Top left: the cave opening; top middle: a timber prop; top right: a wooden ladder; middle left: timber supporting frames; middle right: a wickerwork basket (left) and a wickerwork scoop (right); bottom left: traces of chiseling; bottom right: reddish oxidic zinc minerals and greenish and bluish copper minerals.

Figure 5.5 Cross sections of four mineral samples from Yushi Cave. From left to right: YO1, YO2, YO3, YO4.

Table 5.3 Average bulk compositions of four mineral samples from Yushi Cave (wt%). Analyses on polished sections at areas of ~2 by ~2.5 mm by SEM-EDS. Both unnormalised (top half) and normalised (bottom half) results are presented.

Sample	MgO	Al_2O_3	SiO_2	CaO	FeO	ZnO	CdO	PbO	Total
YO1	0.4	0.4	2.0	0.6	26.9	31.6	0.3	0.1	62.3
YO2	0.5	0.3	8.3	2.4	26.3	24.7	0.2	0.1	62.8
YO3	0.4	0.1	5.3	0.4	0.3	56.9	0.6	0.3	64.3
YO4	1.5	1.1	32.0	3.8	1.9	60.5	0.1	0.4	101.4
YO1	0.7	0.6	3.2	1.0	43.4	50.6	0.4	0.1	100.0
YO2	0.8	0.5	13.1	3.8	42.1	39.3	0.3	0.1	100.0
YO3	0.7	0.2	8.2	0.6	0.4	88.6	0.9	0.4	100.0
YO4	1.5	1.2	31.9	3.8	2.0	59.1	0.1	0.4	100.0

Table 5.4 Mineralogical compositions of four mineral samples from Yushi Cave. Analyses by XRD.

Sample	Mineral
YO1	smithsonite, hemimorphite, hematite, hydrozincite, calcite, quartz, goethite
YO2	hemimorphite, calcite, smithsonite, quartz, dolomite, goethite
YO3	smithsonite, hydrozincite, hemimorphite, quartz
YO4	hemimorphite, calcite, smithsonite, dolomite

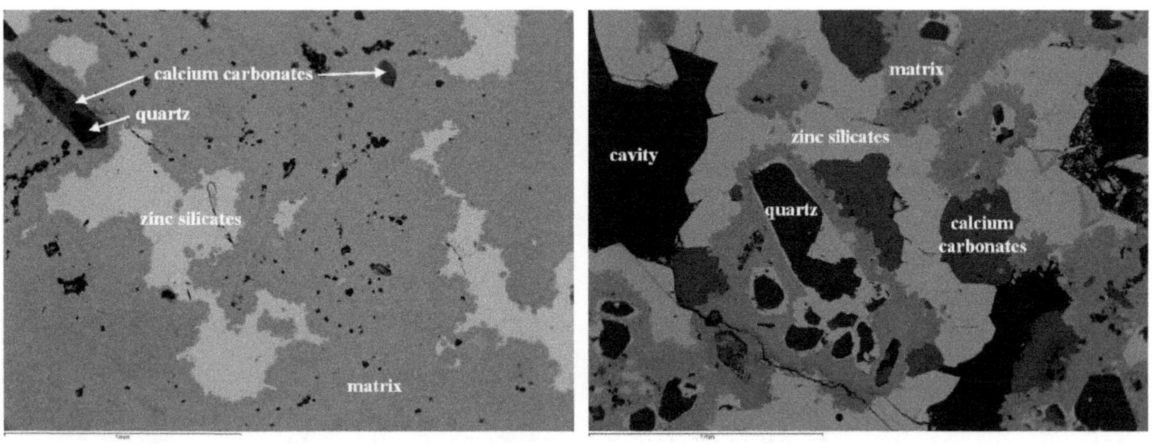

Figure 5.6 BSE images of ore YO1 (left) and YO2 (right), showing zinc silicates, quartz and calcium carbonates within the matrices.

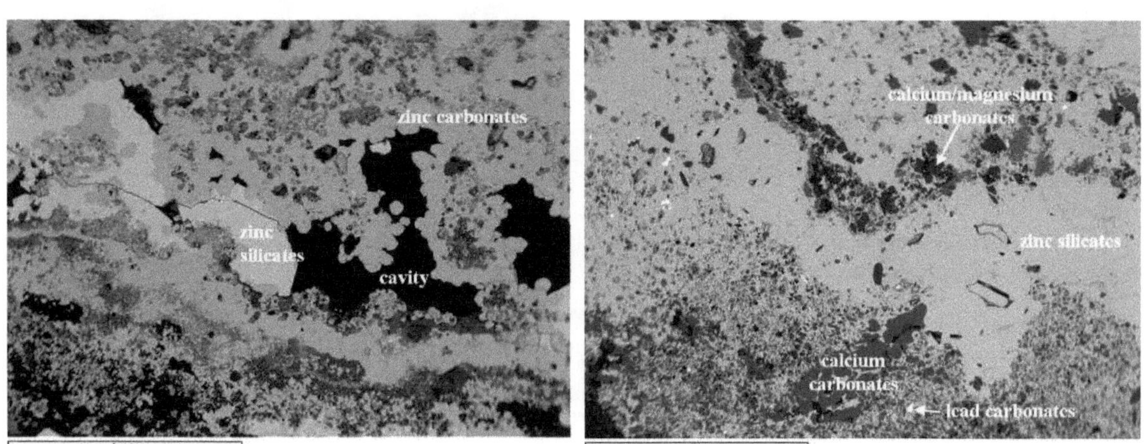

Figure 5.7 BSE images of ore YO3 and YO4, showing zinc silicates within zinc carbonates in YO3 (left) and calcium carbonates, calcium/magnesium carbonates and lead carbonates within zinc silicates in YO4 (right).

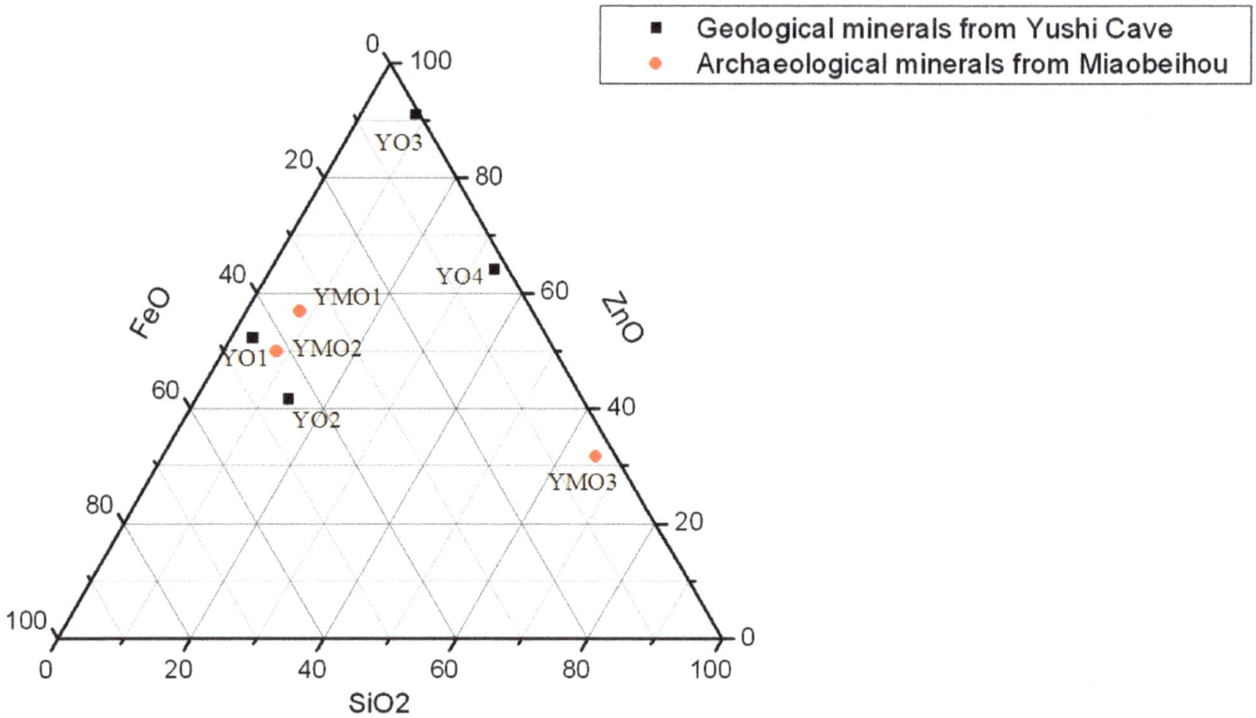

Figure 5.8 Ternary diagram of ZnO-SiO_2-FeO showing the compositions of the geological minerals from Yushi Cave (YO1, YO2, YO3 and YO4) and the archaeological minerals from Miaobeihou (YMO1, YMO2 and YMO3).

5.3 Dafengmen (SD)

5.3.1 Site description

The Dafengmen site is about 7-8 km west of the Laochangping region, situated in a long and narrow valley between mountains. Dafengmen 大风门 literally means 'large wind gate', a name which reflects its windy location. When we visited the site one day in high summer, August 2010, it was rainy, foggy, windy and chilly. Abundant retort fragments and slag were scattered all over the valley area. Trees and plants were rarely visible in this area, probably because of the severe pollution problems caused by the metallurgical debris affecting both the earth and water. A small stream flowed across the heaps of retorts and slag (Figure 5.9).

Owing to the adverse weather, we did not walk further to explore the size of the site, as there was a risk of getting lost in the mist. The size of the site, however, has been estimated by a modern mining and smelting company which planned to utilise these smelting remains to extract zinc for the production of zinc oxide. It is estimated that the Dafengmen site is around 2 km in length and 250 m in width. The heaps of retorts and slag are 1.5-3 m thick, weighing approximately 250,000 tonnes.

Figure 5.9 Top: retort and slag heaps at Dafengmen. Note the small stream flowing through the area; bottom: retort fragments littering the site.

5.3.2 Results

5.3.2.1 Retorts

Numerous retorts were identified in the 2004 field survey. The best preserved pot was about 40 cm in height, with a volume estimated at about 2L (Figure 5.10 left). During the 2010 field survey, only fragments of different parts of retorts were discovered. The retorts encountered in the field were fragments, including the bottom parts of the pots, usually with metallurgical residues remaining inside (Figure 5.10 middle and right), the pot walls, the top parts of the pots, the condensers and one iron lid. Although no pockets were found, horizontal ceramic partitions with a hole on one side must have been added to separate the two zones (for distillation and condensation) and to collect the zinc. As seen from the fragments, the retorts from Dafengmen were also made of pots, condensers, pockets and lids, but they display different shapes, sizes and material makeup from those from the river sites in Fengdu. The analyses of the pots, the condensers and an iron lid from Dafengmen are presented below.

Pots

The pots are elongated, conical but with small flat bottoms. Their circular bases are 4-5 cm in diameter; the bodies open gradually to the largest diameters at the rims; the rims are straight sided with internal diameters of around 10 cm. The sizes of the pots cannot be accurately determined due to the lack of complete examples, but they can be as tall as 40 cm as the relatively intact pot found in 2004 shows (Figure 5.10 left). The wall thickness increases from the rims (~0.8 cm) to the bases (~1.5 cm). They were made by wheel throwing as seen from the shallow spiral markings on the surfaces. All the fragments studied were used, displaying rusty colours on the surfaces. The colours of cross sections of the pot fragments are mostly bluish-grey, resulting from the reducing conditions inside the pots. Sometimes the outer parts are yellowish-brown (SDB11), indicating a more oxidising condition surrounding the pots (Figure 5.11).

Just as the pots from Dafengmen are different in shape to those from the river sites in Fengdu, their micro-analyses show different manufacturing materials and techniques as well as a peculiar contamination pattern. The pot fabrics show high levels of ZnO, varying from 5% to 17%. To estimate the original compositions of unused pots, the ZnO contents were omitted, and the data renormalised to 100%. The renormalised compositions of the ceramic matrices show 44-52% SiO_2, 29-37% Al_2O_3, 5-12% FeO, 3-6% TiO_2 and 3-5% K_2O and low levels of other oxides (Table 5.5). Their bulk compositions are generally indistinguishable from the matrix compositions, apart from higher levels of ZnO and slightly lower levels of K_2O.

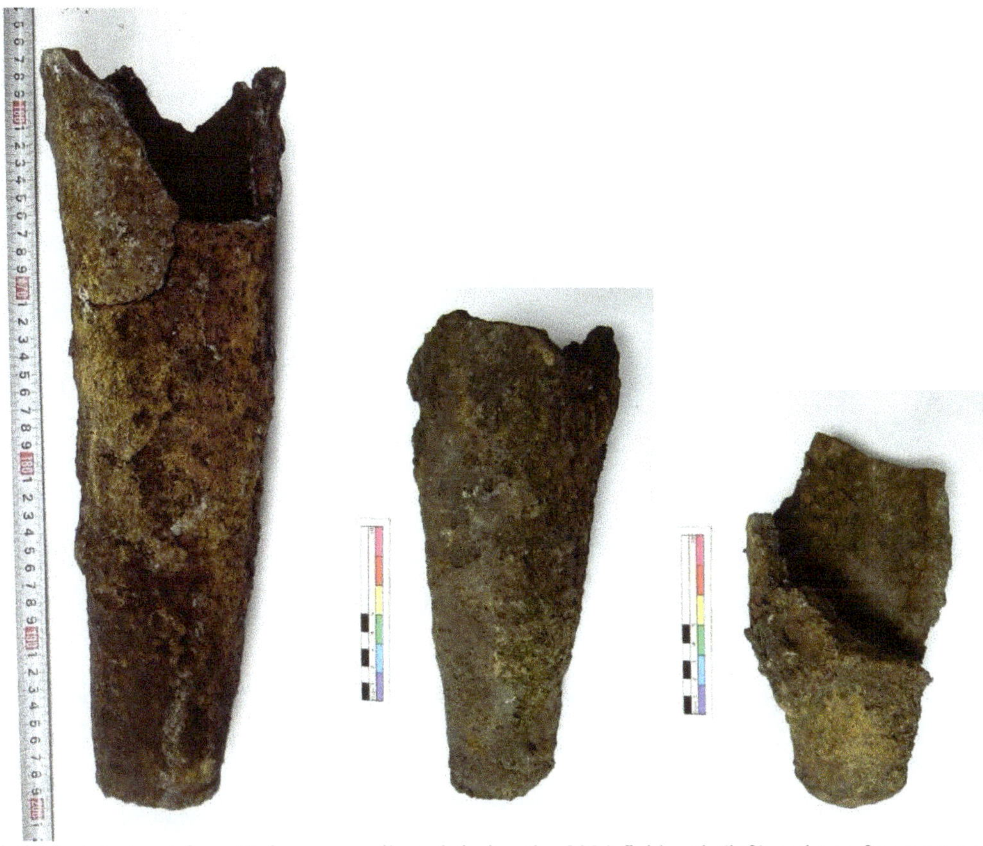

Figure 5.10 A relatively intact pot from Dafengmen collected during the 2004 fieldwork (left) and two fragmentary ones collected during the 2010 fieldwork (middle and right).

Figure 5.11 Cross sections of used pots SDB11 (left), SDT14 (middle) and SDT17 (right). SDT17 is the straight sided rim of the pot. The internal surfaces of these samples are facing up.

Table 5.5 Average matrix (top half) and bulk (bottom half) compositions of six pot samples from Dafengmen (wt%), normalised to 100%. Analyses on polished sections for areas of ~150 by ~200 μm and ~2 by ~2.5 mm respectively by SEM-EDS. For both compositions, the rows at the bottom show the same results after omitting ZnO and renormalising to 100%. '-' means 'not detected'.

Sample	MgO	Al_2O_3	SiO_2	P_2O_5	SO_3	Cl	K_2O	CaO	TiO_2	Cr_2O_3	MnO	FeO	ZnO
SDB11	0.6	34.9	44.9	1.0	-	-	4.0	0.4	3.4	0.1	0.1	5.3	5.3
SDT13	0.4	25.3	42.7	0.3	0.1	-	3.7	0.8	2.7	0.1	0.8	5.9	17.2
SDT14	0.5	29.1	40.7	0.4	0.1	-	2.9	1.7	3.9	0.1	0.4	4.0	16.2
SDT15	0.6	25.6	44.3	0.2	-	0.1	4.0	0.6	2.9	0.1	0.2	10.9	10.5
SDT17	0.5	30.2	43.2	0.2	0.1	-	4.3	0.9	4.1	0.2	0.1	7.4	8.8
SDT18	0.5	30.8	41.1	0.2	0.2	0.1	4.6	1.2	5.7	0.3	0.4	9.2	5.7
SDB11	0.6	37.0	47.4	1.0	-	-	4.2	0.4	3.6	0.1	0.1	5.6	
SDT13	0.5	30.5	51.8	0.4	0.1	-	4.4	0.9	3.2	0.2	0.9	7.1	
SDT14	0.6	34.8	48.5	0.5	0.1	-	3.5	2.0	4.7	0.2	0.4	4.7	
SDT15	0.6	28.6	49.5	0.2	-	0.1	4.5	0.7	3.2	0.2	0.2	12.2	
SDT17	0.5	33.2	47.3	0.2	0.1	-	4.8	1.0	4.5	0.2	0.1	8.1	
SDT18	0.5	32.6	43.9	0.2	0.2	0.1	4.9	1.2	6.0	0.3	0.4	9.7	
SDB11	0.6	31.6	44.8	0.6	0.1	0.1	2.3	0.5	3.1	0.2	0.2	8.6	7.3
SDT13	0.3	22.9	35.7	0.3	0.5	-	1.9	0.5	2.5	0.1	0.5	9.4	25.4
SDT14	0.5	25.8	31.2	0.4	0.3	0.1	1.6	0.5	2.9	0.2	0.4	6.7	29.4
SDT15	0.4	26.5	41.3	0.2	0.1	0.1	2.9	0.4	3.1	0.1	0.2	8.2	16.5
SDT17	0.3	25.5	35.7	0.3	0.2	0.1	2.5	0.5	3.5	0.1	0.1	9.6	21.6
SDT18	0.4	29.3	36.8	0.2	0.2	0.1	2.9	0.5	3.1	0.3	0.3	7.9	18.0
SDB11	0.6	34.2	48.4	0.6	0.1	0.1	2.5	0.5	3.3	0.2	0.2	9.3	
SDT13	0.5	30.7	47.7	0.4	0.6	0.1	2.5	0.7	3.3	0.2	0.7	12.6	
SDT14	0.6	36.5	44.4	0.5	0.5	0.1	2.2	0.7	4.1	0.2	0.6	9.6	
SDT15	0.5	31.7	49.4	0.3	0.1	0.1	3.4	0.5	3.7	0.2	0.2	9.9	
SDT17	0.4	32.5	45.6	0.4	0.3	0.1	3.1	0.7	4.5	0.1	0.1	12.2	
SDT18	0.5	35.7	44.8	0.3	0.3	0.1	3.5	0.7	3.8	0.3	0.3	9.7	

Table 5.6 Average compositions of aluminium-rich inclusions in the pot samples from Dafengmen (wt%), normalised to 100%. Analyses on polished sections by SEM-EDS. '-' means 'not detected'.

Sample	MgO	Al_2O_3	SiO_2	P_2O_5	SO_3	K_2O	CaO	TiO_2	Cr_2O_3	MnO	FeO	ZnO
SDB11	0.4	40.8	49.1	0.3	0.1	3.8	0.3	3.6	0.2	0.1	0.9	0.4
SDT13	0.4	39.0	47.8	0.2	0.1	4.9	0.9	3.7	0.2	0.2	1.6	1.0
SDT14	0.5	41.0	44.8	0.9	0.1	4.9	1.0	4.3	0.3	0.1	1.4	0.9
SDT15	0.3	39.3	49.2	0.2	0.1	4.9	0.6	2.4	0.1	0.1	2.1	0.6
SDT17	0.5	35.9	48.4	0.2	0.1	6.1	0.9	4.0	0.2	-	3.0	0.7
SDT18	0.5	36.5	45.6	0.5	-	5.2	0.8	3.9	0.3	0.1	5.8	0.7

The fabrics contain abundant ill-sorted inclusions up to 2 mm in length. Most of the inclusions are rock fragments, with edges dissolving into the matrices (Figure 5.12 top left); some of them developed bloating pores as a result of exposure to high temperatures (Figure 5.12 bottom left). These rock fragments contain high levels of SiO_2 (46-49%) and Al_2O_3 (36-41%), and a few percent of K_2O (4-6%), TiO_2 (2-4%) and FeO (1-6%) (Table 5.5). Their high Al_2O_3 concentrations indicate they are probably fragments of aluminous rocks, i.e. shales or mudstones composed of kaolinite, $Al_2Si_2O_5(OH)_4$, and other minerals bearing K_2O, TiO_2, FeO, CaO and MgO. They show similar compositions to kaolinitic clay matrices, which have elevated levels of FeO but lower levels of Al_2O_3 due to the higher levels of ferruginous impurities present in the clay. The clay is therefore probably the weathering product of aluminous parent rocks. Rock fragments were likely procured together with the clay, serving as temper. Aluminous rocks and associated clay are generally found in coal-bearing strata from the Permo-Carboniferous eras in north China and also in southwest China, often directly underlying the coal seams. Bauxite, an aluminous rock, is often found underlying the coal seams, according to geological survey reports of coal deposits within the Fengdu and Shizhu regions. The bauxite ore in the Fangdou Mountain coal deposits contains ~37% SiO_2, ~32% Al_2O_3 and ~11% FeO (Sichuan Geology Bureau 1961), highly comparable to the raw materials used for making pots at Dafengmen.

In addition to the inclusions of rock fragments, there are a significant amount of ferruginous concretions (up to 2 mm large), some of which contain metallic iron prills. Iron prills can also be seen in the matrices, indicating highly reducing conditions maintained during the firing (Figure 5.12 top right). Metallic iron was also identified in the ceramic fabrics of nineteenth-century Indian steel making crucibles (Freestone and Tite 1986, 18). In some areas, silica formed from melted quartz grains which subsequently recrystallised intergrown with tiny crystals rich in iron, zinc and titanium (Figure 5.12 bottom right).

It is worth noting that the pot fabrics are very porous (Figure 5.12). Most of the pores have elongated fractures surrounding large inclusions. The pot fabrics were heavily permeated by zinc vapour: both the clay and the inclusions were chemically altered by zinc oxide, and zinc oxide reacted with the ceramic fabrics, forming zinc-rich layers within cracks and pores. The XRD results of used pot SDT 17 show that its main crystalline phase is gahnite, zinc spinel ($ZnAl_2O_4$) and willimite (Zn_2SiO_4). It also contains small amounts of zinc ferrite ($ZnFe_2O_4$), quartz and mullite. The bluish colour of the pots resulted from the formation of grains of zinc spinels, which were also found in the blue refractory fragments related to zinc production in Bristol (Day 1988; Dungworth and White 2007).

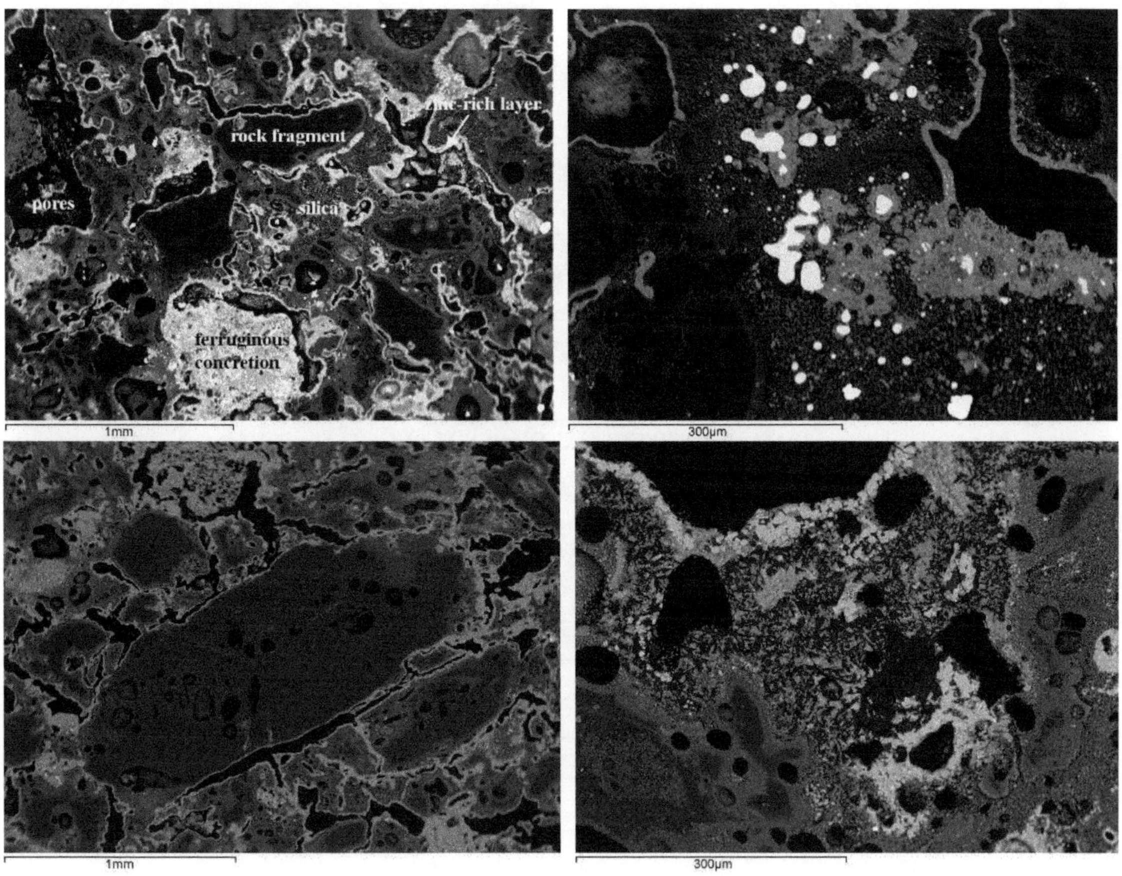

Figure 5.12 BSE images of the pot fabrics from Dafengmen. Top left: the pot fabric of SDT13, including rock fragments, ferruginous concretions, silica, pores and the zinc-rich layers formed along the pores; top right: the iron prills in SDT13; bottom left: the rock fragments with bloating pores in SDT14; bottom right: the silica with zinc-rich crystals intergrown in SDT15.

Condensers

The condenser fragments found at Dafengmen usually appear detached from the pots. One relatively well preserved fragment is about 6 cm in height and over 10 cm in diameter. Whitish zinc crusts adhere to the yellowish and brownish fragments internally and externally. There are large mineral inclusions visible in the fractures of two samples, SDT3 and SDT5 (Figure 5.13). The condensers show various shades of yellow, brown and grey, while the cross sections of the three samples studied are yellow or brown in the inner parts and grey in the outer parts, resulting from the oxidising and reducing conditions, respectively (Figure 5.14).

The condenser fabrics have extraordinarily high levels of ZnO (53-55%). After omitting the ZnO contents and renormalising the data, the condensers contain 63-68% SiO_2, 14-16% Al_2O_3, 8-12% FeO, 2-6% MgO, 1-3% K_2O, ~1% CaO and 0.5-1.4% SO_3 (Table 5.7). They appear to have been made of a different clay from the pots, with lower levels of Al_2O_3, K_2O, TiO_2 and higher SiO_2 and MgO. Compared to the condensers from Fengdu, they are richer in FeO, MgO and SO_3.

The makeup of the condenser fabrics is not easily established due to the heavy zinc contamination. A few large grains of quartz and feldspar are scattered in the fabrics. Coal fragments were only detected in SDT1, while rock fragments, composed mainly of silt-sized quartz and feldspar grains, were found in SDT2 and SDT3. Zinc silicates are occasionally present, which might have derived from the crushed slag or gangue crushed and recycled as temper (Figure 5.15).

Figure 5.13 Left: the internal surface of SDT1; middle: a large inclusion in SDT3; right: a large inclusion in SDT5.

Figure 5.14 Cross sections of condenser SDT1 (left), SDT2 (middle) and SDT3 (right). The internal surfaces of these samples are facing up.

Table 5. 7 Average bulk compositions of three condenser fabrics from Dafengmen (wt%), normalised to 100%. Analyses on polished sections for areas of ~2 by ~2.5 mm using SEM-EDS. The rows at the bottom show the same results after omitting ZnO and renormalising to 100%.

Sample	MgO	Al_2O_3	SiO_2	P_2O_5	SO_3	Cl	K_2O	CaO	TiO_2	MnO	FeO	ZnO
SDT1	1.0	7.1	29.8	0.3	0.5	0.1	1.1	0.4	0.4	0.1	4.3	54.9
SDT2	1.3	7.1	30.0	0.3	0.7	0.1	1.2	0.5	0.4	0.1	5.8	52.5
SDT3	2.6	6.4	30.1	0.2	0.2	0.1	0.5	0.3	0.4	0.1	3.6	55.5
SDT1	2.3	15.8	66.2	0.6	1.1	0.1	2.4	0.9	0.9	0.2	9.5	
SDT2	2.7	14.9	63.2	0.7	1.4	0.2	2.5	1.1	0.9	0.2	12.2	
SDT3	5.8	14.3	67.6	0.5	0.5	0.3	1.1	0.8	0.9	0.1	8.1	

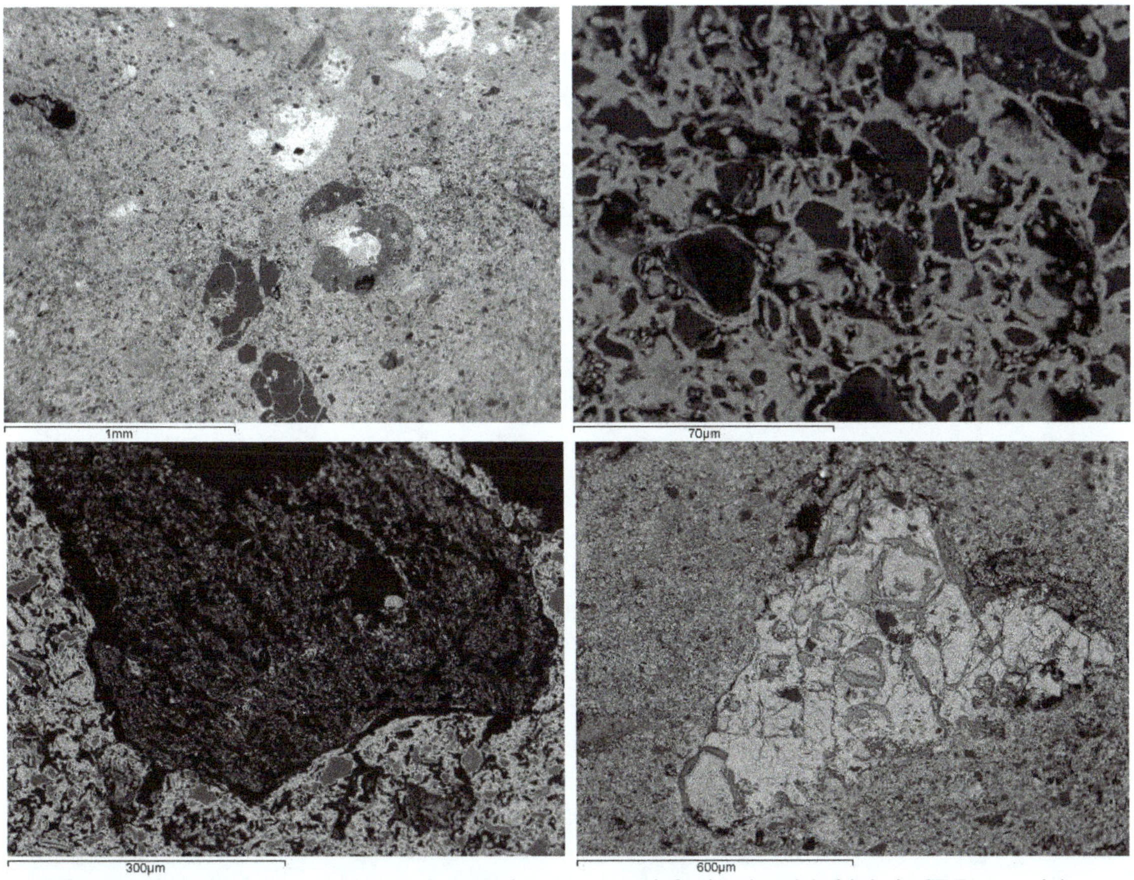

Figure 5.15 BSE images of the condenser fabrics from Dafengmen. Top left: the zinc-rich fabric in SDT1; top right: small quartz grains in the zinc-rich matrix in SDT2; bottom left: coal in SDT1; bottom right: zinc silicate in SDT3.

Iron lid

The one iron lid fragment found is approximately 2 mm thick and 12 cm in diameter (Figure 5.16), i.e. it is quite a good fit for the opening of the condensers. A small sample cut from the edge of the lid was analysed.

The iron lid shows ferrite grains with a large number of slag inclusions, mostly elongated and parallel to the surfaces. The slag inclusions consist of wüstite, FeO, and a glassy silicate matrix. In some of the large inclusions, the wüstite phases show dendritic structures (Figure 5.17). The inclusions are mainly composed of FeO (69-72%), P_2O_5 (13-15%) and SiO_2 (10-12%), with around 1% of Al_2O_3, K_2O, CaO and MnO (Table 5.8). Some of the slag inclusions were surrounded by a layer of corrosion (Figure 5.17 bottom right), which is generally composed of FeO with a few percent of SO_3.

The lid was heavily corroded on both surfaces and the edge. It is surprising that no detectable zinc was found within the surface corrosion, since the lid should have been in contact with zinc metal during distillation, forming some intermetallic compounds (as in the modern galvanising process). The corrosion layer on the edge contains about 1% zinc, with a few tiny particles of cadmium sulphide embedded, containing 13-17% iron and 3-6% zinc. More analyses on the surfaces of the lid would provide more representative data.

The presence of ferrite grains and abundant two-phase slag inclusions suggests that the lid was made of bloomery iron rather than the more common cast iron usually found in ancient China. The lid was shaped by hammering bloomery iron into a thin disk, as seen from the elongated slag inclusions. The extremely high levels of P_2O_5 in these slag inclusions indicate that the lid would have been produced from iron ores rich in phosphorus.

Figure 5.16 An iron lid fragment from Dafengmen.

Table 5.8 Average compositions of the slag inclusions (bulk and matrix), wüstite and corrosion layers of the slag inclusions in the iron lid from Dafengmen (wt%), normalised to 100%. Analyses on polished section by SEM-EDS. '-' means 'not detected'.

Slag inclusion	Na_2O	MgO	Al_2O_3	SiO_2	P_2O_5	SO_3	K_2O	CaO	TiO_2	MnO	FeO
Bulk	0.1	0.3	1.5	11.0	13.8	0.4	0.8	1.5	0.1	0.6	70.1
Matrix	-	0.4	0.5	16.0	17.9	0.1	0.7	1.2	-	0.8	62.3
Wüstite	-	0.1	0.8	0.3	0.4	0.1	-	0.1	0.1	0.2	98.0
Corrosion	0.3	-	0.1	1.4	2.8	5.9	0.2	1.1	-	-	88.2

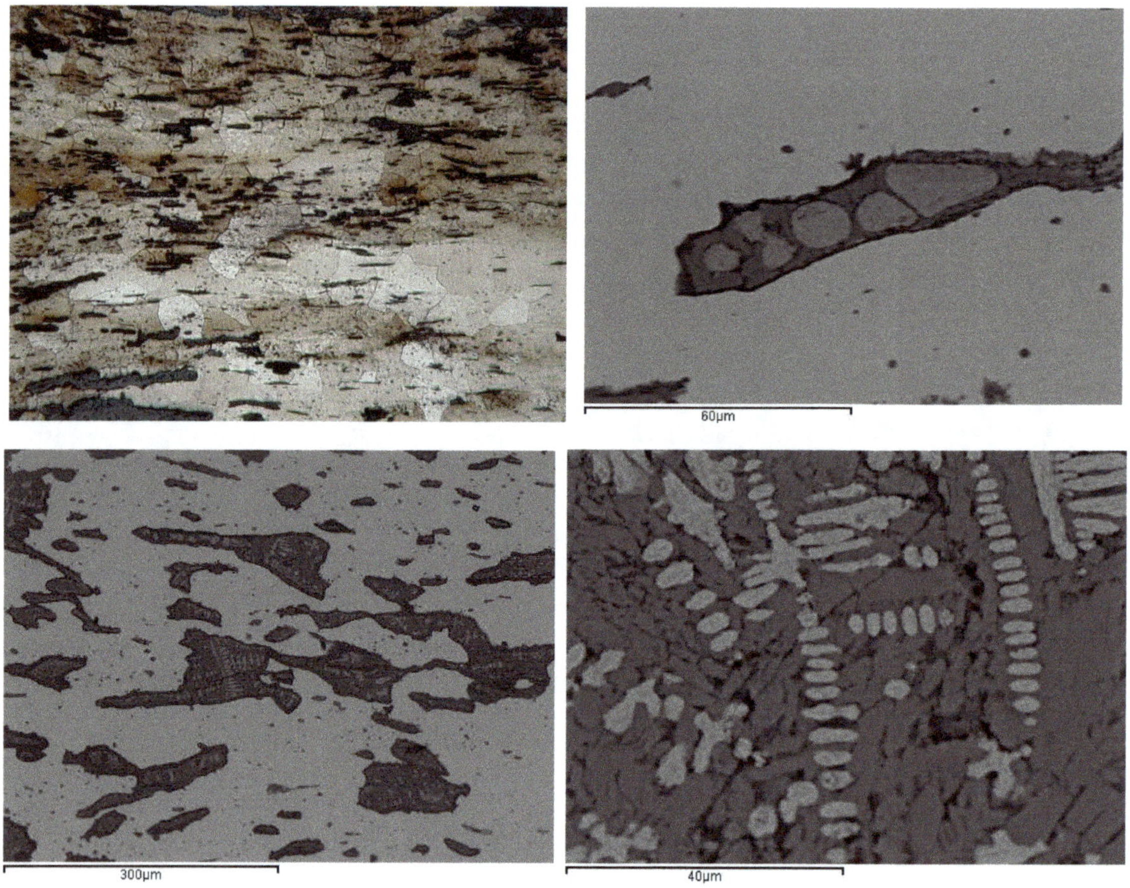

Figure 5.17 Top left: metallography of the iron lid, showing the ferrite grains and elongated slag inclusions. The width of picture is 2 mm. Top right: BSE image of the iron lid, showing an elongated slag inclusion. Wüstite (white grey) and silicate matrix (dark grey) are surrounded by a corrosion layer; bottom left: BSE image, showing large slag inclusions; bottom right: BSE image, showing dendritic wüstite (white grey) within the silicate matrix (dark grey) in a large slag inclusion.

5.3.2.2 Slag

Most of the slag samples are rusty lumps attached to fragments of the bottom parts of retorts. Occasionally white inclusions are seen trapped within the slag (Figure 5.18 left). One slag sample (SDB12) was found *in situ* within the cylindrical pot wall (Figure 5.18 right).

The five slag samples from Dafengmen contain variable concentrations of SiO_2 (18-35%), FeO (17-26%), ZnO (9-32%), BaO (5-19%), SO_3 (3-12%), Al_2O_3 (5-8%), CaO (2-6%), MgO (1-4%), and below 1% K_2O and TiO_2. It should be noted that their BaO and SO_3 contents are rather high; in addition, they contain around 1% PbO (except SDB9) and minor amounts of CuO, As_2O_3 and Sb_2O_3 (Table 5.9). They are different from the slag samples from the Fengdu sites not only in terms of the chemical composition but also in microstructure.

Figure 5.18 Left: rusty slag with white inclusions in SDB7; right: top view of slag in SDB12.

The slag samples are composed mainly of primary crystals rather than glassy phases. The assemblages of crystals in one small area differ significantly from other areas even in a small sample of about 2 cm in size. The common primary crystals include feldspar (mostly barium feldspar), olivine and pyroxene (Table 5.10 and related figures). These crystals are larger in size than those in the slag samples from the Fengdu sites, denoting slower cooling inside the furnace.

Together with these crystals, there is also a considerable amount of zinc sulphide. Some of the large grains are most likely the residues of sulphidic zinc ores (SDB7 and SDB9) (Figure 5.19 top right), while the small ones probably formed during the process (Figure 5.20 left). There are other residues of ore minerals, such as barite in SDB7 and SDB10 (Figure 5.19 bottom right) and zinc silicates in SDB6 (Figure 5.19 bottom left). The zinc silicates appear as internally recrystallised iron-bearing zinc silicates, similar to those found in the Muxiexi slag samples (see Section 4.3.2.2). Therefore, it can be inferred that the macroscopically visible white inclusions within the slag of SDB7 are potentially residual grains of zinc silicates from the ores.

In most of the samples only a few metallic iron prills were found, mostly in the glassy phases, where they appear together with zinc sulphides (Figure 5.20 left). In SDB12 there is a large lump consisting of numerous iron prills and zinc oxide (~12% FeO) together with zinc- and iron-rich olivines and zinc-rich pyroxenes (Figure 5.20 right), resembling those in sample CMB1 from Muxiexi. The formation of the lump is either due to insufficient reducing gases or the large size and low porosity of the original zinc ore fragment. The coexistence of metallic iron and zinc oxide confirms that iron oxides were reduced more easily than zinc oxide. There are more secondary iron oxides deposited in pores, cracks and the holes left by the corrosion of metallic iron as well as on surfaces.

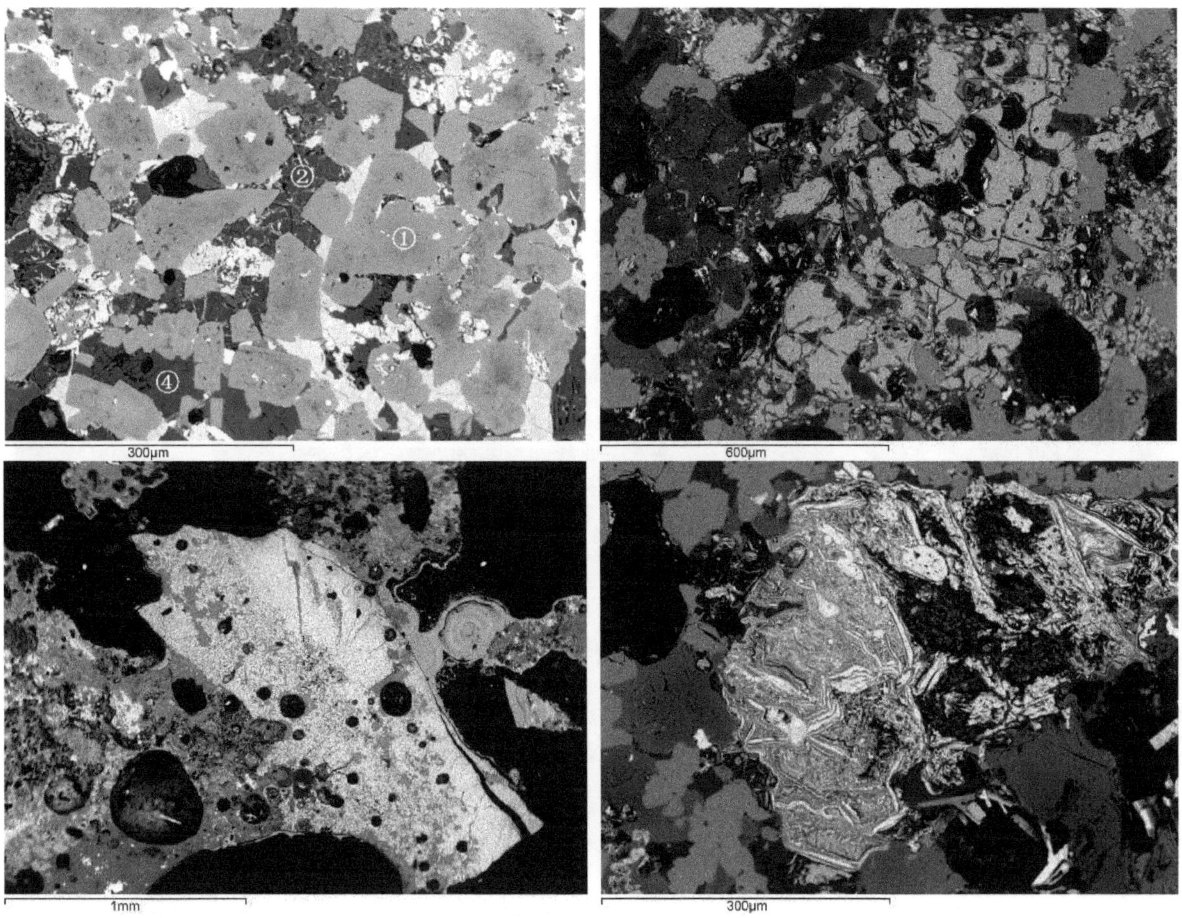

Figure 5.19 BSE images of the slag samples from Dafengmen. Top left: barium feldspar (1), olivines (2)(3) and pyroxene (4) in SDB6; top right: residual grains of zinc sulphides in SDB9; bottom left: residual grains of zinc silicates in SDB6; bottom right: residual grains of barite in SDB7.

A few inclusions of metallic lead, lead oxide and lead carbonate are embedded in SDB6, SDB7 and SDB9. The lead particles in SDB6 contain up to 6% silver (Figure 5.21 left). Among the samples, SDB9 contains the highest level of PbO: there is a significant amount of lead oxide surrounded by lead carbonates and in some cases metallic lead is embedded within the lead oxide (Figure 5.21 right). Some small metallic copper prills were identified within the lead oxide; they contain 1.5-2.5% zinc, 1.8-1.9% antimony and 0.6-0.8% iron. These lead and copper phases may have derived from lead and copper minerals in zinc ores, probably cerussite and malachite. The presence of lead minerals in zinc ores is expectable, as noted in the case at the Baishaling white lead factory, where sometimes black lead was produced (Appendix 4.2). So is the presence of copper minerals, which were found inside Yushi Cave (Figure 5.4 bottom right). As demonstrated in the Ellingham diagram, copper and lead oxides are more easily reduced to metal than zinc oxide, which would explain the presence of these metals in zinc distillation retorts. Lead particles also formed in traditional zinc distillation retorts, e.g. in Xiaoshuijing, Hezhang, Guizhou. Indeed, heavier lead particles from the distillation residues were recovered deliberately by panning crushed slag in water. Each retort could produce 0.07-0.08 kg of lead particles, in addition to 0.56-0.69 kg of raw zinc (Xu 1986).

In contrast with the slag samples from the Fengdu sites, there are no coal or charcoal residues identified in these samples, except SDB10. This suggests that a relatively lower proportion of carbon-rich material was added to the charge.

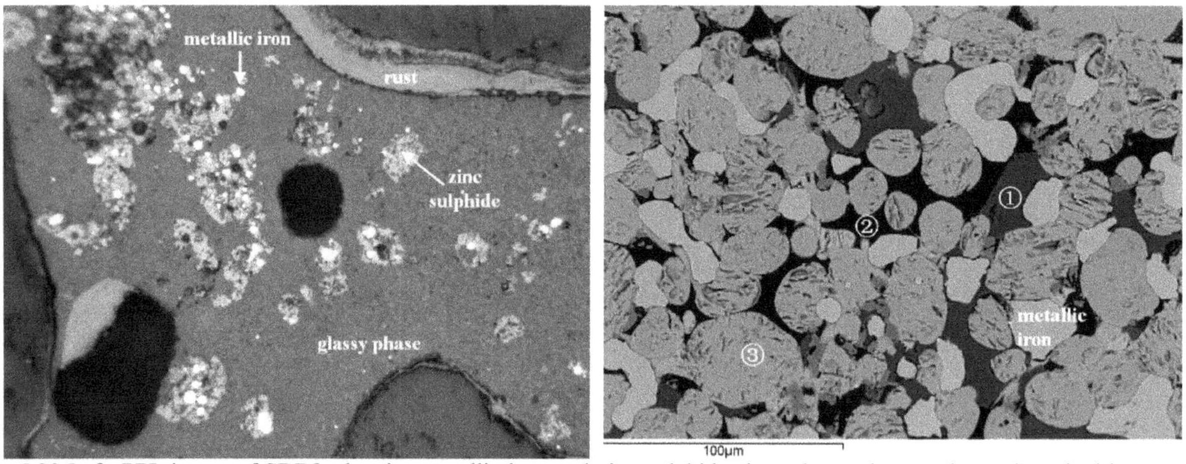

Figure 5.20 Left: PPL image of SDB9, showing metallic iron and zinc sulphides in a glassy phase and rust deposited in pores. The width of the picture is 2 mm. Right: BSE image of SDB12, showing metallic iron prills, zinc oxide (3), olivine (1) and pyroxene (2) in a large lump.

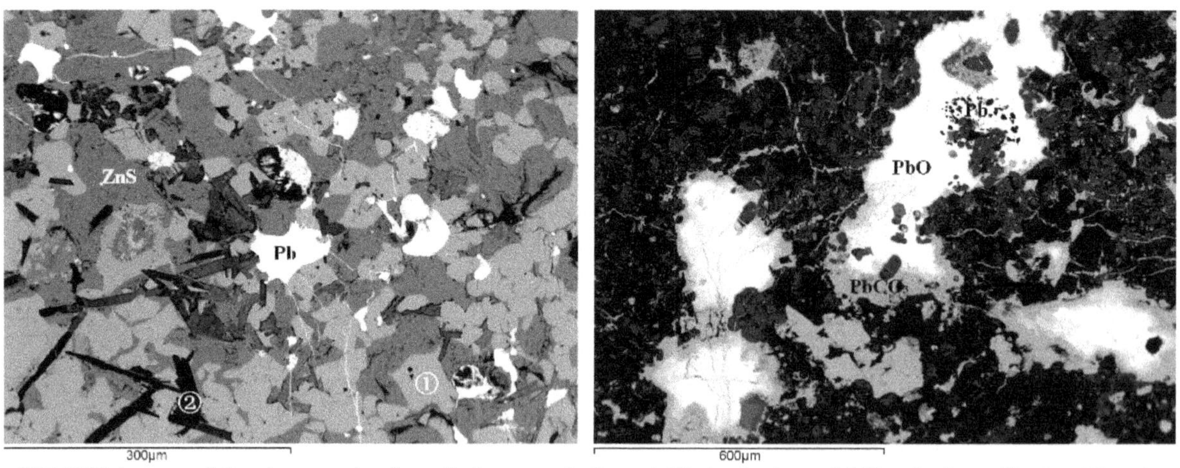

Figure 5.21 BSE images of the slag samples from Dafengmen. Left: metallic lead, zinc sulphides, barium silicates (1) and a silica-rich phase (2) in SDB6; right: lead, lead oxide and lead carbonates in SDB9. The lead particles are impressed with silicon carbide grains (black), which were introduced from grinding papers during sample preparation due to the extreme softness of lead metal.

Table 5.9 Bulk compositions (top half) and crystalline phases (bottom half) of five slag samples from Dafengmen (wt%), normalised to 100%. Analyses on polished sections for areas of ~2 by ~2.5 mm and ~400 by ~600 μm respectively by SEM-EDS. '-' means 'not detected'.

Sample	MgO	Al_2O_3	SiO_2	P_2O_5	SO_3	Cl	K_2O	CaO	TiO_2	MnO	FeO	CuO	ZnO	As_2O_3	Sb_2O_3	BaO	PbO
SDB6	1.7	4.6	18.3	0.1	11.6	0.1	0.1	2.9	0.6	0.1	17.2	0.2	21.8	0.5	0.2	18.9	1.1
SDB7	1.9	8.0	30.8	0.1	6.8	0.1	0.9	3.3	0.6	-	22.3	0.4	9.2	1.5	0.3	13.1	0.7
SDB9	1.9	5.4	24.5	0.2	6.5	0.1	0.3	2.0	0.5	0.1	26.3	0.9	13.5	0.4	0.2	12.4	4.8
SDB10	1.3	6.6	35.3	0.2	2.6	0.1	1.0	2.1	0.4	0.1	24.6	0.1	16.8	0.3	-	7.2	1.3
SDB12	3.7	4.5	21.0	0.2	5.1	0.1	0.2	6.0	0.2	0.1	19.6	0.2	32.4	0.3	0.2	5.4	0.8
SDB6	1.7	6.5	20.5	0.2	12.8	-	0.2	2.4	0.4	-	8.6	0.3	19.4	0.2	0.1	26.3	0.4
SDB7	2.5	9.0	36.8	0.2	3.0	-	1.4	3.6	0.9	0.1	21.0	0.3	7.6	0.9	0.2	12.2	0.3
SDB9	2.1	6.6	27.5	0.3	9.5	0.1	0.3	2.5	0.5	0.0	13.7	1.3	11.9	0.3	0.2	21.6	1.6
SDB10	1.6	9.6	42.8	0.2	2.0	0.1	1.5	3.7	0.4	0.1	13.8	-	17.4	0.2	0.2	5.9	0.5
SDB12	1.8	3.7	22.2	0.1	8.2	-	0.2	2.6	0.2	0.1	14.0	-	41.7	0.2	0.2	4.7	0.1

Table 5.10 Compositions of some phases in the slag samples from Dafengmen (wt%), normalised to 100%. Analyses on polished sections by SEM-EDS. '-' means 'not detected'.

Figure	Spot	Phase	MgO	Al_2O_3	SiO_2	K_2O	CaO	TiO_2	FeO	CuO	ZnO	BaO	PbO
5.3.11 top left SDB6	①	Barium feldspar (Ba, Zn, K)(Al, Si)$_4$O$_8$	0.2	25.5	34.1	2.1	-	0.3	0.4	-	3.0	34.5	-
	②	Olivine (Zn, Fe)$_2$SiO$_4$	0.2	2.7	39.9	-	0.3	0.3	6.7		49.5	0.5	-
	③	Olivine (Ba, Fe, Zn, Mg)$_2$SiO$_4$	6.8	0.5	30.8	0.2	1.0	0.4	10.3	-	11.6	38.5	-
	④	Pyroxene $CaSiO_3$	5.2	0.8	39.8	0.3	33.7	0.2	3.9	-	11.8	4.3	-
5.3.12 right SDB12	①	Olivine (Zn, Fe, Mg)$_2$SiO$_4$	4.4	0.2	28.4	-	-	-	14.2	-	52.9	-	-
	②	Pyroxene $CaSiO_3$	1.2	1.8	37.4	0.3	32.2	-	4.8	-	20.9	1.6	-
	③	Zinc oxide (Zn, Fe) O	0.8	0.4	0.2	0.1			17.0		81.6		
5.3.13 left SDB6	①	Olivine (Ba, Zn, Fe)$_2$SiO$_4$	0.7	0.4	20.4	-	0.2	-	4.8	0.2	20.7	52.8	-
	②	Silica-rich	5.2	0.2	46.2	0.1	2.1	-	6.9	0.8	10.0	12.9	15.5

5.4 Discussion

5.4.1 Retort design

The pots forming the main parts of the retorts from Dafengmen are not jar-shaped like those from the Fengdu sites, which were probably directly adapted from domestic jars with few, if any, modifications in shape. They are taller and slimmer without drum bellies, although still flat-bottomed. While retaining a volume similar to that of the jar-shaped pots (about 2 L), the shape of the Dafengmen pots can be argued to be technically more advantageous for zinc distillation. Firstly, it would be easier to adjust temperature gradients between the pots and the condensers when using taller pots, or rather, pots with a higher ratio of height to rim diameter. Furthermore, such pots have a higher surface area to volume ratio, which could facilitate heat absorption and even heat distribution within the charge (Rehren 2003). These are thought to be the reasons why most of the twentieth-century traditional processes utilised tubular retorts, although their bases were mostly pointed rather than flat (see Section 2.3). The tall pots are thicker at the bottom than at the top, which would increase structural stability and strength to hold the charge.

Turning to the composition of the ceramic fabrics, the pots were made of kaolinitic clay and rock fragments. Compared to the pot fabrics from the Fengdu sites, the Dafengmen pots seem to be much more refractory, with higher levels of Al_2O_3 (over 30%). On the other hand, the moderate concentrations of fluxing oxides, K_2O and FeO in particular, would decrease their refractoriness to a certain extent (Figure 5.22). However, the partial reduction of the iron oxide to metal would decrease the fluxing action of iron oxide. The high TiO_2 levels would further increase the refractoriness. Although the degree to which their refractoriness was decreased by the fluxing oxides is difficult to estimate, the fabrics can be deemed quite refractory. The presence of elongated cracks around the inclusions and open voids could help arrest and stop crack propagation, thus increasing the toughness and thermal shock resistance. However, the high levels of zinc oxide documented in the pots from Dafengmen would suggest that they were less resistant to the penetration of zinc vapour and thus losses of metal, contrasting sharply with the zinc-resistant pots from the Fengdu sites.

Kaolinitic clays were commonly employed for making fine-grained high-fired ceramics in ancient China, including white stonewares during the Neolithic period and the Shang Dynasty and white porcelains of the Xing and Ding Kilns during the Tang and Song Dynasties (Kerr and Wood 2004). They are also known to have been used for metallurgical activities in medieval times and later, for example, the famous Hessian crucibles (crucibles made in the German region of Hesse), crucibles made in Stamford, Lincolnshire (Martinón-Torres *et al.* 2008; Freestone and Tite 1986), and the steel-making crucibles of Central Asia (Rehren and Papachristou 2003) (Figure 5.23). The clay, also called fireclay, found in coal-bearing strata, was also used for lining furnaces and making crucibles in eighteenth and nineteenth-century Britain (Highley 1982). In each of these applications, the key aspect that maximises the performance characteristics of pyrotechnical ceramics made from kaolinitic clays is a high-temperature prefiring in the potter's kiln. When fired at temperatures of over 1100 °C, kaolinitic clays develop mullite crystals, which can greatly improve strength, thermal shock resistance, thermal and chemical refractoriness, thus rendering a superior material for high-temperature operations (Martinón-Torres *et al.* 2008). This is why the Hessian crucibles were of much better quality than those from Stamford. It appears that the main secret behind the Hessian crucibles was their high-temperature prefiring at 1300-1400 °C, which generated continuous vitrified matrices with mullite developing in the fabrics. The Stamford crucibles, although sharing similar compositions to the Hessian crucibles, were low fired with no mullite developing until they were used (Martinón-Torres and Rehren 2009).

From the discussion above it can be inferred that the good thermal refractoriness of the Dafengmen pots, contrasting with a seemingly poor chemical refractoriness, may be due to the fact that they were not prefired at temperatures high enough to develop vitrified matrices in order to be resistant to zinc vapour. Furthermore, only a small amount of mullite developed. Unfortunately, the prefiring temperatures cannot be estimated as no unused pots have been found at Dafengmen, and thus this explanation remains as an untested hypothesis. A further aspect which may account for the higher zinc levels in these pots is the large proportion of rock fragments, generating too many open pores. Employing such porous retorts for zinc distillation led not only to zinc losses into the pot fabrics, but more losses into the air, thus decreasing the yield. In this regard, the higher surface area to volume ratio of the pots may been a negative factor for zinc smelting, contrary to the assumption stated above. If the pots were reused, the cumulative fluxing effect of zinc could also lead to an earlier collapse, i.e. fewer reuses.

It is worth noting that the Dafengmen pot fabrics, which were made of the kaolinitic clay and rock fragments rich in fluxing oxides, were rarely employed for the production of ceramics in ancient China, based on the published compositional data of Chinese ceramics (Li 1998). However, kaolinitic clays related to coal measures were exploited to make retorts in some traditional zinc smelting processes (see Section 2.3). In Hezhang in 1986, the kaolinitic clay found in the lower layers of coal measures was mixed with common clay rich in fluxing oxides and a sandy temper found in the upper layers of coal measures. The resulting materials contained similar levels of Al_2O_3 (32%) to that of the Dafengmen pots but lower levels of fluxing oxides (5.1% Fe_2O_3, 0.7% CaO and 2.3% MgO) and TiO_2 (1.0%) (Table 2.3).

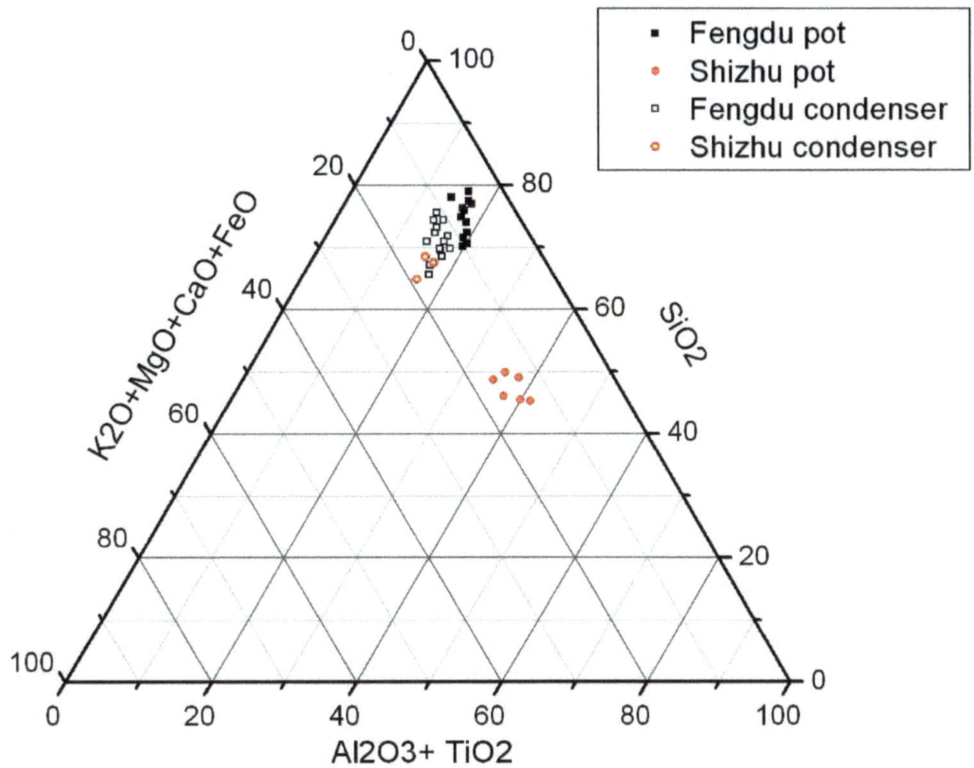

Figure 5.22 Ternary diagram of SiO_2-(Al_2O_3+TiO_2)-(K_2O+MgO+CaO+FeO) showing the bulk compositions of the pot and condenser fabrics from Fengdu (three sites) and Shizhu (Dafengmen).

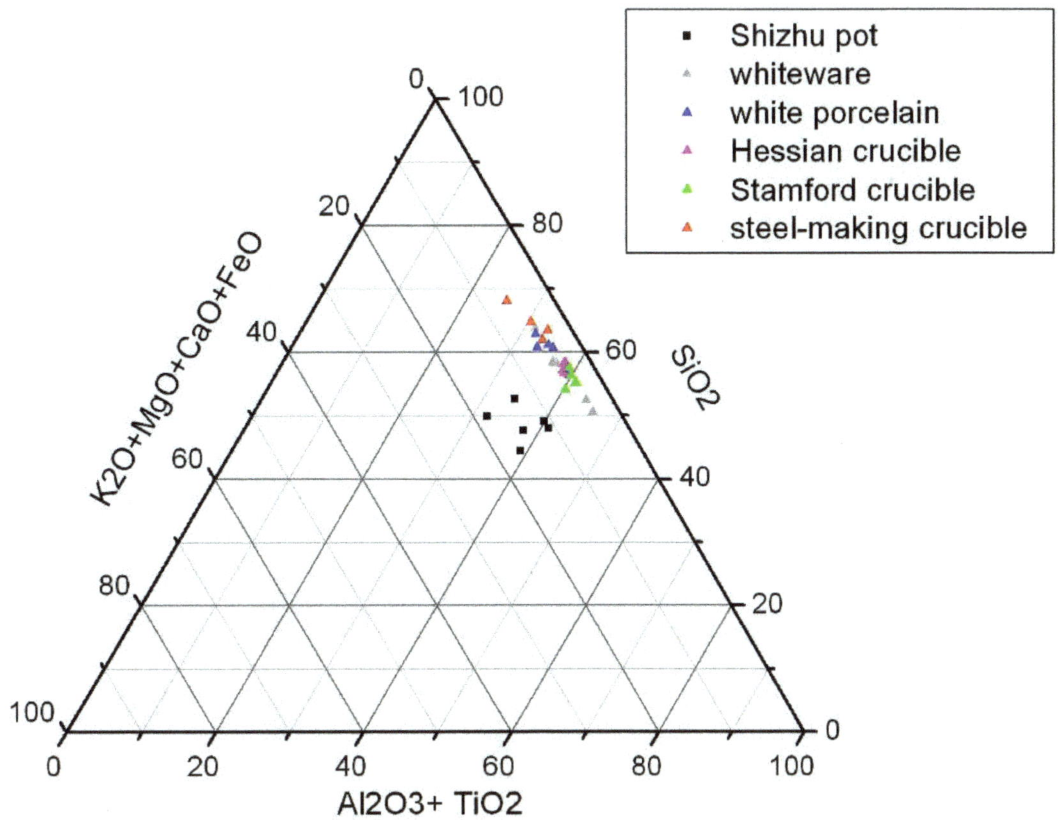

Figure 5.23 Ternary diagram of SiO_2-(Al_2O_3+TiO_2)-(K_2O+MgO+CaO+FeO) showing the matrix compositions of the Dafengmen pots and some kaolinitic ceramics: Neolithic and Shang whitewares (Kerr and Wood 2004), Ding white porcelains (Kerr and Wood 2004), Hessian crucibles (Martinón-Torres and Rehren 2009), Stamford crucibles (Freestone and Tite 1986) and Central Asian steel-making crucibles (Rehren and Papachristou 2003).

Overall, it can be argued that the pots were highly specialised ceramics produced for zinc smelting in terms of formal and material characteristics. The long-shaped pots made of the kaolinitic clay and rock fragments rich in fluxing oxides were specifically chosen to make retorts for zinc distillation. They were probably manufactured in specialised ceramic workshops and fired in kilns. This is supported by the fact that there was a 'earthenware jar factory' at Dafengmen as recorded in a local history document (Ran and Ran 1991). However, the pots were made somewhat cursorily, as indicated by the larger size of the inclusions and the likely lower firing temperatures. On the whole, they seem technically less developed than those from Fengdu.

Turning to the condensers, these are formally and materially similar to those from the Fengdu sites. They contain slightly higher concentrations of fluxing oxides and lower amounts of SiO_2 (Figure 5.22). Their raw materials were not well prepared, as seen from the very large inclusions that were most likely not deliberately added but naturally present in the clay. The condensers were probably not prefired, and during use they were exposed to temperatures lower than 800 °C. The absence of pockets suggests that either the pockets were too fragile to survive or they were reused to charge the retorts to recover some zinc (see next section). The condensers and pockets were likely made at the zinc workshops rather than in specialised ceramic workshops.

Finally, the lid was made of bloomery iron and hammered into a thin disk. Bloomery iron was among the first iron smelting products in China, but soon cast iron technology became dominant, which favoured mass production of iron and steel objects (Han and Ke 2007; Wagner 2008). Iron objects with a great number of slag inclusions have been frequently discovered from the Yuan, Ming to Qing periods, and were probably produced with coal (Chen Jianli pers. comm. 2011). The sulphur content of the slag inclusions also suggests the use of coal. The use of bloomery iron in making the lid indicates that the production took place on a small scale, while the presence of abundant slag inclusions suggests that the technology was rather rudimentary. It is thus presumed that the bloomery iron lid was produced with local iron ores using coal as fuel at or near the zinc smelting site. Although the quality of the iron was not exceptional due to the abundant slag inclusions, it was sufficient for its use as a lid for a zinc distillation retort. As iron has good thermal conductivity, a thin iron lid could transfer heat fast into the air and keep the condensation zone at adequate temperatures for condensation. More lid samples should be analysed to fully understand their manufacturing technology and performance.

5.4.2 The charge

The analytical results show that oxidic zinc ores, coal/charcoal and recycled zinc-impregnated ceramics made up the charge at Dafengmen.

Compared to those from Fengdu, the slag samples from Dafengmen contain lower levels of SiO_2, Al_2O_3, K_2O, MgO and CaO, higher levels of BaO, ZnO, SO_3 and PbO, and similar levels of FeO and other oxides (Figure 5.24). The ratios of SiO_2 to Al_2O_3 concentrations in the slag samples are not similar to those in the pot fabrics, but rather to the condenser fabrics (Figure 5.25). Thus most of the SiO_2 and Al_2O_3 could have come from used condensers, and also used pockets which might have had similar ratios of SiO_2 to Al_2O_3 to the condenser fabrics. Part of the SiO_2 and Al_2O_3 could also have derived from zinc ores and coal. Ores with hemimorphite minerals would have introduced more SiO_2, but this source of SiO_2 did not contribute much to the slag.

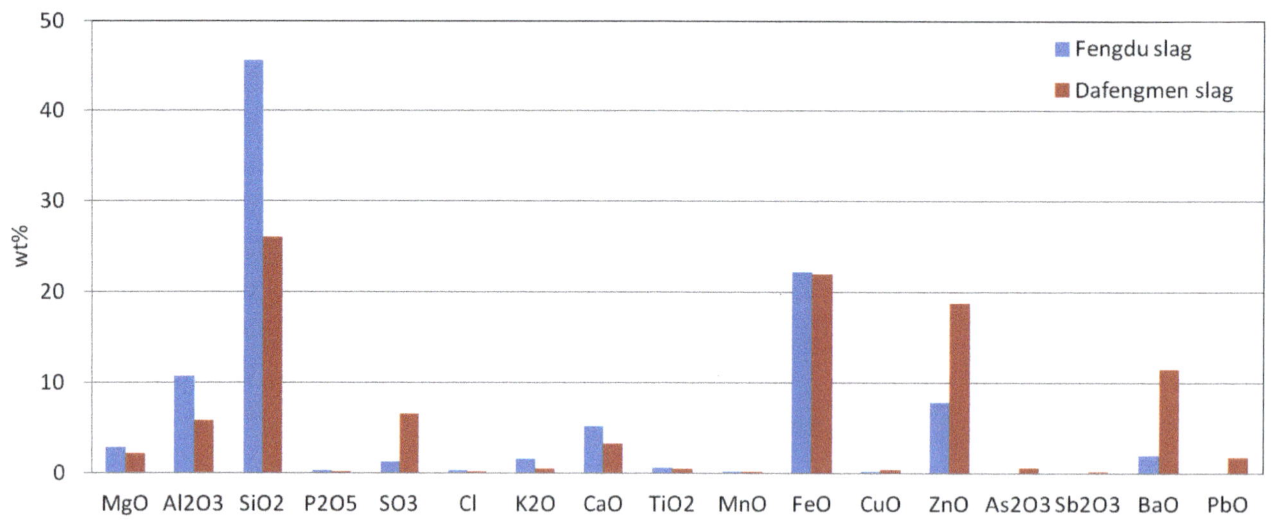

Figure 5.24 Bar chart comparing average bulk compositions of the slag samples from the Fengdu sites and Dafengmen.

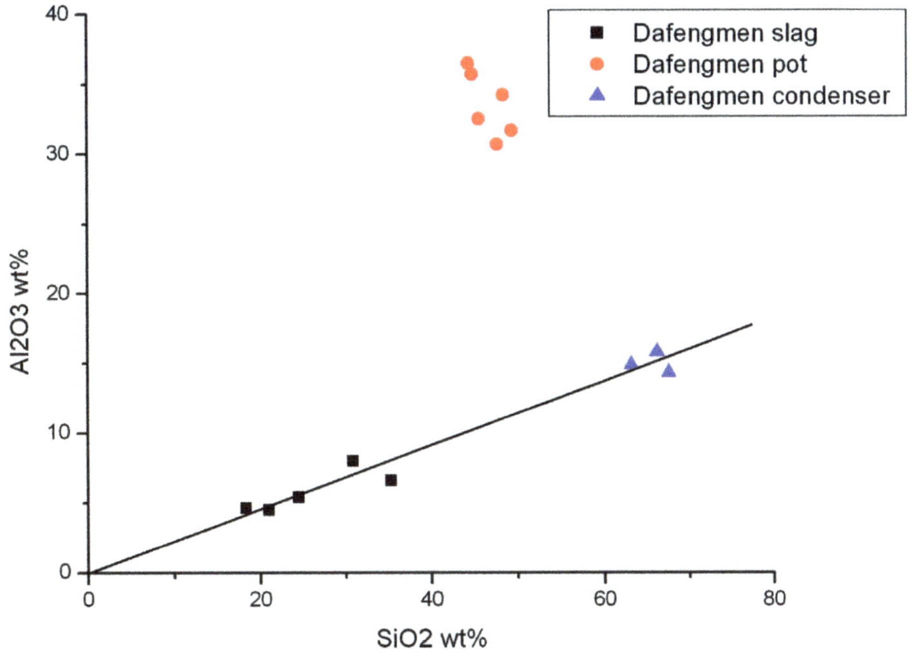

Figure 5.25 The SiO_2 and Al_2O_3 contents in the slag, pot and condenser samples from Dafengmen. Note the linear correlation of SiO_2 and Al_2O_3 of the slag and condenser samples.

The enrichments of FeO, ZnO, SO_3, PbO and minor amounts of CuO and As_2O_3 clearly derived from the ores, which were dominated by iron-rich oxidic zinc ores but also contained sulphidic zinc minerals (sphalerite) and oxidic lead/copper minerals (probably cerussite and malachite). In addition, the ores contained gangue minerals (barite, calcite and dolomite), leading to enriched levels of BaO, CaO and MgO. It should be noted that the abundance of sphalerite and barite minerals could generate a lot of sulphur, which would readily recombine with metallic zinc to form zinc sulphide. This resulted in the great losses of zinc into the slag, which contains 9-33% ZnO. In addition, as with iron impurities, the lead/copper minerals were also easier to be reduced to metals, which would consume more reducing agents. Lead is a harmful element in zinc distillation as it tends to form lead silicate which is very corrosive to the retorts and fusible slag which would hamper the volatilisation of zinc (Ingalls 1903, 212). However, most of the lead from the minerals in the slag samples was reduced to metallic lead, so it was not so harmful. In general, therefore, the zinc ores used at Dafengmen appear of worse quality than the ores used at the Fengdu sites. Most of the impurities would have stayed in the slag, while the reduced lead would have partly volatilised and condensed with the zinc. Although no metallic zinc lumps have been found at Dafengmen, the zinc products are expected to have been richer in lead than those from the Fengdu sites.

The reducing agents might have been coal and charcoal, but there are very limited amounts remaining in the slag. It may be inferred that all the carbon was consumed during the process, and perhaps an insufficient quantity of reducing agents was present in the charge. The ash of coal and charcoal would have contributed further CaO, MgO, K_2O and FeO to the slag, while some sulphur could also have come from the coal.

The internal heterogeneity and abundant residual grains in the slag suggest that the charge was never fully molten during the process, so it is difficult to estimate the operating temperatures from the slag itself. As an alternative, it it possible to estimate it from the degree of vitrification of the pots. Refiring experiments showed that the Fengdu pots had been exposed to temperatures around 1200 °C (see Section 4.2.2.4). Conversely, the Dafengmen pot (SDT11) bloated until failture when refired at 1200 °C, indicating that operating temperatures at Dafengmen must have been lower. The appearance of small amounts of mullite in the pots indicates that the processes may have reached at 1100 °C. After heating, the charge probably cooled slowly, explaining the large crystal sizes if the slag phases.

5.4.3 Final remarks

The technology of zinc distillation at Dafengmen was generally similar to that documented at the Fengdu sites, involving broadly the same *chaîne opératoire* from retort production, mining and ore preparation, reducing agent and fuel preparation, furnace construction to smelting. Some of the technical details of zinc production at Dafengmen are different, however, as outlined above and further discussed in the next chapter.

The yield per retort at Dafengmen must have been lower than that at the Fengdu sites because of the lower grade of zinc ores and greater zinc losses. From the estimated 250,000 tonnes of retort fragments and slag left in the Dafengmen valley, the scale of production can be deemed as factory scale. As the site was dated to the Qing period, it may have been one of the localities of the white lead factory documented in the Shizhu gazetteers (see Appendix 4.2 and 4.3). If this is correct, the zinc factory at Dafengmen should have been privatised and run by merchants, while supervised and taxed by the government, who would have set up local offices nearby to administer mining and smelting affairs. The factory was either composed of a number of production units centred around the furnaces, just like the Fengdu sites, or spatially divided into different working areas. Without further historical evidence and archaeological information, the workshop layout and labour division within the factory remain unclear.

Chapter 6

Zinc Production in Chongqing and Beyond

The archaeological, historical and geological evidence, combined with the present archaeometallurgical evidence based on the analyses of the production remains, allow a tentative account of the technologies of zinc distillation practiced at the three Ming river sites in Fengdu and the Qing site in Shizhu. In this chapter, the differences in the technologies of zinc distillation at these two groups of sites are first reviewed. Considering the resource procurement (zinc ores, coal, charcoal, clay and water) and transport of raw materials and products, the choice of location is assessed for both groups of sites. Applying a modern theory of choice of industrial location, the possible reasons for the differences in technology and organisation are discussed with reference to the social, political and economic constraints affecting the production of zinc in the region for the different periods.

Notwithstanding the differences between sites, however, the overall technological similarities strongly suggest that all the zinc production sites examined can be broadly considered to belong to the same technological tradition. The second section of this chapter therefore attempts a perspective of this broader picture. An outline of the main components of this technological tradition is identified through commonalities in the various case studies presented and the historical and ethnographic sources examined. This is followed by a detailed comparison of the Chinese and Indian zinc distillation technologies.

6.1 Zinc smelting sites in Fengdu and Shizhu: a comparison

6.1.1 Technical aspects

The zinc smelting technologies as documented at the two groups of sites in Chongqing, although dated to different periods, bear clear similarities: the use of iron-rich oxidic zinc ores; a combination of coal and charcoal as reducing agents inside the retorts but with only coal as fuel outside them; the overall design of retorts with pots, condensers, pockets and lids forming two chambers; and the rectangular furnaces. As will be discussed in Section 6.2, these can all be deemed as elements of the same technological tradition. However, the high-resolution reconstruction allowed by the analytical study also highlights some important differences in technical aspects, which are discussed first:

1. Ore grade and preparation

Zinc smelters in Fengdu and Shizhu used the same types of oxidic zinc ores, but of different grades. The ores smelted at the river sites in Fengdu were mainly zinc carbonates with significant amounts of iron oxides, small amounts of barite and possibly calcite and dolomite. Zinc silicates were occasionally incorporated and only few zinc sulphides were present. The grades of the ores used at the three Fengdu sites studied show no significant differences. In contrast, the ores used at the site in Shizhu were also mainly zinc carbonates rich in iron oxides, but contained greater amounts of zinc sulphides, zinc silicates, barite, lead and copper minerals, i.e. oxidic ores collected near sulphidic ore deposits. Thus, while acknowledging that the minerals analysed may not be wholly representative of the ores employed, it would appear that the ores used at Dafengmen were of worse quality than those at the Fengdu sites. Interestingly, it is unlikely that the better oxidic ores had been exhausted by the Qing period, as even the geological ores collected from the Laochangping region in modern times still constitute relatively pure zinc carbonates and medium-grade zinc carbonates rich in iron but low in other impurities (see Section 5.2.2). It therefore seems that the Dafengmen smelters used ores from specific zinc deposits that had been near exhausted, and they did not try to explore and exploit new deposits – an enterprise that would require capital investment and entail a high risk of failure (Golas 1999). They also made no or few efforts to improve their quality, either by properly beneficiating the ores to get rid of unwanted gangue materials or by roasting them to remove sulphur prior to the smelting process.

2. Reducing agent to ore ratio

Coal and charcoal were both used as reducing agents at the zinc smelters in Fengdu and Shizhu, but their residues were less frequently found at the later site than the earlier sites. Generally, an excess of carbon was charged into the retorts at the river sites to ensure efficient reduction of zinc oxide. No such excess of reducing agents was charged at Dafengmen, resulting in insufficient reduction of zinc oxide, with a subsequent decrease in yield due to loss of zinc into the slag. There are two possible reasons for the lower proportion of reducing agents in the Dafengmen retorts. One is that the craftspeople at Dafengmen were not aware of the importance of charging an excess of reducing agents. This is unlikely because this need would have been easily recognised through experience. The other more likely possibility is that they deliberately charged smaller amounts of reducing agents in order to lower the cost.

3. Retort design and performance

The pots of the retorts used at the river sites in Fengdu were made of quite refractory clay tempered with quartz sand, which improved their toughness and thermal shock resistance. They were prefired to high temperatures of 1100-1200 °C to produce extensively vitrified fabrics with closed porosity, which were resistant to the attack of the charge and zinc vapour. As a result, their levels of zinc oxide were very low. Overall, they were fairly good quality pots which could have been reused several times. Conversely, the pots used at Dafengmen were made of highly refractory kaolinitic clay tempered with parent rock fragments. The use of these raw materials could have resulted in pots of superb toughness, thermal shock resistance and both thermal and chemical refractoriness. However, the pots were too porous to be resistant to the zinc attack, most likely owing to their lower prefiring temperatures and to the large numbers of ill-sorted rock fragments. Therefore, they had much higher levels of zinc oxide, which acted as flux to render the pots less refractory. Thus these pots were technically less appropriate and would have collapsed more quickly preventing multiple reuse.

The choice of raw materials for the condensers and pockets was less crucial for zinc smelting than the pots as they both were exposed to much lower temperatures. They were made of less refractory clay tempered with coal ash fragments from the burnt coal fuel. The large mineral inclusions found in the condensers from Dafengmen might indicate a less careful preparation of raw materials than performed at the river sites.

All in all, even though the taller, slimmer shape of the Dafengmen retorts indicates an improvement, their overall material properties seem less adequate for their ultimate application. This is not only due to possible regional differences in clay sources, but also, crucially, to the way in which these materials were processed.

4. Technical efficiency

Ideally, one might assess technical efficiency by calculating the proportions of zinc metal produced relative to the amount of zinc contained in the ores. However, we do not know the typical zinc content in the ores, the quantity of ores that went into each retort load, or how much zinc was produced per retort. As an alternative proxy, one can look at how much zinc was lost, i.e. zinc that was contained in the ores but did not make it to the final ingot. This is an approach that, while not without its problems, has been used in other archaeometallurgical studies, assessing copper or iron losses in the slag (e.g. Rehren *et al.* 2007; Pryce *et al.* 2010).

For zinc distillation, there are three kinds of zinc losses. One is the loss of zinc into the slag which was caused by the incomplete reduction of zinc minerals in the ores. An estimate of the amount of zinc lost in this way may be obtained by measuring the zinc contents of the slag, and the quantity of slag produced per smelt. The loss of zinc to the slag is affected by several parameters, such as: the purity of ores; an excess of sulphur (either from the ores or the coal) combining with the zinc; ore grain sizes being too large or too small; an insufficient amount of reducing gases in the retort; or perhaps the process temperatures or duration not being appropriate. Our study indicates that the zinc ores used at Dafengmen contained more types of impurities than those at the Fengdu sites. They contained more sulphur-bearing minerals (zinc sulphide and barite), which generated more sulphur to trap zinc in the slag. They also contained relatively higher amounts of iron oxides and other impurities (lead/copper minerals), which would consume reducing agents. The charge may also have had larger ore grain sizes, as indicated by the large lump in SDB12. Furthermore, there was a smaller proportion of reducing agents inside the Dafengmen retorts, which would make the reduction less efficient. In addition, the process temperature at Dafengmen may have been lower than that at the Fengdu sites.

Secondly, zinc vapour would be lost as it penetrated into the ceramic fabrics of the retorts, such as pots, condensers and pockets. Although the taller and slimmer Dafengmen pots were made of highly refractory clay, they were made very porous and probably prefired in a relatively low temperature, so they were not resistant to the zinc attack. If such pots were reused, the cumulative fluxing effect of zinc contamination would lead to collapse. It thus appears that the Dafengmen pots could not be reused, otherwise they would have collapsed. As such, the higher levels of zinc in the Dafengmen pot fabrics could also indicate more zinc loss into pots per smelt. Since condensers were also not likely to be reused, their zinc levels can also be informative. As no pockets were found at Dafengmen, zinc losses into these retort parts cannot be quantified.

Thirdly, an excessive temperature or an insufficiently reducing atmosphere in the condenser could result in the oxidation of zinc and the escape of metal as a vapour into the air; however, this zinc loss cannot be easily evaluated from the production remains.

Obviously, any comparison of the zinc losses between sites using the above parameters can only be indicative. We do not know how much slag was left inside each retort per smelt and how many times a given retort was used. More importantly, we should not forget that efficiency is a relative parameter that should be assessed in each particular context – for example, by considering the costs of raw materials, fuel and labour, the value of products, etc. In addition, the sample analysed so far is relatively small and quite variable. However, even if only superficially compared, the remarkably higher zinc levels in the slag, pots and condensers from Dafengmen stand out next to those at the Fengdu sites (Figure 6.1), and indicate higher zinc losses at the Qing site. This is even more significant if one bears in mind the extremely large numbers of such artefacts left behind by the zinc makers. It thus seems compelling that, on a technical basis, the later, Qing technology was less efficient than that documented at the Ming sites.

6.1.2 Choice of location

In order to assess how these multiple crafts associated to zinc production were spatially organised, and how production was related to consumption, this research will rely on an economic model on industrial location, borrowed from the discipline of modern economic geography. The theory of industrial location was first systematically formulated by the famous German economist Alfred Weber in the early twentieth century (Weber 1929), who posted that an industry should be located where it can minimise its costs and thus maximise its profits.

Weber analysed various factors which could influence the choice of location and argued that costs of raw materials and fuel, transport costs and labour costs were the three most important general locational factors affecting all industries. To simplify the theory, the differences in prices of the same raw material or fuel in different locations are abstractly expressed as the differences in transport costs. The application of the theory of industrial location can be divided into three stages. The first stage only considers the first locational factor, transport costs. The transport orientation law pulls the industries to the optimal point with the least transport costs of raw materials, fuel and products. In the second stage, the second locational factor, labour costs, is taken into consideration. The labour costs drive the location of industries if the savings in labour costs in a particular location with cheaper labour are more than the increase in transport costs incurred when moving away from the location with the minimum transport costs. The third stage investigates the influence of agglomeration on the location determined by the transport and labour costs. The agglomeration of a large number of enterprises decreases the costs as they can provide assistance to each other through shared technology and facilities, thereby further relocating the industries (Weber 1929). This theory has been applied to various industries, including modern zinc smelting (Cotterill 1950).

In Weber's transport orientation law, the two fundamental factors which determined transport costs were the weight to be transported and the distance to be covered. Other factors included the type of transport system and the extent of its use, the nature of the region and its kinds of roads, and the nature of the goods themselves. There were ubiquitous materials, i.e. raw materials widely spread, and localised materials (raw materials only located in geographically well-defined localities). The localised materials were further classified into two groups: pure materials, which imparted total weight to the products, and weight-losing materials, of which only part of or no weight went into the products. The ubiquitous materials were not deemed important factors if they did not enter the products. The transport orientation law uses the 'material index', the proportion of the weight of localised materials to the weight of products, to judge the impacts of different raw materials on the choice of location. When the material index is below 1, the industry is attracted towards the location of consumption; when the material index is above 1, the industry is attracted towards the location of the materials. When there is more than one kind of weight-losing material (material index >1), then the industry tends to be located nearer the materials with more weight loss, i.e. higher material index, to minimise the overall transport costs (Weber 1929).

To apply Weber's transport orientation law to zinc smelting, it is necessary to analyse the nature of the various raw materials required, including zinc ores, coal, charcoal, clay and water.

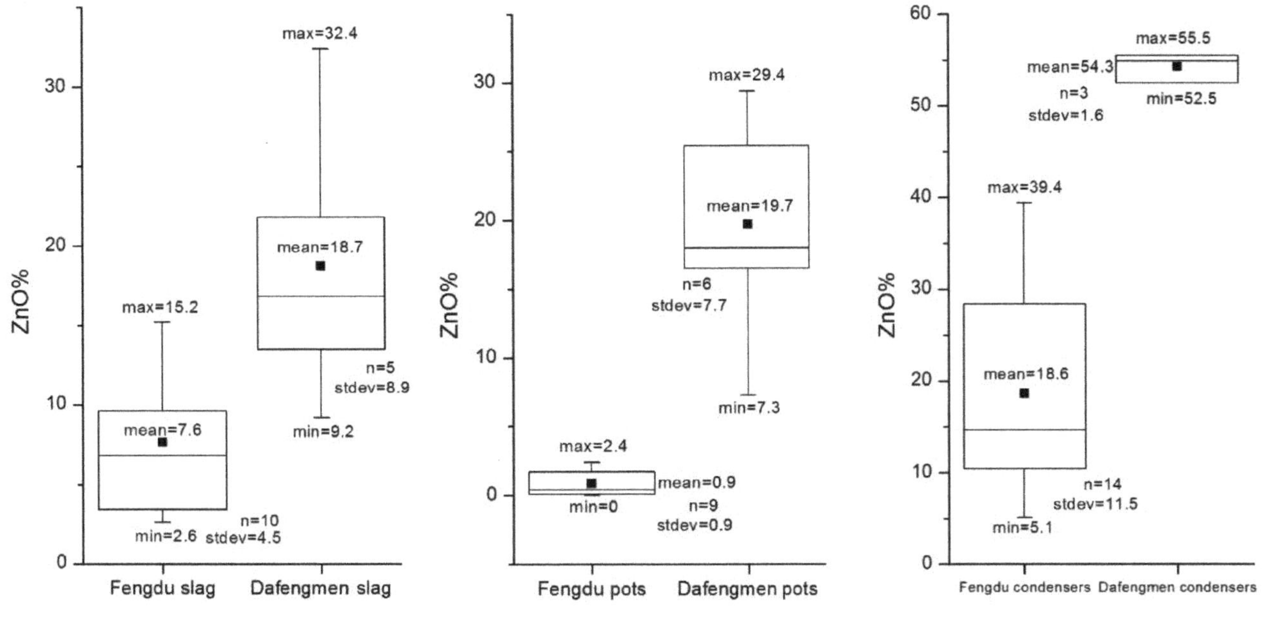

Figure 6.1 Box and whisker charts of zinc oxide levels (wt%) of the slag (left), pot (middle) and condenser (right) samples from the Fengdu sites and the Dafengmen site. All the zinc oxide values were bulk compositions of the slag, pots and condensers. (max=maximum value; min=minimum value; mean=average value; n=number of samples; stdev=standard deviation)

Charcoal, clay and water may be considered as ubiquitous materials, which can therefore be neglected when applying Weber's theory. Only relatively small amounts of charcoal were needed as part of the reducing agents, and these could be produced from the local trees growing in the regions near the river sites and the site in the mountain. The making of the pots of the retorts required refractory clay and suitable temper. Although we do not know where these pots were produced, we can only assume that the clay would have been procured from nearby deposits for each locality. The ordinary clay for making condensers and pockets was more widespread. To process the clay, a large amount of water is needed: this would most likely to have been provided by the Yangtze River at the river sites, or by the small streams or lakes at Dafengmen.

Zinc ores and coal are localised, weight-losing materials, and therefore may have been the two main materials that influenced the choice of location. The coal consumed both as fuel and as a reducing agent was much heavier than the zinc ores. In the traditional processes, the weights of coal consumed were normally two to three times those of the zinc ores (see Section 2.3). This indicates that the material index of coal was higher than that of zinc ores. Coal would not only be much heavier, but also much bulkier because of its lower density than the zinc ores. In addition, part of the zinc ores entered the product while coal did not. Therefore, according to Weber's recommendations, the transport costs of zinc ores, coal and zinc products, would be minimised when the smelters were located in or near the coal deposits. This is consistent with historical records: the Shizhu gazetteers explicitly points out that zinc production was restricted by coal availability, and the frequent relocation of the white lead factory was partly due to the procurement of coal (see Appendix 4.2 and 4.3).

In Ming and Qing China, land transport of bulk goods was mainly by animal-drawn carts, animal pack, wheelbarrows and coolies, among which the costs of coolies were the highest. Water transport was much more economical, although not always faster, than land transport (Twitchett and Fairbank 1998, 603; Golas 1999). Apart from the Yangtze River, the Fengdu and Shizhu region has numerous small rivers and streams, although they are not large enough for water transport. Even the largest river, the Dragon River, a tributary of the Yangtze River, was not suitable for shipping (RGALHS 2009). *Revised Local Gazetteer of Fengdu*, compiled in the early twentieth century (Huang and Lang 1992), records that a coal mine at Qingbaidang, Xinjian township, 30 *li* (c. 17 km) away from the capital town of Fengdu, supplied coal to sell in the capital town. It took seven days to transport the coal from the mine to the capital town via the Dragon River, as there were dangerous rapids and shoals in the river. The transport costs doubled the price of coal in the capital town (Huang and Lang 1992). The mountainous nature of the region also made the land transport tough. The mountain roads were narrow, unpredicatable and steep, so the bulk goods could only be carried by coolies on foot (Peng and Wang 2007). A coolie could only carry around 40 kg at a rate of 40 km per day, which is considerably lower than other forms of transport (Twitchett and Fairbank 1998, 603). When the poor road conditions and coolie transport in the region are taken into consideration, the transport costs emerge as a key locational factor for zinc smelting sites.

By plotting zinc smelting sites, zinc mines and coal mines on the regional maps (Figure 6.2), the choice of location can be assessed in more detail. On this basis, the location of Dafengmen seems ideal in terms of transport, as it was situated within the coal deposits of the Qiyao Mountain Coalfield, which spanned from the southwest to northeast of Qiyao Mountain. The coal seam (mainly bituminous coal and anthracite) at Xiaofengmen, near Dafengmen, was in the weathering zone, about 0.5 m thick, and therefore suitable for opencast mining (Sichuan Geology Bureau 1961).

As for the proximity to ores, the site was about 7-8 km away from the Laochangping lead-zinc deposits. Local elders reported that coolies used to carry zinc ores from Laochangping to Dafengmen, with a break for resting by Xijiaoxi, 'a small river for washing feet'. It took three days for two round trips. The Shizhu gazetteers are not explicit as to whether the Baishaling white lead factory was located near the ores or the coal, but it makes apparent that local officials were aware of the importance of coal sources for zinc smelting. The factory was moved to many localities partly in response to the need to be close to coal deposits (see Appendix 4.2 and 4.3).

Like the Baishaling factory, the zinc produced at Dafengmen should have been partly paid as tax to the government and partly sold to the government – with the rest left for further commerce. Both government parts were probably sent to the Sichuan provincial mint, Baochuanju 宝川局, which was located in Chengdu, 350 km west of Chongqing (Wang 1990). Sichuan province was the second largest copper production area after Yunnan province; its copper not only supplied its provincial mint, but was also procured by the central mints in Peking and several other provincial mints. Baochuanju was the most productive provincial mint and the coins cast there were circulated beyond the province (Wang 1990). It began to mint coins in 1668, but stopped minting already in 1670. The office reopened in 1732 and 8-40 furnaces were in operation. In 1756, the office used 1,400,000 catties of copper, 1,162,000 catties of zinc, 182,000 catties of lead and 59,360 catties of tin; in 1781, it used 600,000 catties of copper, 498,000 catties of zinc, 78,000 catties of lead and 25,440 catties of tin (Huang 2008, 2349). As recorded in *Local Gazetteer of Shizhu* [1775], 300,000 catties of zinc produced by the Baishaling factory were sent to Baochuanju in 1770 (see Appendix 4.2). Thus Shizhu was an important zinc production area for Baochuanju. To transport zinc from the Laochangping region, the best way would probably involve first land transport by coolies to the Yangtze River, followed by water transport via the Yangtze River and its tributary and land transport to Baochuanju. The rest of the metal for commerce could be sold easily to merchants around the Yangtze River after the first leg of transport.

Turning to the river sites at Fengdu, the choice of location seems far from ideal based on the least cost transport theory. The nearest and largest zinc mines were the Laochangping lead-zinc deposits, about 50 km away (see Section 5.1.1). As noted before, the stela inside Yushi Cave (Wang 1991) suggests that the old mine had been exploited at least since the Wanli period, i.e. contemporary with the mass production of zinc in Fengdu. In addition, the ores exploited in the cave are similar to the mineral fragments discovered at Miaobeihou. Thus it is quite likely that the zinc smelting sites in Fengdu used ores from the relatively distant Laochangping region.

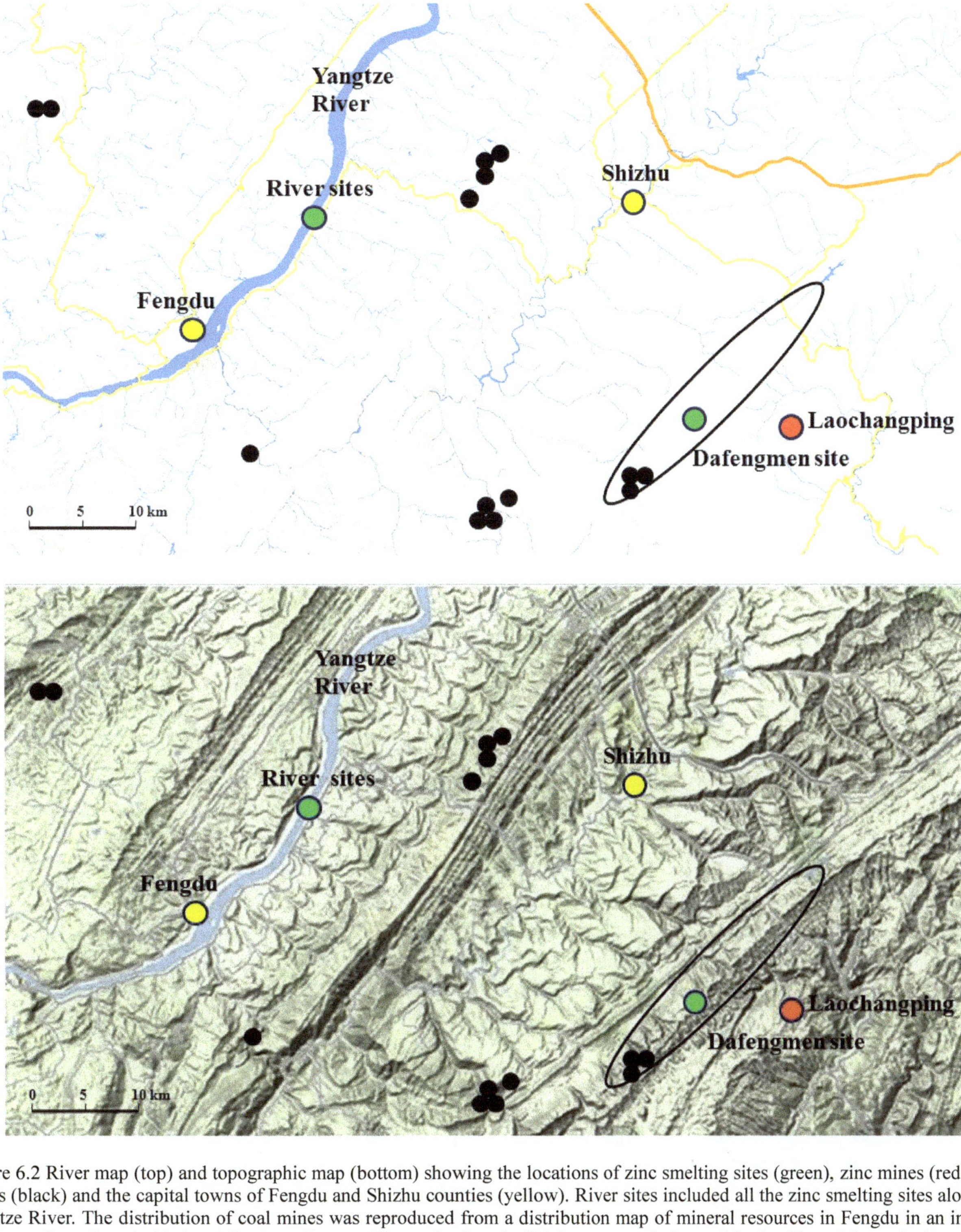

Figure 6.2 River map (top) and topographic map (bottom) showing the locations of zinc smelting sites (green), zinc mines (red), coal mines (black) and the capital towns of Fengdu and Shizhu counties (yellow). River sites included all the zinc smelting sites along the Yangtze River. The distribution of coal mines was reproduced from a distribution map of mineral resources in Fengdu in an internal document from Fengdu National Territory Bureau. The coal mines in Shizhu are mostly located in the Qiyao Mountain (the circled area). From Google Maps.

There are several coal deposits reported in modern Fengdu (Figure 6.2). The Qingbaidang coal mine, as mentioned above, was probably the nearest coal deposit to the Fengdu sites. It could have also been exploited for the zinc smelters during the Ming period. There were other coal mines in Fengdu, but they were exploited only for local use because they were situated far from the city and not connected by rivers (Huang and Lang 1992). Coal from other sources, potentially even further away, could have arrived at the Fengdu sites via the Yangtze River, which was one of the largest watercourses for transporting goods at that time (Luo 1991) – but the distance would obviously add to its costs.

Although there is no direct historical evidence to inform us about the distribution of the zinc produced at the river sites, a review of the use of zinc in Ming China indicates that its main use was for the manufacture of brass coins, which were cast in the central mint in the capital as well as in provincial mints in most provinces (see Section 2.2.1). Thus the large-scale zinc factories in Fengdu were probably set up to supply minting metals for the Sichuan provincial mint in Chengdu, which was active in the Wanli period. Overall, it can be argued that the locations of the river sites were pulled by consumption, with much higher overall transport costs. However, it seems clear that there must have been other factors that drove them away from the localities deemed 'ideal' from a transport cost perspective.

6.1.3 Explaining the differences between the two groups of sites

Based on the comparisons of the technology and location of production presented so far, it has been argued that the earlier zinc smelters by the Yangtze River were not ideally located in terms of their higher transport costs, but they were technically more efficient in terms of ore grade, reducing agent to ore ratio, quality of retorts and zinc losses. In contrast, the later site at Dafengmen was ideally located with lower transport costs, but the technology employed was less efficient. Possible reasons for these differences lie in the different social, political and economic contexts in which the two groups of sites operated.

As discussed above (see Section 4.4.2), the zinc sites in Fengdu may date more specifically to the late Ming period, when brass coins began to be manufactured on a large scale by alloying copper and zinc. In the early Ming Dynasty, mining and smelting were strictly controlled by the state by recruiting specialist households, corvée, conscript and convict labour. With the development of a commercial economy, the state-run industries declined in the mid-Ming period, while private mining and smelting boomed. The government still supervised the industries and reaped benefits by collecting tax. By the late Ming period, most of the industries were operated by private entrepreneurs (Bai 1956; Xu and Wu 1985) and it is therefore likely that the Fengdu zinc workshops were also run privately. Whether the approximately 20 sites so far documented were individual small workshops or just units belonging to a larger factory is unknown due to the lack of historical and archaeological evidence. The close agglomeration of these workshops and the technical similarities among them seem to indicate that they probably shared the same sources of ores, coal and pots. This mass production of zinc must have been supervised and taxed by the government.

Why were these Fengdu workshops located far away from the zinc mines, where there was also available coal? This seemingly strange choice is probably explained by the specific political and social environment in which they operated. Laochangping was then situated in the region under the *Tusi* reign, but also near neighbouring Fengdu. Local conflicts could have arisen between the Han people and the *Tujia* people for access to ores, as had happened for lead and silver smelting in the region in the fifteenth century (see Section 5.1.2). In order to control the production and collect tax, the Han government perhaps moved the smelting workshops out of the mining area to a safer, easier to control, location. A safer area could also potentially attract more private capital investment. Another possible reason for the location near the river could have been the lack of skilled labour for zinc smelting in Shizhu. This would be consistent with the poor reputation of the *Tujia* people as craft specialists (see Section 5.1.2). It is possible that the sophisticated zinc distillation technology could have been mastered by the Han people. Thus locating the sites by the Yangtze River would have been advantageous not only in its facilitation in the transport of products, but also in the control of products and access to capital and skilled labour.

Turning to the Dafengmen site, dated to the Qing period, this might constitute one of the zinc factories mentioned in the Shizhu gazetteers. The site should have been in operation after the abolition of the *Tusi* system and the establishment of the *Liuguan* system in Shizhu in 1761. The change of administrative system resulted in the influx of Han people and economic prosperity, thereby bringing in skilled labour and capital. By analogy with historically documented Baishaling white lead factory, it is likely that private merchants invested in this enterprise, which would have been supervised and taxed by the government.

Although the location of Dafengmen seems more suitable from a transport point of view than that of the river sites in Fengdu, the technology of zinc distillation was less technically efficient than those practiced at the river sites. Craftspeople at Dafengmen smelted zinc ores with more impurities, mixed the ores with a lower proportion of reducing agents in less-than-ideal retorts, leading to greater losses of zinc. It seems that they were trying to save on materials and labour costs. This would be consistent with the recorded decline of zinc production at Baishaling due to the growing availability of cheap zinc from Guizhou on the market. Guizhou was the largest zinc production area in Qing China and Guizhou zinc supplied the central mint and also provincial mints (Ma 2010a; Ma 2010b; Ma 2011). If zinc produced locally could not compete with the cheap Guizhou zinc, then the production could not bring in profits and even result in serious losses. Under such circumstances, it appears that the smelters would have tried to manage the production by keeping down the costs. The result was a distillation process that, even if taking place in a theoretically ideal location, was not technically efficient with great metal losses. Therefore, the zinc yield at Dafengmen would have been lower than that of the sites in Fengdu – but in both cases the technology was shaped both by material and sociocultural parameters.

6.2 The technological tradition of Chinese zinc distillation

Notwithstanding the regional variants exemplified by the archaeological sites in Chongqing and the traditional processes, a number of shared elements of Chinese zinc distillation make it a consistent technological style through the Ming and Qing Dynasties to the twentieth century. This style can be characterised by four key technical features related to zinc ores, fuel/reducing agents, retorts and furnaces, which explain the overall success of this technology, and also provide a detailed picture that will facilitate comparisons with its Indian counterpart.

6.2.1 The procurement of oxidic zinc ores

According to the earliest account of the zinc smelting process recorded in *Tian Gong Kai Wu*, the zinc ores used to produce zinc in the late Ming period were *luganshi*, which had been used for the cementation process since the Five Dynasties, and for medical applications since the Yuan Dynasty (see Section 2.1). The first detailed description of *luganshi* is in the medical compendium *Ben Cao Gang Mu*, where *luganshi* is described in some detail:

> *Luganshi* got its name because of its fierce fumes in furnaces and its sweet taste. It exists in every mining and smelting place, particularly richly in Sichuan and east Hunan, but those from Taiyuan, Zhezhou, Yangcheng, Gaoping, Lingqiu, Rongxian and Yunnan[6] have the best quality. It is the precursor of gold and silver. It has different sizes, like lamb brains, loose as *shizhi*[7] and sticky to the tongue. Those growing in gold pits are yellow and of the best quality; while those growing in silver pits are white, blue, green or pink. (Needham 1974, 213-214)

As seen from the text, *luganshi*, literally 'furnace sweet stone', was most likely named from the fierce fumes rising from the furnaces for making cementation brass and its sweet taste. According to this source, *luganshi* was widely exploited in provinces of Sichuan, Hunan, Shanxi, Guangxi and Yunnan. Similar locations of *luganshi* were also recorded in *Tian Gong Kai Wu*: *luganshi* 'was produced primarily in the Taihang Mountains of Shanxi province, followed by Jingzhou [in Hubei province] and Hengzhou [in Hunan province]' (Sun and Sun 1966, 247). The *luganshi* described by Li Shizhen was lamb brain-shaped and appeared in variable sizes and colours ranging from white, yellow, blue and green to pink. In addition, *luganshi* was 'loose', indicating its porous nature. Based on these parameters, it can be inferred that the *luganshi* described is quite pure smithsonite ($ZnCO_3$), which is typically botryoidal, reniform and stalactitic, either compact, porous or earthy. Smithsonite is mostly white, often greyish, and sometimes variably coloured by the presence of traces of impurities, such as copper and iron. The best quality *luganshi* from several areas in Shanxi, Guangxi and Yunnan could possibly refer to pure smithsonite with these characteristics. The impressions when tasted – 'sweet' and 'sticky to the tongue' – were the properties often mentioned in the pharmaceutical works (Golas 1999, 233).

Luganshi is generally considered by historians of mineralogy and chemistry to denote zinc carbonates, namely smithsonite ($ZnCO_3$) or hydrozincite [$Zn_5(CO_3)_2(OH)_6$] (Zhao and Zhou 1998, 359), which easily decompose to zinc oxide and then reduce to metallic zinc. Another zinc mineral, the silicate, hemimorphite [$Zn_4Si_2O_7(OH)_2 \cdot H_2O$], usually occurs together with smithsonite. Both minerals are similar in shape and colour and cannot be distinguished easily, so *luganshi* may also have included hemimorphite (Xia *et al.* 1980, 282). This is also the case in European contexts, where the 'calamine' of historical sources was found to include both smithsonite and hemimorphite (Rapp 2009, 179). However, the discovery of residual zinc silicates in some of the slag samples studied in this research shows that hemimorphite was not as easily reduced as the zinc carbonates and would remain in the residues. It was therefore not an ideal zinc ore.

As seen from the analyses of the archaeological ore and slag samples from the Fengdu sites (see chapter four), the main zinc ore mineral used at Fengdu was smithsonite, which was intergrown with iron oxides. Such iron-rich zinc ores were also reported by some Europeans in the eighteenth century[8] involved in the Chinese zinc trade and interested in its production (Bonnin 1924). Sphalerite minerals might also have been present in the ores, but they could not be reduced and would remain in the residues, as shown by residual sphalerite identified in two

[6] Taiyuan, Zhezhou, Yangcheng, Gaoping are all in south Shanxi province, while Lingqiu is in northeast Shanxi province. Rongxian is in Guangxi province.

[7] *Shizhi* 石脂 is kaolin [$Al_2Si_2O_5(OH)_4$], a soft, earthy, irregularly shaped stone, used as a medicinal material in ancient China.

[8] In 1756, Carl Gustave Ekeberg, a supercargo of the Royal Swedish East India Company, described zinc ores: 'its ore is ash-coloured, bluish, slightly glittering like iron-ore and very heavy, varying however according to its percentage' (Bonnin 1924, 10). In 1775, Gustave v. Engestrom, in *Memoirs of Stockholm*, stated that a zinc ore from China: '...was of a white colour, interspersed with red streaks of calx of iron...It contained in various specimens from 60 to 90 per cent. of zinc: the remainder was iron and a small proportion of clay' (Bonnin 1924, 8).

slag samples from Dafengmen (see Section 5.3.2.2). Gangue minerals, such as calcite, dolomite, cerussite and malachite, might also exist in the ores, as documented at Dafengmen (see Section 5.4.2). Unlike brass cementation and medical applications, zinc distillation did not require very pure smithsonite – low grade smithsonite with a relatively high impurity content could still be used, since most impurities are involatile and would therefore remain trapped in the residues, never reaching the condensers. The downside would be a higher consumption of fuel and reducing agents relative to the yield obtained per retort.

Overall, both archaeological evidence and historical evidence suggest that smithsonite was the main ore mineral used, that specialists were capable of discerning better ores, but that impurity-bearing ores were also employed.

6.2.2 The use of coal as fuel and reducing agents

Based on historical, archaeological and ethnographic evidence collected in this research, it appears that the Chinese zinc distillation process employed coal as the main source for both fuel and reducing agents. Coal had been in use as fuel for industrial and household purposes since the Han Dynasty and was widely used in the Song Dynasty (Golas 1999, 195; Li *et al.* 2007). By the Ming Dynasty, craftspeople had already gained much experience in coal mining and coal's uses, as recorded in *Tian Gong Kai Wu*:

> There are three kinds of coal: 'bright coal', 'broken coal' and 'coal dust'. The large pieces of bright coal are about the size of a one *dou* [c. 10 L] measuring box. The coal is produced in Yan, Qi, Qin and Jin [i.e. the provinces of Hebei, Shandong, Shaanxi and Shanxi]. It is kindled with a little charcoal, and can burn throughout a whole day and night without using bellows. The fragments found with the large pieces can be used as fuel after being mixed with clean yellow earth and water and made into briquettes. There are two kinds of broken coal, which are produced in Wu and Chu [i.e. in the middle and lower Yangtze region]. That which burns with a high flame is called 'rice coal' and is used for cooking. That which burns with a low flame is called 'iron coal' and is used for smelting and forging metals. This coal is moistened with water before being placed in the furnace, and the bellows must be used to bring it up to red heat. Coal should be continuously added as it burns. Coal dust is as fine as wheat [flour] and is called 'automatic wind'. It is mixed with mud and water to form briquettes. When placed in the furnace and ignited, it is like bright coal in that it will burn constantly throughout the day and night. Half of this kind of briquettes is used for cooking and half for smelting, calcination, and the manufacture of cinnabar. As for burning stones to produce lime, alum and sulphur, all three kinds of coal can be employed. (Sun and Sun 1966, 205)

This passage shows that the Ming people had already classified several kinds of coal based on size, fumes and combustion performance, and had mastered how to process each category to meet its specific purpose. These different categories roughly correspond to modern classifications (Golas 1999, 190). In China, anthracite and bituminous coal are the most useful and widely used coal. Anthracite contains a high carbon content and little volatile matter, and its flame is clean; bituminous coal contains lower carbon, but more tar and impurities. 'Bright coal' seems thus to correspond to anthracite; most 'coal dust' is likely to have been bituminous coal. The 'rice coal' with a high flame would be bituminous coal, while 'iron coal' with a low flame was probably anthracite or high-quality bituminous coal. Iron smelting in blast furnaces fuelled with coal had been widespread since the Song Dynasty, as evidenced from the high sulphur levels of iron artefacts and coins (Huang 2006; Wagner 2008); coke, produced from the destructive distillation of coal, had been also in use, but it was not mentioned by Song Yingxing. Small pieces of coal, smaller fragments of 'bright coal' and 'coal dust', were made into coal briquettes with clay and water. Such coal briquettes were in use as early as the Han Dynasty (Henan Relics Group 1962; Zhengzhou Museum 1978). By the Ming Dynasty, coal briquettes were widely used as fuel for various purposes, including smelting.

For zinc smelting, the fuel outside retorts and the reducing agents inside them were two different kinds of coal, and required different preparation methods. The coal fuel for zinc smelting was in the form of coal briquettes, as recorded in *Tian Gong Kai Wu* (Sun and Sun 1966, 247). Coal briquettes were also used in the traditional zinc smelting processes (see Section 2.3). There are several advantages of making coal briquettes (Wu 2002b). Small fragments of coal, even coal dust, could be fully utilised. Coal of low quality, such as bituminous coal, could also be used as it was not in direct contact with the charge. The clay bound coal together and its sintering made coal briquettes strong enough to support the retorts and upper coal briquettes. In addition, the space left between coal briquettes could help the air supply and circulation in the furnaces, preventing coal dust choking the furnaces. Larger chunks of anthracite would provide the same ventilation effect but would perhaps burn too hot as well as being weaker. Such coal briquettes were named 'automatic wind coal', as they could burn all day long without using bellows. It is not certain whether coal briquettes or raw coal was used at the zinc smelting sites studied, but the coal gangue materials left on the external surfaces of some retorts seem to indicate the use of raw coal rather than coal briquettes as anticipated. If this is the case, it will be interesting for future work to discern whether these constitute parallel technological choices or whether, alternatively, there was at some point a transition from one to another.

As for the reducing agents in direct contact with the charge, anthracite and high-quality bituminous coal were preferred in the traditional processes for their high carbon contents and little ash and sulphur (see Section 2.3). Coal residues were discovered inside the retorts from the Fengdu and Shizhu sites, but their specific types could not be identified. Based on their low sulphur levels detected by SEM-EDS, they are likely to have been anthracite. Coal can be used as a reducing agent for zinc production because of the nature of zinc distillation: impurities in the coal will not go to the zinc metal but rather stay with the residues within the retorts. Low sulphur coal, anthracite, was probably preferred to avoid the formation of zinc sulphides and thus the loss of zinc. In addition, anthracite was to some extent more advantageous than charcoal because of its much higher heating value (Rostoker and Bronson 1990, 62), which means that a smaller volume of reducing agents were included along with the charge within the restricted space of a retort, thereby allowing a larger proportion of ores and a higher yield per retort. Charcoal was also identified together with coal inside some of the Fengdu retorts. This could suggest that charcoal had been used as a reducing agent before the use of coal. The Fengdu process could perhaps represent a transition from the use of charcoal to coal. However, it is also possible that charcoal was added in order to kindle the anthracite which is difficult to ignite.

Overall, the use of coal as fuel and a reducing agent for zinc distillation could help keep production costs low, as compared with the more expensive use of charcoal (Golas 1999, 194). Metal production in ancient China was affected not only by a shortage of wood for fuel, but also by competition from domestic consumption and other industries. In a mining and smelting region, large quantities of wood were required not only for the production of charcoal for smelting, but also for timber supports and firesetting in mines and for ore roasting (Craddock 1995, 193). In the Laochangping region, where a multiplicity of metallurgical operations took place, zinc smelting with coal could avoid the competition with copper and lead smelting over the use of charcoal (see Appendix 4.2).

The procurement of coal was of great importance to zinc smelting. In order to minimise the transport costs, zinc smelters were often set up near coal mines, such as Dafengmen (see Section 6.1.2). 'The transport of ores to coal' was a universal phenomenon for zinc production in Guizhou province in the Qing Dynasty (Chen Hailian 2014). In the Qianlong period, zinc ores from Sidongshan, Rongxian county, Guangxi province, were transported to the nearby coal mines for smelting. Later the ores were also transported to the coal mines in neighbouring Luocheng county (Renmin University of China 1983, 372).

6.2.3 Retorts for zinc distillation

Arguably the most crucial technical challenge for Chinese zinc distillation was designing installations with two chambers – the reaction zone and the condensation zone – in order to make zinc distillation work. Owing to the extreme volatility of zinc and its readiness to be oxidised in contact with air, both zones have to maintain highly reducing conditions. Ordinary smelting furnaces did not work for zinc distillation because zinc, once reduced, would evaporate and oxidise to zinc oxide when in direct contact with the heating gases within. The use of crucibles, which could be externally heated by the combustion of the fuel surrounding them, would provide a much more closed, highly-reducing chamber for the reduction of zinc ores by reducing agents without exposure to the hot gases. By separating the reduction from the heating, closed crucibles could guarantee strongly reducing condition required for the process (Rehren 2003, 213). Zinc smelting was generally restricted to closed crucibles, called retorts, with the condensation zone created with internal or external condensers. The retorts tended to be tubular or elongated, which would facilitate the cooling in the condensers by keeping them away from the heating around the reaction zone.

The earliest retorts known so far in China are the jar-shaped ones found in Fengdu (see chapter four), which are similar to those illustrated in *Tian Gong Kai Wu* (Figure 1.4). Such retorts may have initially been adapted from domestic jars, to which condensers, pockets and lids were added in order to form two chambers. In the Qing Dynasty, the pots tended to be longer, with condensers also built onto the tops of the pots, as was the case in Shizhu (see chapter five). As with the twentieth-century traditional processes, retorts of variable sizes and forms were employed in different areas. Besides retorts with added condensers, retorts with external condensers and internal condensers were also used. Retorts with elongated condensers were used in the first half of the twentieth century, but the pointed-based retorts with internal condensers became the dominant type and widely used in the second half of the twentieth century (see Section 2.3).

The material properties of the main parts of the retorts, the pots, were of great importance for the overall performance of the retorts. Good pots required good strength, toughness and thermal shock resistance, the ability to endure high temperatures and aggressive chemical environments as well as being impervious to zinc contamination. Ideal pots should be made of refractory clay with adequate coarse temper, and fired to high temperatures. As discussed in chapter four and chapter five, the pots from Fengdu were of excellent technical quality, while those from Shizhu were less so, but still good enough to be used despite higher zinc losses and fewer reuses. The raw materials for making pots in the two regions were most likely to have been procured locally. The Fengdu pots may have been made of a more refractory clay than the domestic jars, which generally did not require a refractory clay. The Shizhu zinc smelter used kaolinitic clay associated with coal deposits, which was also the case in traditional zinc smelting in Guizhou and Yunnan, where fireclays from the upper and lower coal measures were the main raw materials for making pots (Xu 1986; Craddock and Zhou 2003). For making condensers and pockets, which were exposed to much lower temperatures, the selection of clay and temper was not so crucial. Ordinary clay was tempered with coal ash, which was available in great abundance in the zinc smelting furnaces fuelled with coal. The condensers and pockets could be easily removed at the end of each firing and the pots could be reused if no serious damage occurred.

6.2.4 Rectangular furnaces

Unfortunately, *Tian Gong Kai Wu* fails to mention the furnaces or properly illustrate them (Figure 1.4). The furnace foundations excavated in Chongqing show that the furnaces are rectangular. The upper parts of the furnaces are usually missing, leaving only furnace bars and lower parts of walls, from which it can be inferred that they resembled the trough-shaped furnaces used in the traditional practices (see Section 2.3). The traditional furnaces usually contained three retorts on each furnace bar, while the Qing document *Dian Nan Kuang Chang Tu Lue* states either two or four (Wu 2002a, 184). For the zinc smelting furnaces in Chongqing, the number of retorts placed on each furnace bar remains unknown. Considering that the width of these furnaces was similar to that of the traditional furnaces but that the archaeological retorts are generally smaller than the traditional retorts, there could have been a similar number or perhaps a few more retorts on each bar. The length of the furnaces was variable, from 4 m to over 10 m in both the archaeological and traditional cases. Thus, such rectangular furnaces could accommodate from a few dozens up to over a hundred retorts per firing. The yield of one furnace per firing could be controlled by adjusting the length of the furnace, i.e. the number of furnace bars, when the furnace was built.

As noted above, the furnaces were fuelled by coal briquettes, which could burn for a long time by themselves. No evidence of forced draught, such as bellows and tuyères, has been found in the archaeological records in Chongqing or the field surveys of the traditional processes. However, the latter often reports that there were air passages at the bases of the furnaces to provide air for the combustion. In actuality, the air passages were the spaces between each pair of furnace bars. Limiting the number of retorts on each bar as well as packing coal briquettes could easily regulate the air supply. *Dian Nan Kuang Chang Tu Lue* states that the furnaces were powered with leather-made bag bellows (Wu 2002a, 184), but these might have been operated only at the beginning rather than throughout the whole firing.

The rectangular furnaces have no obvious parallel in ancient metallurgy. The zinc smelting furnaces were loaded with retorts, charged with ores and reducing agents and the reaction occurred within the retorts with the external heat supplied by the fuel in the furnaces. Such crucible smelting processes separated the heating from the reduction, so the heating could take place in oxidising atmospheres such as that produced by the combustion of fuel by natural draught. Such processes could be readily carried out in pottery kilns (Rehder 2000, 104). It has been suggested that the zinc smelting furnaces at Zawar were possibly adapted from the traditional Indian updraught pottery kilns (Craddock *et al.* 1998, 48). Similarly, the less common type of furnaces of the traditional Chinese zinc smelting processes, 'climbing hillside' furnaces, operating in Bijie in 1994-95, resembled traditional Chinese dragon pottery kilns (Craddock and Zhou 2003, 275), which were widely used in southeast China and Sichuan province (Kerr and Wood 2004, 347). However, no pottery kilns are known to resemble the typical Chinese trough-shaped zinc smelting furnaces.

A potential source of inspiration for the design of the trough furnaces may have been the linear furnaces used for iron/steel smelting in Sri Lanka, Burma, Cambodia and Japan (Juleff 2009). The V-shaped Japanese *tatara* furnaces, in particular, are very similar to the trough-shaped furnaces used in the traditional zinc smelting processes. These linear furnaces are characterised by the use of multiple tuyères embedded into one or two furnace walls with long sides and the use of natural or induced draught. Unlike the zinc smelting furnaces, they were directly charged with ores and fuel and the reaction occurred within them. However, some modifications could have been made to the linear furnaces to adjust to crucible furnace smelting and zinc distilling, including the arrangements of furnace bars to hold the retorts and regulate the air supply and the adjustment of temperatures in the upper parts of the furnaces. It is thus possible that the rectangular zinc smelting furnaces could have developed from the Pan-Asian linear furnace tradition, adapted for the specific technology of zinc distillation. Similar rectangular furnaces were also used in crucible iron/lead smelting processes (see Section 8.3) and also in melting copper alloys for lost-wax casting (Figure 6.3).

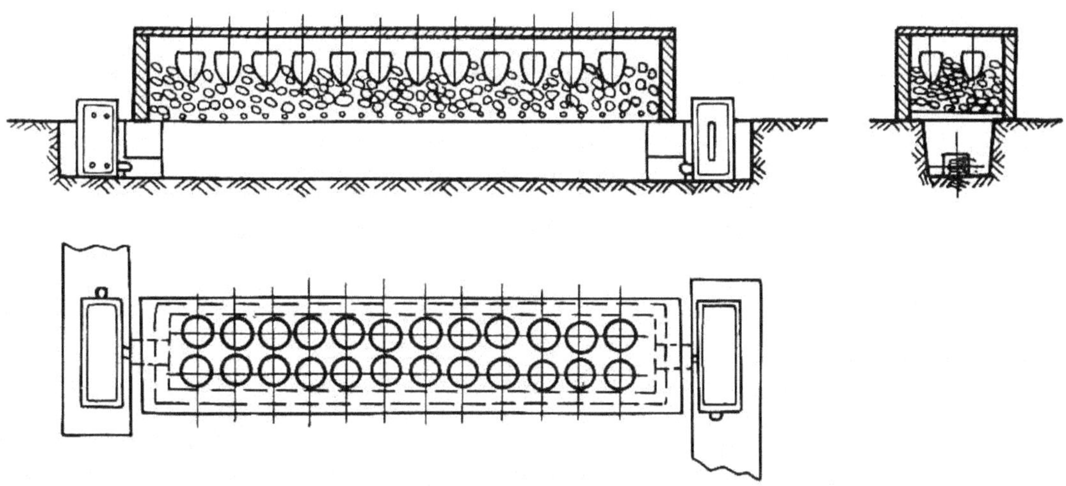

Figure 6.3 A crucible furnace for melting brass for traditional lost-wax casting in modern China (Hua 1999, 543). The furnace was built on a 0.5 m deep hole, with furnace bars and four brick walls. Two rows of crucibles were embedded within the charcoal in the furnace. The furnace was blasted using two wooden bellows situated at the short ends of the furnace.

6.3 Zinc distillation in India and China

The four technical features mentioned above make Chinese zinc distillation unique in the world history of distillation of metals (Craddock 1985). A very different approach to zinc distillation evolved in India, which had begun earlier than its Chinese counterpart (Craddock *et al.* 1998). The relationship between the Chinese and Indian zinc distillation technologies has been addressed before, mostly in the context of discussion of whether they were independent developments. As early as the 1910s, Hommel had no doubts that the Chinese process originated in India, but he did not provide any convincing evidence to support his claim (Hommel 1912). Conversely, most scholars have tended to support an indigenous origin, especially since the reconstruction of the Indian process based on the Zawar remains. Given the stark differences between the early Indian process and the Chinese traditional process, it was concluded that 'the Chinese methods owe nothing to India but are clearly based on the principle of the traditional Mongolian still' (Craddock *et al.* 1998). Mei (1993) agreed that the two processes derived from different technological traditions with their own technological roots, which he thought lay in their local traditions of mercury distillation. However, the Indian process, reconstructed from scientific studies of ancient production remains at Zawar (Freestone *et al.* 1985b; Freestone *et al.* 1985a; Freestone and Tite 1986; Freestone *et al.* 1991; Craddock *et al.* 1998), was indirectly compared with the traditional Chinese zinc smelting processes practiced during the twentieth century. New archaeological data from Chongqing provided in this research and their scientific studies allow for the first time direct and specific comparisons of the two processes. As detailed above for the Chinese case, the Indian process can be characterised by the four technical features associated with ores, fuel/reducing agents, retorts and furnaces (Table 6.1).

The ores used at Zawar were sulphidic zinc ores, specifically sphalerite which occurred with galena and pyrite, and which was separated by hand picking. The mechanical separation of galena was obviously sufficient as the slag contains only 1.2% PbO on average, but pyrite was probably not so easily removed as seen from the high levels of FeO in the slag (14.1% on average). The beneficiated ores were roasted thoroughly to remove sulphur, resulting in low levels of sulphur in the slag (about 0.5%). As the ores contained dolomite as gangue minerals and lime could have been added to the charge to act as an initiator, the slag samples have high levels of CaO (12.4% on average) and MgO (5.0% on average). The Chinese process, in contrast, used oxidic zinc ores with little (Fengdu sites) or some zinc sulphide (Dafengmen). Since no roasting took place prior to smelting, any sulphidic minerals in the ores would remain in the slag; so the slag samples from Dafengmen contain higher levels of sulphur, 2.6% on average. The oxidic ores are rich in iron and contain calcite and dolomite as gangue minerals, so the slag samples from the Fengdu sites and Dafengmen contain about 20% FeO and a few percent MgO and CaO (Table 6.2).

The Zawar retorts were composed of aubergine-shaped pots and conical condensers with long tubes (Figure 6.4). The pots were hand made with a common clay tempered with coarse crushed quartzite, a local rock type. The clay was less refractory than that used for making the Fengdu pots, as indicated by the relatively higher levels of fluxing oxides (Figure 6.5). The quartzite inclusions, much larger than the inclusions in the Fengdu pots, provided the pots with good strength and thermal shock resistance. The pots sometimes appear vitrified and bloated, and even distorted (Figure 6.4). Refiring experiments of the pots suggest that the maximum temperatures reached were about 1200 °C, at which point the ceramic was on the verge of collapse (Craddock *et al.* 1998, 41). This suggests that the Zawar pots, although obviously functional, were technically less proficient and could not have been reused. The prefiring temperatures of the pots cannot be estimated as no unused pots were found and analysed. It is surprising that the ZnO content

(2.2%) of the only used pot analysed (Freestone *et al.* 1985b, 233) was slightly higher than those of most of the Fengdu pots studied (0.9% on average), but much lower than the Dafengmen pots (19.7% on average). The relatively low chemical contamination perhaps suggests that the Indian pots were prefired to quite high temperatures to develop dense fabrics with low porosity, or that the potassium-rich vitrified layer on the surface could act as an effective glaze to stop the penetration of zinc vapour into the ceramic fabrics. However, more Zawar pot samples need to be analysed to make such a conclusion. The condensers, also made of a common clay, show relatively fine, silty fabrics (Figure 6.4). Like the Fengdu and Dafengmen condensers, they were exposed to lower temperatures and probably broken up to extract the zinc within them (Freestone *et al.* 1985b, 231).

Table 6.1 Differences between Chinese and Indian zinc distillation on the basis of the archaeological finds.

Difference	Chinese zinc distillation	Indian zinc distillation
Ore	Smithsonite (hemimorphite)	Sphalerite
Fuel	Coal	Charcoal
Reducing agent	Coal, charcoal	Charcoal, organic matter, cow dung
Retort	Pot, condenser, pocket and lid	Pot and condenser
Furnace	Rectangular	Square
Distillation way	Ascending	Descending

Table 6.2 Average compositions of the slag samples from the Fengdu sites, Dafengmen and Zawar (wt%). Some major/minor oxides and trace elements of 22 slag samples from Zawar were analysed by AAS (unpublished data from Paul T. Craddock and Ian C. Freestone). Only five oxides (MgO, CaO, FeO, ZnO and PbO) can be compared to those of the samples from the Fengdu sites and Dafengmen. The sulphur level in the slag from Zawar is given in (Craddock *et al.* 1998, 35).

Site	MgO	SO_3(S)	CaO	FeO	ZnO	PbO
Fengdu sites	2.8	1.2 (0.5)	5.1	22.2	7.8	-
Dafengmen	2.1	6.5 (2.6)	3.2	22.0	18.8	1.7
Zawar	5.0	1.2 (0.5)	12.4	14.1	5.3	1.2

Figure 6.4 Cross section of an Indian retort from the teaching collection of UCL IoA. The aubergine-shaped pot (now distorted from firing) and a conical condenser were luted together; the long tube of the condenser is missing. Note the large inclusions of quartzite in the pot and the fine texture of the condenser fabric.

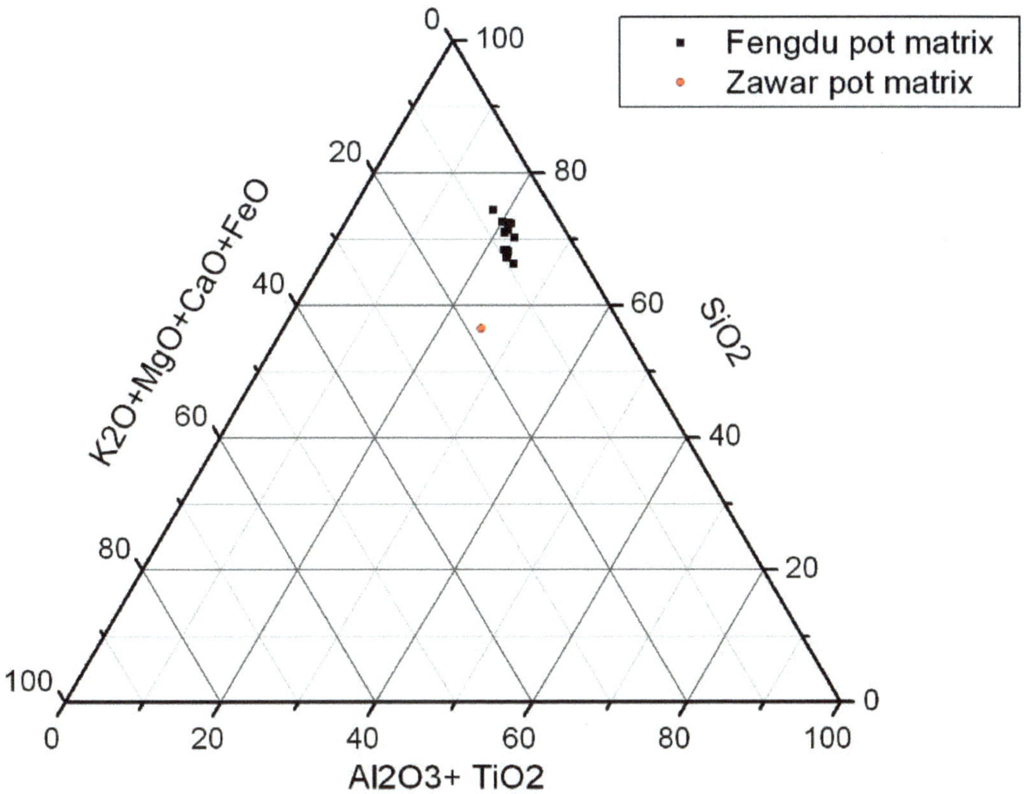

Figure 6.5 Ternary diagram of SiO_2-(Al_2O_3+TiO_2)-(K_2O+MgO+CaO+FeO) showing the matrix compositions of the Fengdu pots and a Zawar pot.

The Zawar furnaces were also different from the Chinese rectangular furnaces. They had the overall shape of a truncated pyramid, and internally divided by perforated bricks into two parts: the upper firing chamber and the lower cool chamber (Figure 1.2). The retorts were thus placed with the condensers facing down and loaded into the upper chamber with the condensers sitting in the holes of the perforated bricks. The pots were heated with the fuel in the upper chamber, at high temperatures (about 1200 °C) comparable to those recorded in the Chinese process in Fengdu. Highly reducing conditions were maintained within the retorts, as indicated by the metallic iron prills in the slag. As the zinc formed it descended and condensed in the condenser tubes, and was collected in vessels placed in the lower chamber. Each furnace accommodated 36 retorts in a 6 by 6 arrangement, while several furnaces were built together side by side, forming a rectangular bank (Craddock *et al.* 1998).

Overall, the Indian style of zinc distillation differs remarkably from the Chinese process in terms of the four defining technical features. The most significant distinction was in the arrangement and orientation of the distillation equipment. The Zawar process was based on distillation by descending, a long-established distillation tradition practiced in mercury production and also zinc distillation as described in earlier alchemical texts (Craddock 1985; Craddock *et al.* 1998). The Chinese process, in contrast, adopted the traditional Chinese distillation by ascending (see Section 8.3). However, as the analyses reveal, more subtle aspects of the *chaînes opératoire* also set both traditions apart.

The estimated zinc yield of one retort at Zawar, 0.1 kg for small ones and 0.3-0.4 kg for the large ones, is lower than that of an ordinary Fengdu retort (0.6-0.7 kg). At Zawar, one furnace block could produce 4 kg of zinc per day for small retorts, and 11-14 kg of zinc per day for the large retorts. Zawar furnaces with several blocks could produce dozens of kg of zinc, an estimate which is likely to be lower than or comparable to the yield of one Fengdu furnace, estimated at 60 kg for a furnace with 30 furnace bars. As estimated from the production remains, 24,000-32,000 tonnes of zinc were produced at Zawar over about 400 years from AD 1400-1800, i.e. an average of 60-80 tonnes of zinc annually (Craddock *et al.* 1998). Such large-scale production was probably directly run by the Mewar state with a single effective overall administration, as evidenced by the presence of a massive fortress (Craddock 2009b, 456). Zinc produced at Zawar was used for the manufacture of various products, including brass vessels, coins, zinc oxide, etc.

While Zawar appears to have been the only zinc production area in India, large-scale zinc production was simultaneously taking place in many provinces of Ming and Qing China. Although the production scale at the Fengdu sites and the factories in Shizhu cannot be estimated from the remains, the Shizhu gazetteers record that at least 179 tonnes of zinc was produced annually at the Baishaling factory in 1770, while at least about 50 tonnes of zinc was produced annually in the late eighteenth and early nineteenth centuries (see Appendix 4.2 and 4.3). According to this source, the annual yield of zinc in Chongqing alone could be comparable to that at Zawar, not to mention the fact that Chongqing was only one of the several production areas in China. In Qing China, Guizhou province was the largest production area with an overall output of over 523,000 tonnes of *yuan*[9] from 1725 to 1828, i.e. an average of about 5,000 tonnes annually (Ma 2011), which was much higher than the annual yield at Zawar. Zinc production in Ming and Qing China was generally organised as factories capitalised and run by private merchants, who were in turn supervised and taxed by the government. Although Chinese zinc was used for making brass vessels and Paktong, and considerable quantities were also exported, its production was primarily intended for minting (see Section 2.2). It was the great demand for coinage that led to the boom of zinc production in the late Ming and Qing Dynasties.

Overall, from a modern technical perspective that is only concerned with costs and yields, it can be argued that the Chinese process appears to be more efficient than the Indian process in three main aspects. Firstly, the Chinese process relied on the use of smithsonite, which could be charged directly to the retorts, whereas the Indian process used sphalerite, which had to be carefully roasted before charging. Secondly, the Chinese process used mineral coal extensively for both fuel and as a reducing agent, which would have been much cheaper than the charcoal used in India. Thirdly, the Chinese retorts could be reused several times, while the Indian retorts were only used once. However, even the generally more efficient Chinese process was not always as efficient as it might have been, as evidenced by the process practiced at Dafengmen (see Section 6.1). The different choices of ores, reducing agents, fuel and retort design of the two processes were not only influenced by material constraints in the two regions, but also by their different social, economic, political and cultural environments.

In summary, the new archaeological and archaeometric data from Chongqing confirm previous suggestions about the relationship between the Chinese and the Indian zinc distillation processes: they were two traditions of zinc distillation, arguably different enough to suggest that they originated and developed independently from each other.

6.4 Summary

Based on the analyses of production remains from the two groups of sites in Chongqing presented in the previous two chapters, there were obvious differences in technical details and organisation of production. The discussion presented in this chapter explained how these differences responded to a variety of diachronic and regional economic, political and environmental constraints, thus linking the *chaînes opératoire* to their specific contexts. Notwithstanding the differences between sites, however, the overall technological similarities strongly suggest that all the zinc production sites examined can be broadly considered to belong to the same technological tradition that lasted from Ming and Qing China throughout the twentieth century. The Chinese tradition of zinc distillation was totally different from the Indian tradition, reinforcing previous suggestions of their different origins.

[9] *Yuan* 铅 includes *baiyuan* 白铅 (zinc) and *heiyuan* 黑铅 (lead), but most of *yuan* produced in Guizhou was zinc.

Chapter 7

The Significance of Zinc Production in Ming and Qing China

The previous chapter discussed the zinc distillation technology in Chongqing and summarised the key features of the Chinese zinc distillation tradition. Zinc production in the late Ming and Qing Dynasties was on an industrial scale in many provinces. According to Ming documents, *luganshi* deposits were known and mined in several provinces, such as Shanxi, Guangdong, Sichuan, Hunan and Yunnan, from the early fifteenth to late sixteenth centuries. However, it is unclear whether these ores were exploited to produce metallic zinc or to make cementation brass. As shown in the previous discussion on the use of zinc in coinage, *luganshi* deposits exploited before the Wanli period (AD 1573-1620) were most probably used for the cementation process. From the beginning of the seventeenth century to the 1680s, *luganshi* deposits were exploited for zinc production in the provinces of Shanxi, Henan, Hubei and Hunan. During the eighteenth century, zinc was produced in many provinces, among which Guizhou emerged as the largest zinc mining and production centre (Chen and Souza forthcoming).

The beginning of large-scale zinc production was strongly associated with the minting of coins. The zinc produced quickly replaced *luganshi* for making brass coins. It was proposed by Zhou that zinc production developed from the brass cementation process, as the cementation process was technologically inefficient and was thus soon replaced by the speltering process. The change of technology was resulted from the great social demand for brass for minting in the late Ming period (Zhou 2007b; Zhou 2008). This viewpoint appears to be fundamentally sound, but it lacks of a sufficient discussion. Therefore, the first section of the chapter tries to explore how and why the speltering process replaced the cementation process in the late Ming period, followed by a discussion on the role of zinc in coinage. The industrial production of zinc in Ming and Qing China played an important role not only in the domestic territory, but also in the world beyond. The second section of the chapter tries to investigate the significance of Chinese zinc in the world, with India and Europe as two cases.

7.1 Chinese zinc at home

As shown in the review of the use of brass and zinc in Ming and Qing China (see Section 2.2), brassmaking did not flourish until the use of brass for coinage in the Jiajing period (AD 1522-1566). It is generally believed that the earliest brass coins were made of cementation brass, while from the Tianqi period (AD 1621-1627) brass coins were made by the speltering process (Zhou 2007a). There was a transition from the cementation process to the speltering process not long after the Jiajing period, but how and why this transition happened has not been fully explored. This exploration is attempted here by revisiting relevant historical documents and analyses of brass coins, combined with recent experiments and studies on the European cementation process. In addition, the monetary policy in Ming China will be considered in order to understand the transition and the role of zinc in coinage.

It should be noted that this section focuses on brass coins because, on the one hand, minting was the main use of zinc and brass in the periods studied, and, on the other hand, brass coins can be dated to specific emperors' reigns allowing diachronic changes in their alloying recipes to be resolved.

7.1.1 Two brass making processes: cementation vs. speltering

As already noted, two main procedures could be used to produce brass for coinage: cementation and speltering. Both processes were two-step operations, but in different ways: the cementation process involved brass making by cementation followed by brass melting, while the speltering process included zinc distillation and the subsequent alloying of copper and zinc (Figure 7.1). Brass making by cementation was described in *Tian Gong Kai Wu*:

> Place one hundred catties of 'self-bellowing' coal [briquettes] in the furnace. Put ten catties of copper into an earthen jar, followed by the addition of six catties of *luganshi*. Heat the jar inside the furnace until the resultant metal melts. (Sun and Sun 1966, 247)

The description is so simplified that the addition of reducing agents was not mentioned. The cementation brass produced was remelted in official mints to cast coins. The technology of zinc distillation has been reconstructed from the production remains in Fengdu, which complemented the vague texts and illustration in *Tian Gong Kai Wu*. The alloying of the two metals, carried out in mints, was briefly described in *Tian Gong Kai Wu*: melt six catties of copper in a jar, and then add and melt four catties of *woyuan* 倭铅 (Sun and Sun 1966, 247).

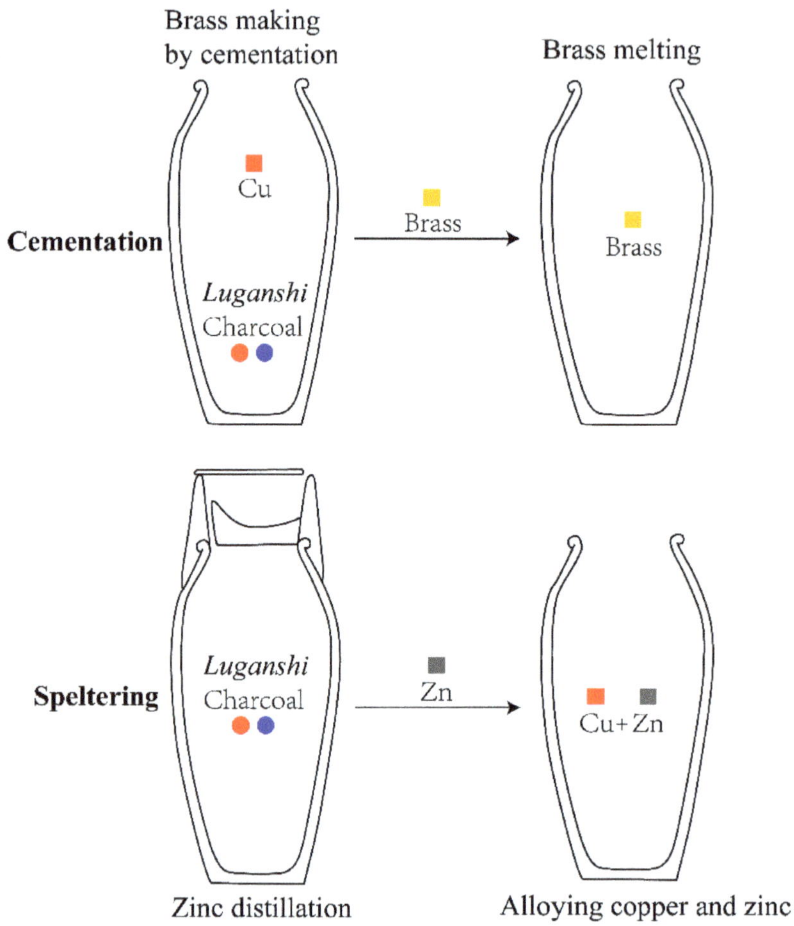

Figure 7.1 Schematic diagram of the cementation and speltering processes for making brass coins.

1. Brass coins

From the early Ming Dynasty, leaded bronze was the main alloy of coinage. A few coins of the Hongzhi period (AD 1488-1505) were found to contain significant levels of zinc, up to 16%. However, from the Jiajing period (AD 1522-1566), all the coins were made of brass. The first large-scale issue of brass coins in the Jiajing period is recorded in a monetary regulation: 'casting 600,000 coins used 47,272 catties of *erhuo huangtong* 二火黄铜 and 4,728 catties of *shuixi* 水锡' (Shen 1989, 982-983). The *shuixi*, literally 'water tin', referred to metallic tin (Zhou 1992), while *huangtong* denoted cementation brass made with copper and *luganshi.* Most Jiajing brass coins were found to contain 12-20% zinc, which are typical zinc levels of cementation brass (Figure 7.2a). Their tin levels range from 3-10%, which are slightly lower than that recorded in the monetary regulation, i.e. ~10% tin, if we assume that the tin was evenly alloyed with the amount of brass recorded. They also contain variable levels of lead, up to 15%. During this period, a *luganshi* mine was exploited in Lin'an, Yunnan province, to make cementation brass with copper for minting, as recorded in the geographic treatise *Du Shi Fang Yu Ji Yao* 读史方舆纪要 (Essentials of Geography for Reading History), written by Gu Zuyu 顾祖禹 in the early Qing Dynasty (Gu 1993, vol.115, 165).

In the Longqing period (AD 1567-1572), only low numbers of coins were issued. Only six coins were analysed and found to contain higher levels of zinc (19-22%) than those from the preceding period.

In the Wanli period (AD 1573-1620), many sources differentiate between *erhuo huangtong* 二火黄铜, literally 'two-fire brass', and *sihuo huangtong* 四火黄铜, literally 'four-fire brass'. The monetary regulation also records that casting 10,000 *jinbeiqian* 金背钱 (gold-back coins) used 85 catties 8 taels of four-fire brass and 5 catties 11 taels of water tin, while the casting of 10,000 *huoqiqian* 火漆钱 (fire-lacquer coins) used the same weights of two-fire brass and water tin (Shen 1989, 983). Based on their analyses, the Wanli brass coins can be roughly grouped into two types in terms of the zinc levels. The first group of coins contain similar zinc contents (6-24%) to the Jiajing coins, while the second group of coins contain higher zinc contents, from 25% to 37% (Figure 7.2b). These two groups were considered to correspond well with two-fire brass and four-fire brass (Zhao *et al.* 1988) – an interpretation that will be further discussed below. Their tin levels are less than 5%, the standard proportion in the alloying recipe of the Wanli coins. Their lead contents are also variable, up to 12%.

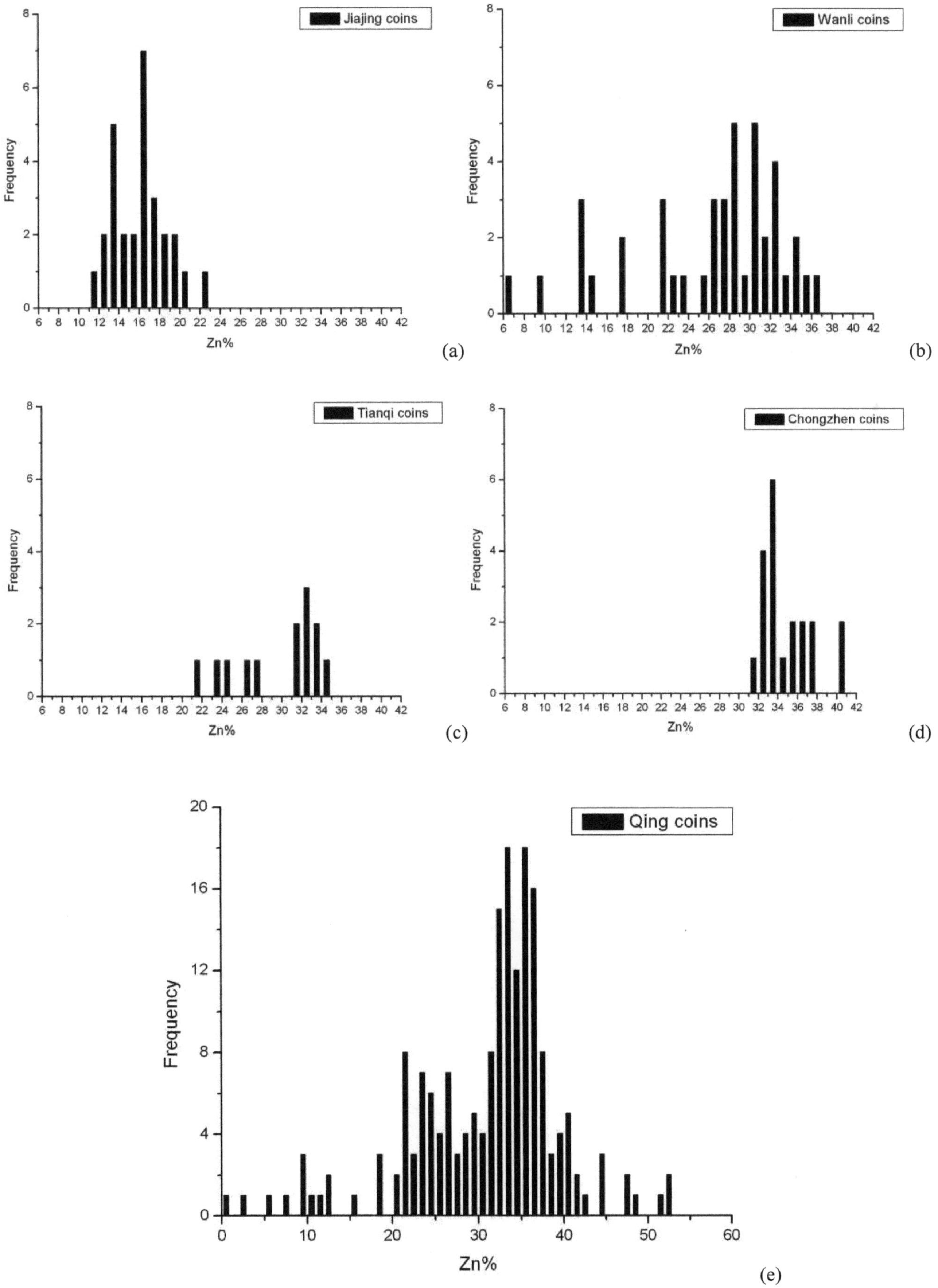

Figure 7.2 Frequency distribution histograms comparing the zinc levels in brass coins of the Jiajing (a), Wanli (b), Tianqi (c) and Chongzhen (d) periods and the Qing Dynasty (e). Data from (Zhou 2004; Wang *et al.* 2005).

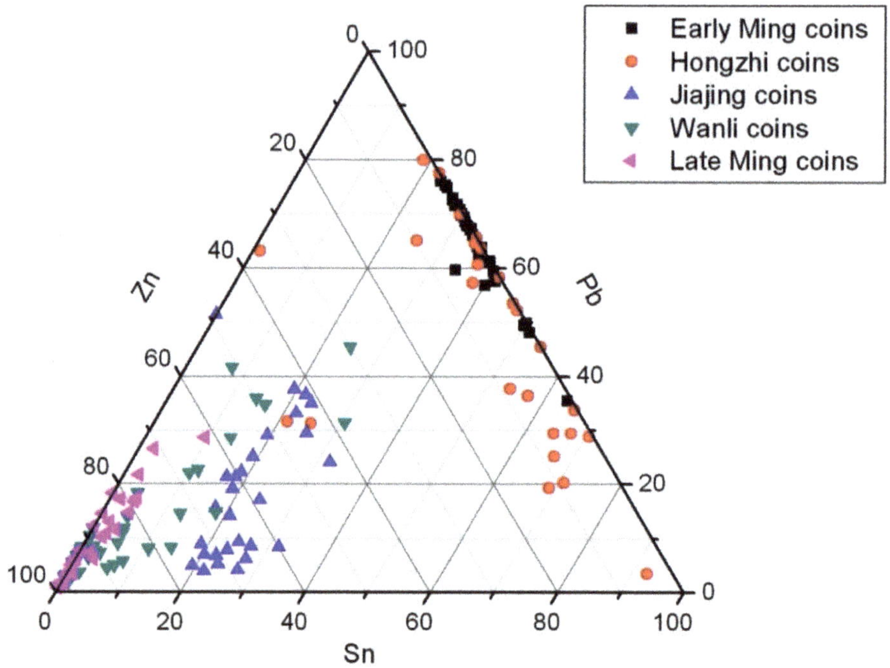

Figure 7.3 Ternary diagram of the zinc, tin and lead contents of the coins of the early Ming, Hongzhi, Jiajing, Wanli and late Ming periods, showing the transition from leaded bronze to brass.

Table 7.1 The official coin alloy standards during the early and middle Qing Dynasty (Vogel 1987, 12). Zn' denotes the zinc contents after considering a 25% zinc loss during alloying.

Year	Cu (wt%)	Zn (wt%)	Pb (wt%)	Sn (wt%)	Zn' (wt%)
1644-1683	70	30			24.3
1684-1726	60	40			33.3
1727-1739	50	50			42.9
1740-1793	50	41.5	6.5	2	34.7
1794-1799	60	40			33.3
1800	52	41.5	6.5		34.7
1801-1804	54	42.75	3.25		35.9
1805	56	37.75	4.75	1.5	31.3
1806-1852	54	42.75	3.25		35.9

Since the Tianqi period (AD 1621-1627), brass coins were cast by alloying copper and metallic zinc. The minting documents after 1621 no longer mention brass and tin, but instead frequently refer to the proportions of copper and zinc. The proportions of zinc documented were mostly 30% or 40%, and even as high as 50% (Zhou 2001). Considering a 25% zinc loss during alloying[10], the zinc levels of the resulting brass coins should have been 24.3%, 33.3% and 42.9% respectively. The zinc contents of the Tianqi coins are mostly around 31-34% and also distributed from 21-28% (Figure 7.2c), indicating that they could have been made with brass of the prescribed zinc proportions of 40% and 30%, respectively. The zinc contents of the Chongzheng coins peak at around 32-34% (Figure 7.2d); they were mostly made of brass with the prescribed zinc proportion of 40%. However, the patterns are not so clear since only 13 and 20 coins were analysed for the Tianqi and Chongzhen periods, respectively. Their tin and lead levels are low.

The ternary diagram of the zinc, tin and lead proportions of the Ming coins shows that the alloys change from leaded bronze to brass in the Jiajing period, a period from which the zinc levels increased with the tin and lead levels decreasing (Figure 7.3).

Monetary regulations of the Qing Dynasty also document the proportions of copper and white lead, zinc. The most common proportion of zinc was 40%, followed by 30% and 50% (Table 7.1). A large number of brass coins were analysed and their zinc concentrations range from a few percent up to 53%, with two peaks centred around 33-37% and 21-27%, which correspond well with the

[10] The 25% zinc loss during alloying is recorded in *Tian Gong Kai Wu*. It is suggested that 15-20% zinc loss is a more reasonable figure based on experiments (Zhou 2004, 436; Wang *et al.* 2005, 88).

prescribed zinc proportions of 40% and 30%, respectively. There are also a few coins made of brass with prescribed zinc proportions of 10%, 20%, 50% and even 60% (Figure 7.2e).

2. Two-fire brass and four-fire brass

It has been proposed that 'two-fire' and 'four-fire' denoted the times *luganshi* was added to the crucible charge during the cementation process (Zhao and Zhang 1987). In order to investigate how the two types of brass were made, Ma Yue conducted laboratory experiments, trying to obtain brass with 15%, 20%, 25% and 30% zinc. She found that the method of adding zinc ores (smithsonite) onto copper, as recorded in *Tian Gong Kai Wu*, led to great losses of zinc, even if some graphite was added as a reducing agent. After four additions of zinc ores, the final brass only contained a few percent of zinc. However, if charging mixed copper or brass from the previous smelt and zinc ores, covered by a graphite layer, the brass contained 5.4%, 10.6%, 15.1% and 20.3% zinc after one, two, three and four smelts respectively (Table 7.2). Nevertheless, the actual zinc loss was greater than the 15% originally expected, so the final brass was less rich in zinc than the predicted values (Ma 2008). Although the experiments were not very successful, they did demonstrate the possibility that the zinc contents of brass could be elevated by multiple steps of charging copper/brass and more zinc ores.

Table 7.2 Results of the experiments by charging mixed copper or brass from previous smelts and zinc ores (Ma 2008). The temperature in the furnace reached 1100 °C, which was maintained for 8 h.

Name	Copper/brass from previous smelt (g)	Zinc ores (g)	Zn (wt%) of products
One fire	86.06	42.49	5.4%
Two fire	47.5	14.16	10.6%
Three fire	49.25	14.16	15.1%
Four fire	33.5	14.16	20.3%

A different method of arranging the raw materials inside crucibles is found in the twelfth-century European treatise *On Divers Arts*, written by the German artisan monk Theophilus Presbyter. He described the brass making process in detail:

> And when the crucibles are red-hot take some calamine, about which I spoke above, that has been [calcined and] ground up very fine with charcoal, and put it into each of the crucibles until they are about one-sixth full, then fill them up completely with the above-mentioned [crude] copper, and cover them with charcoal. ... Now, when the copper is completely melted, take a slender, long, bent iron rod with a wooden handle and stir carefully so that the calamine is alloyed with the copper. Then with long tongs raise each crucible slightly and move them a little from their position so that they may not stick to the hearth. (Hawthorne and Smith 1979, 143-144)

This process appears more efficient than that described in *Tian Gong Kai Wu* in two ways. Firstly, by putting a mixture of zinc ores and charcoal, a reducing agent, underneath the copper, it would ensure that the zinc vapour formed would rise to meet the copper and could be absorbed by it, potentially reducing zinc loss. Secondly, the charge was covered by a layer of charcoal to prevent the oxidation of zinc vapour, thus limiting the zinc loss. Theophilus's text then goes on:

> Put calamine in them all again as before and fill them with copper and cover them with charcoal. When it is once more completely melted, stir again very carefully and remove one crucible with the tongs and pour out everything into [little] furrows cut in the ground. Then put the crucible back in its place. Immediately take calamine as before and put it in, and on top as much of the copper that you have [just] cast as it can hold. When this is melted as before, stir it and add calamine again and fill it again with the copper you have [just] cast and allow it to melt. Do the same with each crucible. When it is all thoroughly melted and has been stirred for a very long time pour it out as before and keep it until you need it. (Hawthorne and Smith 1979, 143-144)

It is surprising that Theophilus's records appear to describe so well what is likely to resemble the Chinese 'two-fire' and 'four-fire' processes. Charging zinc ores and copper twice obtained two-fire brass, while charging zinc ores and two-fire brass twice generated four-fire brass. The latter process, using the crude brass, allowed for the absorption of more zinc (Craddock and Eckstein 2003, 226). Considering the results of Ma Yue's experiments coupled with the examination of the zinc levels in brass coins of the Wanli period, it is likely that the Chinese craftspeople followed a process similar to that described by Theophilus rather than that recorded in *Tian Gong Kai Wu*. However, none of the texts mention lids, so a considerable loss of zinc was inevitable for the multi-step process.

3. The transition from cementation to speltering

The transition from cementation brass to speltering brass was considered to have occurred suddenly in 1621, the first year of the Tianqi period. This was inferred from the cadmium contents of brass coins. Both the research groups studying brass coins found that the cadmium contents of brass coins after 1621 were significantly higher than those before 1621[11]. Zhou believed that the

[11] Two groups of compositional data of coins obtained by the British Museum and the Chinese Numismatic Museum show similar changes. Cowell and his colleagues found that in brass coins issued before 1621 cadmium was hardly detected and never exceeded 0.002%; after 1621 the cadmium content increased with some later coins containing over 0.01% (Bowman *et al.* 1989; Cowell *et al.* 1993). Zhou and his colleagues found that before 1621 cadmium was around 0.0003%, and after that date it was around 0.002% (Dai and Zhou 1992; Zhou 2004).

coins minted after 1621 were made of metallic zinc, which would have introduced more cadmium into coins than the cementation process (Zhou and Fan 1994). However, variable cadmium contents have been found in Chinese zinc ingots. Those from shipwrecks contain quite high levels of cadmium, 0.04% (the *Witte Leeuw*), 0.06% and 0.4% (the *Diana*), while an ingot made by the traditional process in Guizhou contains undetectable levels of cadmium (<0.002%) (Craddock and Zhou 2003, 285) and the zinc lumps from Miaobeihou again show very low levels of cadmium (see Section 4.2.2.3).

Another problem with the purported 1621 transition is that some of the high-zinc coins of the Wanli period contain extremely high levels of zinc, up to 36.7%, which exceed the widely accepted upper zinc limit for the cementation brass, 28-32%. In this sense, the Wanli coins with over 32% are most likely made by the speltering process. However, the Wanli brass coins with extremely high levels of zinc could also have been produced by the cementation process, possibly the four-fire process. A confirmation of this possibility requires further experiments based on the multi-step process described by Theophilus.

The transition of cementation to speltering brass could have started during the Wanli period (AD 1573-1620). Zinc ingots dated to AD 1585 have been found in Guangdong province (Browne 1916); two shipwrecks were found to have been loaded with Chinese zinc ingots in 1607 and 1613 (Craddock and Hook 1997). Whatever the case, the cementation process was only used for a short period from the Jiajing period. The rapid transition to the speltering process was to some degree favoured by the fact that the speltering process shows many advantages over the cementation process.

7.1.2 Advantages of speltering process over cementation process

The new method of making brass coins using zinc metal had technical and economic advantages over the cementation process, which are discussed in terms of raw materials, efficiency of the process, control of alloy compositions and costs of minting.

1. Raw materials

The requirements for the qualities of raw materials, including zinc ores, copper and reducing agents, used in the two processes are different.

For the cementation process, zinc ores with few impurities would be needed, as the impurities like lead, iron and sulphur would enter the brass during the process. For example, lead from zinc ores would make hammering, annealing and gilding of the brass difficult, as lead is present as isolated prills along with the alpha (α) phase of the brass. The German metallurgist Lazarus Ercker notes in his 1580 treatise on ores and assaying that if calamine was collected from lead smelting furnaces and used, the brass produced had a grey colour and was more prone to crack during annealing (Sisco and Smith 1951, 257). However, for the speltering process, any involatile impurities in zinc carbonate ores would not enter the zinc metal and thereby the brass, so low quality ores could also be used (see Section 6.2.1).

Copper metal with few impurities would be preferred for the cementation process in order to make brass with high zinc contents (Rehren 1999a). Tin, lead and iron in the copper would depress the amount of zinc absorbed in the α phase to different extents. For example, copper containing 1% tin would absorb 2% less zinc than it would without the tin; copper with 1% lead or iron could absorb about 1% less zinc (Craddock 1978, 12; Zhou 2004, 450). Theophilus specifies in his treatise that copper free of lead is better for cementation, especially when the resultant brass is to be gilded (Hawthorne and Smith 1979, 144). Contrastingly, the speltering process could utilise copper with more impurities and also bronze scrap.

Although not mentioned in *Tian Gong Kai Wu*, charcoal was probably the reducing agent for the cementation process. The alchemist Rihuazi did refer to the addition of charcoal in his recipe of brassmaking by cementation in the tenth century (see Section 2.1). However, zinc distillation used both charcoal and coal at the sites studied, while only coal was commonly used in the Chinese traditional zinc smelting processes.

Overall, despite the considerable constraints on the raw materials for making cementation brass, the brass produced would tend to contain some impurities from zinc ores and copper. Conversely, zinc distillation could produce pure zinc, even with iron-rich zinc ores and sulphur-bearing coal. Similarly, the copper to be alloyed with zinc to make brass was not restricted to pure copper.

2. Efficiency of the process

As an extremely volatile metal, zinc is inevitably lost at every step of either brassmaking method. The cementation process could result in a great loss of zinc with open crucibles; the multiple steps to obtain brass with elevated zinc contents would achieve so at the expense of even more zinc losses. However, zinc distillation kept the zinc loss to a low level in a relatively closed system, mainly via the fumes from the opening of the lids, the penetration through the ceramic fabrics of the retorts, and into the slag. Both the melting of cementation brass and the alloying of the two metals in speltering would also lead to more losses of zinc. *Tian Gong Kai Wu* explicitly states the zinc loss as a reason why the cementation process was replaced by the speltering process (Sun and Sun 1966, 247).

As a further logistical complication, brass making by cementation would need stirring, pouring and recharging during the process, requiring a great input of skilled

---

Both analyses used AAS, but the two sets of data are different. 0.0003% seems to be a very low detection limit for AAS.

labour and fuel. Although *Tian Gong Kai Wu* does not refer to the size and shape of the brass making furnaces, it can be inferred from the round furnaces described by Theophilus that only a few crucibles could have been accommodated within them in order to operate these complex tasks (Figure 7.4). Conversely, zinc distillation did not require such intense attendance, so a much greater number of retorts could have been heated in the furnaces at the same time, thus leading to much higher yield per firing.

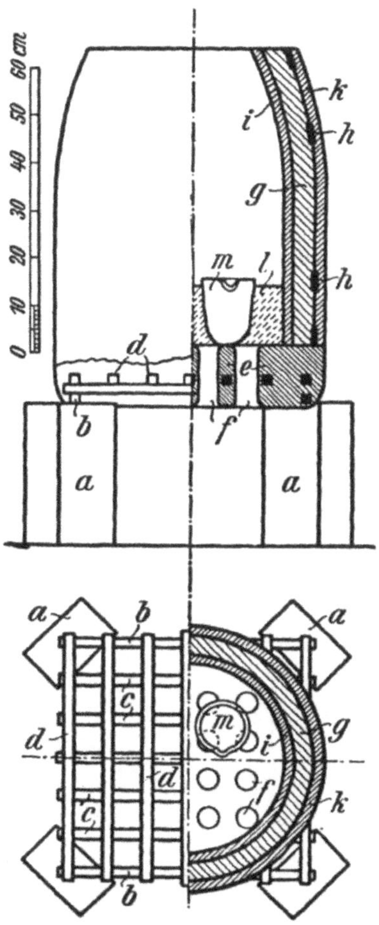

Figure 7.4 The brass making furnaces described by Theophilus (Hawthorne and Smith 1979, 142) (© The Trustees of the British Museum). (a, four stone stands; b, c, d, iron rods; e, clay and dung; f, holes; g, the furnace wall built of stone and clay; h, iron bands to bind the furnace wall; i, k, clay for lining the inner and outer surfaces of the wall; m, crucibles; l, blazing and dead coals.)

3. Control of alloy compositions

It is very difficult to adjust the alloy compositions of brass by the cementation process, because one cannot predict exactly the richness of the zinc ore, how much of it will be reduced, and how much of the resulting vapour will be absorbed by the copper. It is much easier to cast brass of the desired alloy compositions by mixing metallic copper and zinc in the required proportions. The variable properties of brass with different zinc levels had already been realised in the late Ming period. *Tian Gong Kai Wu* lists several recipes of alloying copper and zinc with different ratios, ranging from 9:1 to 4:6 (Table 7.3). According to this text, brass made by alloying copper and zinc with copper to zinc ratios of 7:3 and 6:4 was generally used to cast coins. Allowing for a 25% zinc loss, the actual alloy compositions of brass coins would be around 24.3% and 33.3%, respectively. From the perspective of modern metallurgy, brass containing up to about 37% zinc is single α phase alloys consisting of a solid solution of zinc and copper (Copper Development Association 1961). Such α brass has better strength and ductility due to the increase of zinc levels (Figure 7.5 left). Brass with around 33% zinc shows optimal mechanical properties, suitable for both casting and forging. Most of the speltering brass coins of the late Ming and Qing Dynasties were made of brass with around 33% zinc.

Interestingly, the 9:1 brass (~7.7% zinc) was acknowledged by Song Yingxing as the best alloy for coins, although such brass has lower strength and ductility than 7:3 and 6:4 brass. The 5:5 and 4:5 brass coins, containing 42.9% and 52.9% zinc, respectively, were deemed of the lowest quality. When zinc contents are over 37%, a harder phase, the beta (β) phase, appears, decreasing the ductility and then the strength (Figure 7.5 left). Thus it would appear that, provided that the basic material properties were met, mechanical properties were not the most important criteria to assess the quality of coins. Instead the quality of coins was determined by the proportions of copper: the more copper in coins, the better the quality.

4. Costs of minting

Zinc was a very cheap metal and its use in making brass coins greatly decreased their costs. In the Wanli period, the price of copper was 8 *fen* 5 *li* silver per catty, while two-fire brass was a little cheaper, i.e. 8 *fen* 1 *li* per catty, but four-fire brass was more expensive, i.e. 9 *fen* 4 *li* to 1 mace 5 *li* per catty. Therefore, the earliest brass coins made from two-fire brass and four-fire brass cost about the same or even more than bronze coins, as tin had a similar price to copper, i.e. 8 *fen* per catty. In the Tianqi and Chongzhen periods, the use of cheap zinc, at 7 *fen* 7 *li* per catty, decreased the costs, which was especially significant as copper prices continuously increased, reaching 1 mace 4 *fen* 3 *li* per catty in the Tianqi period (Table 7.4).

Increasing zinc proportions could effectively reduce the amount of copper needed, and greatly reduce the costs, thus rendering minting more profitable. After the Wanli period, the state cast many more coins than in previous periods and also progressively reduced their quality by replacing more copper with cheap zinc. Some local provincial mints even cast coins with copper to zinc ratios of 5:5 and even 3:7 (Peng 1958, 486-487).

Table 7.3 Making brass coins with different ratios of copper to zinc, recorded in *Tian Gong Kai Wu* (Sun and Sun 1966, 165-169, 247).

Cu: Zn	Zn (wt%) (no zinc loss)	Zn (wt%) (25% zinc loss)	Casting coins
9:1	10	7.7	Good quality
8:2	20	15.8	-
7:3	30	24.3	Suitable for casting
6:4	40	33.3	
5:5	50	42.9	Poor quality
4:6	60	52.9	

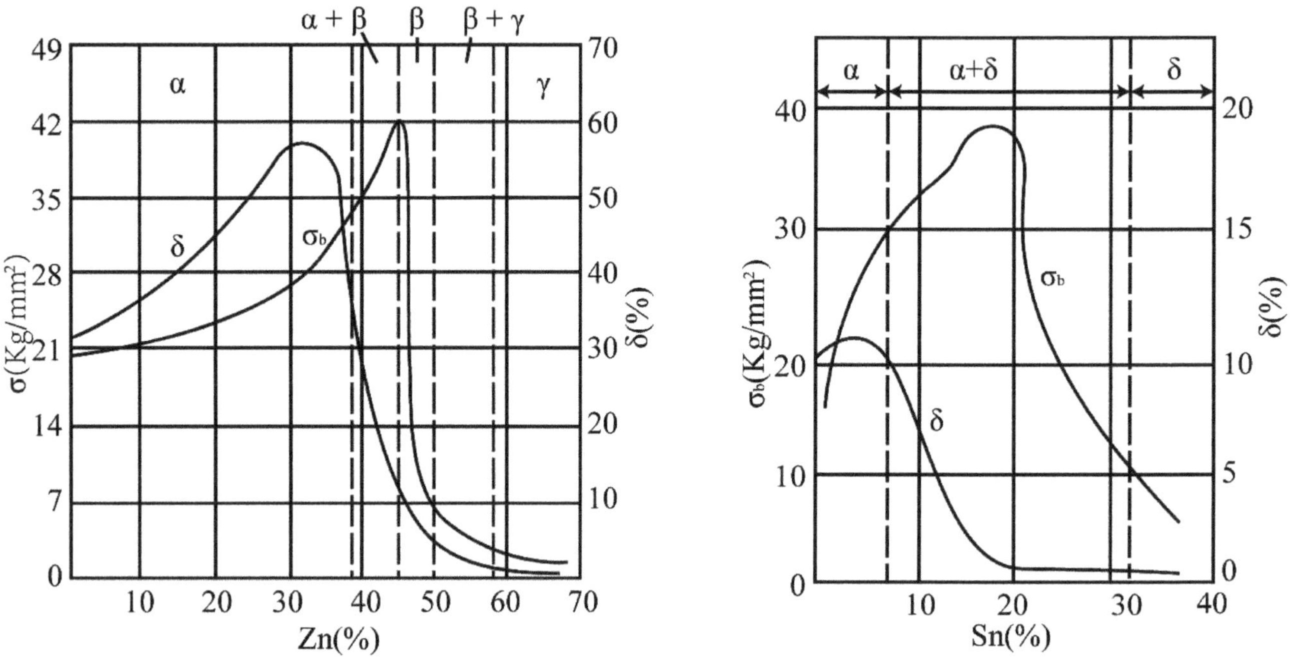

Figure 7.5 The influence of zinc (left) and tin (right) contents on mechanical properties of cast copper-based alloy (σ_b = tensile strength, δ = elongation).

Table 7.4 The prices of minting metals in the late Ming Dynasty. The prices in the Wanli period were recorded in *Gong Bu Chang Ku Xu Zhi* 工部厂库须知 (Instruction on Factories and Storehouses of the Ministry of Works) (Zhou 1992). The prices in the Tianqi period were documented in *Hou Xun Gu Zhu Shi Yi* 侯恂鼓铸事宜 (Minting Issues of Hou Xun) (Peng 1958, 487).

Minting metal	Price (silver per catty) in the Wanli period	Price (silver per catty) in the Tianqi period
Copper	8 *fen* 5 *li* (85 *li*)	1 mace4 *fen* 3 *li* (143 *li*)
Two-fire brass	8 *fen* 1 *li* (81 *li*)	1 mace 2 *fen* (120 *li*)
Four-fire brass	9 *fen* 4 *li* (94 *li*) 1 mace 5 *li* (105 *li*)	
Tin	8 *fen* (80 *li*)	-
Lead	4 *fen* 2 *li* (42 *li*)	-
Zinc	-	7 *fen* 7 *li* (77 *li*)

7.1.3 The role of zinc in coinage

In imperial China, copper-based coins were the fundamental form of currency. The issue of coins was authorised and directly controlled by the state. From the Song to Yuan Dynasties, the monetary system based on coins was difficult to sustain, so the state issued paper currency, a purely nominal form of money, to supplement and sometimes replace coins. Following the monetary policy of the preceding dynasty, the first few emperors of the Ming Dynasty issued large amounts of paper currency, which were used together with coins. To encourage the use of paper currency, the state closed official mints on several occasions and even proscribed the circulation of coins from time to time. However, the unrestricted issue of paper currency led to the increasing devaluation and final failure of paper currency (Peng 1958; Von Glahn 1996).

With the development of the commercial economy, silver emerged as the dominant currency in the mid-fifteenth century. However, silver alone could not meet the demand of the developing private market, when petty retail commerce required a steady supply of coins. The shortage of official minted coins encouraged counterfeiting, producing coins of low quality at reduced weights, driving authentic coins out of the market (Von Glahn 1996). Gradually, silver came to be widely used in the state fiscal administration, including fiscal revenues of land taxes, corvées, salt taxes and tariffs, the court and state expenditures, official salaries and military expenses. Silver also played an important role in monetary transactions amongst people of different social strata, which accelerated commercial development, both domestically and internationally. However, the increasing demand of silver as currency could not be met by the limited exploitation and production of silver at home. The eagerness for silver encouraged maritime trade, via which large amounts of American and Japanese silver entered China to pay for Chinese goods in the late Ming period (Wan 2005).

In the mid-Ming Dynasty, the state, having lost control over both silver and coinage, attempted to enhance the minting and circulation of coins in the hope of improving its adverse fiscal condition (Zhao 1986). Thus great efforts were made to issue high quality coins in the Jiajing and Wanli periods. Several new types of coins were minted, such as 'gold-back' and 'fire-lacquer' coins, which were made with four-fire and two-fire cementation brass, respectively. Minting coins with cementation brass must have been very costly, as indicated by the high prices of two-fire and four-fire brass (Table 7.4). Unfortunately, these official coins could not compete with the counterfeit coins. In the late Ming period, the state underwent a serious financial crisis resulting from the free-spending government and the bloated military. To overcome the crisis, the state minted coins in pursuit of profits, obtained from the margins between the costs of minting and the values of the minted coins (Huang 1974, 250). Under such circumstances, minting brass coins with metallic zinc could help achieve high profits, especially when the zinc content was high. After the Wanli period, the value of coins started to decline until the demise of the Ming Dynasty in 1644. Therefore, the speltering process rapidly replaced the cementation process because it was a more efficient process and a much cheaper way to make coins. The large-scale production of cheap zinc enabled this transition, ultimately leading to the change from leaded bronze coins to brass coins.

Bronze coins had been used in China as early as the Eastern Zhou period, with tin and lead being the major alloying constituents (Zhou 2004; Wang *et al.* 2005). The addition of tin lowers the melting point and improves the mechanical and casting properties of the metal, while the addition of lead could produce a more fluid metal for casting. With limited tin sources and a relatively small stock, mainly in southwest China, tin was much more expensive than lead, and sometimes even more expensive than copper (Zhou 2004, 482). Thus it was preferable to add less tin and more lead in order to reduce the costs of manufacturing bronze coins, especially during the periods of economic decline (Zhou 2004, 480). When cementation brass was first used for minting, it was regarded as a kind of good copper, rather than a useful alloy in its own right, so tin and lead were also added. Later the addition of tin was found to be unnecessary in brass, as brass itself has good mechanical and casting properties (Zhou 1992). As a result, tin contents in late Ming brass coins significantly decreased.

From a strictly technical and economic perspective, brass was an alloy better suited for minting for several reasons. Firstly, brass can be cast more successfully than bronze. It is less subject to segregation and porosity, because the freezing range of brass is narrower than that of bronze, as shown by the liquidus and solidus lines of their binary diagrams (Zhou 2004, 423). In addition, due to the extreme volatility of zinc, the vapour pressure of zinc in molten brass is very high, which can take away the gas within the metal and also prevent it from absorbing air. As tin has low volatility and low vapour pressure, bronze tends to retain some gas, thereby showing more porosity than brass. Secondly, brass generally shows better mechanical properties than bronze. Brass containing up to 37% zinc, the saturation solubility of the α phase, shows good strength and ductility; the ductility of brass with over 37% zinc decreases quickly with increasing β phase, while the strength keeps increasing until the zinc level reaches 45% and then declines rapidly. In contrast, the ductility of bronze starts to decrease when the tin level exceeds approximately 7% because of the appearance of the brittle α and delta (δ) eutectoid phase. Bronze with 20% tin becomes very brittle and its strength also starts to decline (Figure 7.5 right). Thirdly, zinc was a much cheaper metal than tin. When metallic zinc was available for minting, cheap zinc replaced precious tin, which largely lowered the minting costs, like cheap lead, but retained good mechanical and casting properties even with high zinc levels. In other words, smaller amounts of copper were needed to cast coins, as shown by the lower levels of copper in brass coins (Figure 7.6).

Therefore, the use of zinc in coinage not only saved tin, but also saved copper, which favoured the change from bronze coins to brass coins. The occurrence of large-scale zinc production was a result of the social demand for a cheap minting metal in the late Ming period.

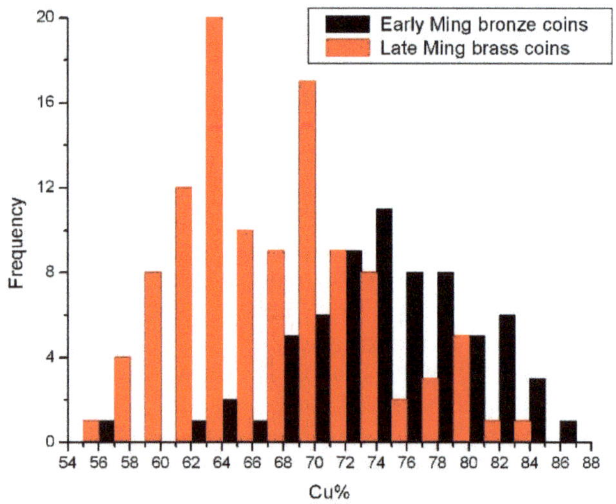

Figure 7.6 The copper contents of the early Ming bronze coins and late Ming brass coins.

7.2 Chinese zinc in the world

Besides the main uses in coinage and copper-based artefacts, Chinese zinc was also frequently exported by European merchants from the seventeenth to nineteenth centuries (see Section 2.2.4). Chinese zinc was first exported under the name of *tutenag* by the Portuguese. A dictionary of Anglo-Indian words and phrases indicates that the root of the word is the Persian *tutiya*, Sanskrit *tuttha*, an oxide of zinc; the last syllable is either an adjective affix, in the same way as *nak* is used in Persian, or it may be *naga* in the sense of lead (Bonnin 1924, 3). The word was spelt by the Europeans in many ways, such as tutenaga, tuttenage, tuthinag, tutanaque, toothanague, and even mistakenly 'tooth and egg' (Bonnin 1924). The Dutch also named the metal spiauter or speauter, while the English named it spelter (de Ruette 1995). *Tutenag* was exported in the form of two types of ingot: flat round cakes directly collected from the tops of the retorts and bars cast by remelting the raw cakes. The zinc bars were more common, weighing from a few kg to over 10 kg (Bonnin 1924).

Chinese zinc was imported to many regions across the world, including Japan, Southeast Asia, India and Europe. The metal played different roles in different regions. The following sections focus on the influence of Chinese zinc in two regions, namely India and Europe, as they are not only of great importance in the history of metallurgy of zinc and brass, but are also the only two regions with sufficient historical and archaeological information for the present discussion.

7.2.1 Chinese zinc in India

Before the industrial scale zinc production developed in China probably at the end of the sixteenth century, zinc production at Zawar, northwest India, had flourished during the fourteenth to sixteenth centuries. In the sixteenth century, Indian zinc was exported by Portuguese merchants to Europe. In 1513, a letter written by an Italian diplomat living in London mentioned that he received letters from Lisbon, Portugal, saying that a cargo of about 400 pieces of tin from India had arrived and the Portuguese claimed that they could have as much as they wanted. 'Indian tin' is regarded to be a mistaken term for zinc, as India never traded tin (Craddock 1995, 318).

In the late sixteenth and early seventeenth centuries, the zinc production at Zawar was seriously interrupted by political turmoil, as the Rajput kingdom of Mewar was attacked by the Moghul Empire. Thus the domestic demand for zinc in India could not be met, resulting in the import of large amounts of Chinese zinc, first by the Portuguese in about 1600. By the early seventeenth century, Chinese zinc had already been in great demand at several ports in northwest India (Prakash 1984). From the seventeenth to nineteenth centuries, India was a dominant market for Chinese zinc (Souza 1991).

Chinese zinc was widely used in India for making brass for domestic utensils, bullets, statuary, and also special uses for coinage and bidri wares. Chinese zinc was employed in the minting of copper coinage, the *bazaruco* first issued by the Portuguese before 1608, which were widely used in the trading communities of the west coast of India (Souza 1986, 179). The introduction of Chinese zinc to minting *bazaruco* caused the debasement and devaluation of the coins, clearly illustrating the massive impact that Chinese zinc had on foreign economics (Souza 1991, 301). *Tutenag bazaruco* coins were minted from 1748 to 1828, with five of them analysed and confirmed to be made of zinc (Yih and de Kreek 1993). Bidri wares were cast with zinc-based alloys containing about 5% copper, inlaid with silver and chemically treated to generate a black patina which contrasted with the silver. During the seventeenth to nineteenth centuries, the production of bidri wares was centred in south India, and progressively spread to the northeast. Limited lead isotope analyses of eighteenth-century bidri wares show that the zinc did not come from Zawar deposits in northwest India or any other zinc deposits in India; it was therefore suggested that the metal might have originated from China (Craddock *et al.* 1998; Craddock 2005).

The import of Chinese zinc into India continued into the nineteenth century. For example, the ship *Diana* is recorded as carrying Chinese zinc from Canton to Calcutta, east India, in 1816 (Ball 1995). In 1826, the import of Chinese zinc into Calcutta ceased as German spelter was introduced into India and eventually supplanted Chinese zinc (Bonnin 1924, 12).

7.2.2 Chinese zinc in Europe

As introduced in Section 1.1, Europe had been making brass by the cementation process with copper and calamine in the Roman period and, after an apparent decline in late Roman times, again extensively from the middle medieval period onwards. In the medieval and early modern periods, cementation brass became the dominant copper alloy and was used for everyday life implements and works of art. Metallic zinc, however, was occasionally obtained from the lead/silver smelting furnaces at Rammelsberg, Germany, from at least the sixteenth century. However, it was not until the first half of the eighteenth century that laboratory experiments by chemists isolated metallic zinc by distillation and the industrial production of zinc was first established in England. Meanwhile, Europe also imported zinc from the East, India and China. As already discussed above, Europe first imported zinc ingots from India during the sixteenth century. In the seventeenth and eighteenth centuries, Chinese zinc dominated the world market and was already frequently exported to India. Europe also imported Chinese zinc ingots, including Sweden, Holland, Denmark and England. The imported Chinese zinc therefore, to some degree, influenced the development of European zinc distillation, both as a scientific quest and as an economic activity.

Before the start of European commercial zinc production from the 1740s, Chinese zinc was sold in small quantities but at high prices in Europe (Day 1973, 62). The use of Chinese zinc in Europe was not explicitly recorded, with only some relevant references indicating that it was used for making solder, as well as high-quality brass, for scientific instruments, jewellery and ornamental castings. From the analyses of European jettons and scientific instruments, it is found that by the 1650s at the latest the use of brass with over 33% zinc had become quite common (Pollard and Heron 1996). Such high-zinc brass was probably made with metallic zinc most likely from the East.

Although oriental zinc did not play a significant role in the European brass industry before the 1740s, it did arouse keen academic interest among European chemists in the metal and its isolation. As the metal was already collected in small quantities from lead/silver smelting furnaces in Germany, several scholars had been shown interest: Erasmus Ebener of Nürnberg is recorded as the first person referring to the metal collected from the furnaces and its use in brass making in the first half of the sixteenth century; Paracelsus, a sixteenth-century Swiss physician and alchemist, is credited with the invention of the name zinc (*zincken*) for the metal from the furnaces. The metal was also known to the famous chemists and metallurgists Agricola, Ercker and Lohneyess (Aitchison 1960, 481; de Ruette 1995, 196). In the 1650s, Johann Glauber, a German chemist, understood that zinc is a volatile metal, and argued that zinc from Germany and the East could turn copper into brass like calamine:

> Zink is volatile metal, or a half ripe metal when it is drawn out of its ore. It is much clearer and brighter than tin, yet not so malleable and fluxile as tin is ... We have it not much growing in Germany, but great quantity of it is every year brought us by the merchants out of the East Indies ... It is a golden but an unripe mineral, it gives red copper a yellow colour and turns it into brass, as lapis calaminaries doth; and indeed that same stone is nothing else but unmeltable zink, and this zink may properly be called a susilc lapis calamnaris (*sic*); for as much as both of them partake of one nature. (translated in Packe 1689, 319).

Experiments with German zinc/spelter and *tutenag* from the East Indies were carried out by the English chemist Robert Boyle, in the late seventeenth century (Boyle 1990, 310-311; Martinón-Torres 2012). Once the volatility of zinc and its ores were recognised, it would not be difficult to isolate zinc with distillation adapted from other areas of metallurgy and beyond.

It appears that Henkel, a German mine owner, first produced zinc from calamine in 1721, but concealed his method. In 1746, Marggraf, a German chemist, found that calamine could be reduced to metallic zinc by distilling with charcoal in the absence of air (Partington 1961, 726). However, the laboratory experiments were not developed to a larger scale on the Continent. It was in England that the first commercial manufacturer was set up. William Champion patented a process producing the metal 'known by the name or names of spelter or toothaneg' in 1738 and established commercial zinc production in Bristol in 1743 (Day 1998). The technology was based on distillation by descending, similar to the Indian process (Figure 7.7).

There are, however, two sources indicating that the first zinc production in Europe was learned from the Chinese process. The German chemist Bergman stated: 'a certain Englishman made a voyage to China for the purpose of learning the art of smelting zinc, but though he became sufficiently instructed in the secret and returned safely home, he carefully concealed it'. It has also been claimed that Champion learned the technology from a sailor who had been to China (Cocks and Walters 1968, 7). However, it appears that the technology of zinc production in China was never explicitly revealed to foreigners, and the Chinese process was completely different from any process adopted in Europe. Even as late as 1751, when zinc production had started in Bristol, a dictionary of trade and commerce still claimed that nobody knew how the imported zinc was produced (Bonnin 1924, 8). In 1756, records of Carl Gustave Ekeberg, a supercargo of the Royal Swedish East India Company, gives details of ores of *tutenag* and also mentions: 'the ore melts easily and emits during smelting and oxidation a thick stinking smoke injurious to the health' (Bonnin 1924, 10-11). It is most likely that Carl also did not understand that zinc was produced by distillation. In 1788, Axel Frederic Cronstedt, in his *System of Mineralogy*, explained the reason why the Chinese zinc production was unknown: 'no European is granted the liberty of entering into the Chinese manufactories' (Bonnin 1924, 9). It was not until the 1790s that Lord Macartney, the first ambassador from Britain to China, first gave a more detailed account of the Chinese process:

> *Tutenag* is the name the Chinese give to zinc, which they extract from a very rich ore of calamine, by pounding the calamine into a fine powder and then mixing it with charcoal dust, and distilling the mixture from earthen jars, in a common distilling apparatus placed on a slow fire. The zinc rises in the form of flowers or sublimate, passes through the worm or pipe of the head of the still, which is luted to the earthen jars, contained the calamine and charcoal powders, and is received under water in a vessel placed for the purpose, into which the extremity of the pipe plunges. The water condenses the flowers of the zinc, which fall in consequence to the bottom, and, when a sufficient quantity has been thus obtained, they are collected out of the vessel, and then put into melting pots, and fused into masses, which are commonly sold under the name of *tutenag*. (Cranmer-Byng 1963, 293)

However, the process described was not the typical traditional Chinese zinc distillation with internal condensers, as the zinc was collected outside the retorts.

Even if European zinc production was not learned from China, Chinese zinc introduced into the European market did attract manufacturers, like Champion, to invest in the new industry, aiming at replacing the expensive oriental zinc. However, Champion's zinc production did not end up with great success. The prices of imported zinc decreased from £260 per tonne in 1731 to only £20-24 per tonne at the London docks after Champion started producing zinc. From 1760 to 1779, about 40 tonnes of Chinese zinc was imported into London at £20 per tonne (Bonnin 1924, 12-13). Champion's total output was only about 200 tonnes of zinc over six years, and his zinc, sold at £40-44 per tonne, could not compete with the cheap foreign zinc (Mosselman 1825). This eventually led to his bankruptcy in 1769 and the production centre moved to the Swansea area, Wales (Cocks and Walters 1968). It was not until the early nineteenth century that zinc was successfully produced on an industrial scale on the Continent, mainly in Belgium and Germany (Day 1998). The success of continental zinc production soon drove the Chinese zinc out of the European market, and ultimately out of the world market.

In summary, Chinese zinc replaced Indian zinc in the world market in the seventeenth century and great amounts of Chinese zinc entered the large Indian market for making various artefacts and coins. Conversely, the European market for Chinese zinc was small because European brass production employed the cementation brass making tradition. However, the import of Chinese zinc attracted keen academic interest in the new metal and inspired laboratory experiments in Europe. Although the Chinese zinc distillation technology was unknown to Europeans, the commercial value of the metal attracted manufacturers in England to invest in its production. Chinese zinc was also exported to many other regions, probably as far as Africa. The seventeenth-nineteenth century Zimbabwean high-zinc brass beads, which were supposed to be made of Indian zinc, are most likely made of Chinese zinc (Thondhlana and Martinón-Torres 2009).

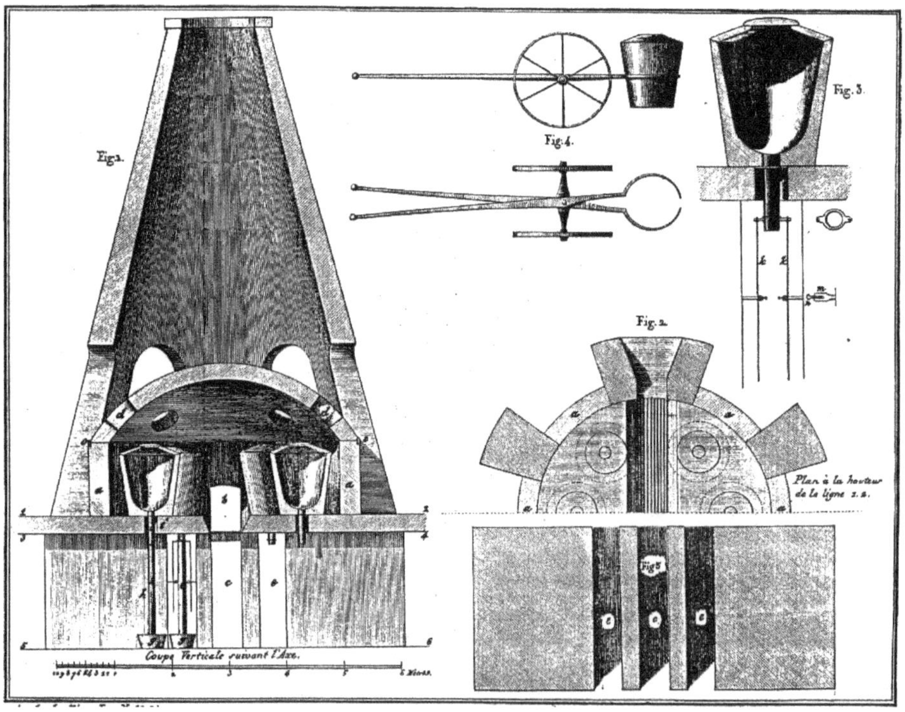

Figure 7.7 William Champion's zinc smelting furnace in Bristol (Mosselman 1825). Six pots were inserted in the upper reaction chamber of the conical furnace fuelled by coal. Zinc formed from the calamine and charcoal charged into the pots, descending via the iron tubes into the containers in the lower condensation chamber.

Chapter 8
Conclusions and Future Work

Chinese zinc distillation stands out in its unique technological characteristics in the history of world metallurgy. In addition to its technological importance, the production of zinc also played a significant role in the economic history of Ming and Qing China and many other regions in the world. This chapter presents a brief summary of the main conclusions of this research, and outlines further potential avenues that may be pursued in future research.

8.1 Zinc smelting in Chongqing and beyond

The studies of production remains recovered at the zinc smelting sites in Fengdu and Shizhu, Chongqing, have shown that large-scale zinc production was in operation following what can be broadly termed the typical Chinese technological tradition of zinc distillation by ascending: oxidic zinc ores with some impurities, reducing agents coal/charcoal and recycled zinc-rich materials were charged into retorts consisting of pots, condensers, pockets and lids, which were fired in rectangular furnaces fuelled by coal. These technical elements constitute the tradition of Chinese zinc distillation, which is significantly different from the Indian tradition and thus appears to have developed as a unique metallurgical strand in the world.

Although rooted in such a tradition, the technologies of Chinese zinc distillation actually show a rather diversified picture in terms of technological choices, which is well exemplified by the range of the traditional processes and archaeological case studies documented in this research. The archaeological and archaeometric evidence of the river sites in Fengdu, dated to the Ming Dynasty, indicates that the technology employed was quite standardised and technically efficient. The jar-shaped retorts were manufactured with quite refractory clay tempered with quartz, formed by wheel throwing and prefired to high temperatures. The resulting fabrics showed superb toughness, thermal shock resistance and both thermal and chemical refractoriness. Charged with iron-bearing oxidic zinc ores and an excess of reducing agents, the retorts would have produced high yields with low zinc losses. In contrast, the long-shaped retorts at Dafengmen, dated to the Qing Dynasty, were made of kaolinitic clay and rock fragments. Although the raw materials were very refractory, they were not prefired to temperatures high enough to maximise the performance characteristics of the fabrics. With oxidic zinc ores containing more impurities and a lower proportion of reducing agents, the process was much less efficient with more zinc loss.

The organisation of production at the two groups of sites was also different in terms of the resource procurement and transport of raw materials and products. Weber's theory of the choice of location aided the interpretation of the relative importance of the various kinds of raw materials to the industrial scale production. The access to the coal sources was deemed of utmost importance. According to the transport orientation law, the Dafengmen site was situated in an ideal location, while the location of the Fengdu sites was much less ideal. However, the Dafengmen site showed a technology that employed potentially good resources (kaolinitic clay for long-shaped pots, ores from nearby lead-zinc deposits and coal from nearby coal deposits), but which did not exploit their full potential by making higher fired, more refined pots and exploiting ores of better grades from the same deposits. This suggests that although the technical knowledge existed in the Qing Dynasty (where to find good design and clays for retorts, and good ores), the smelters probably operated at a time of stringent constraints on resources, labour and capital. Although the Fengdu sites were not well situated in terms of the access to raw materials, the technology was highly efficient and also quite standardised. Furthermore, locating the sites by the Yangtze River facilitated the production with easier access to skilled labours and better control of products in a stable political and social environment.

Overall, the case studies presented in this research provide high-resolution reconstruction of the technology and organisation of zinc production and allow comparisons of regional and temporal variants within their situational contexts (Zhou *et al.* 2012; Zhou *et al.* 2014b). Production remains from more zinc smelting sites in Chongqing and beyond could be analysed and interpreted following the theoretical framework and methodology established in this research, in order to progressively map the variability of Chinese zinc distillation technology.

In Chongqing, zinc smelting sites probably dated to the Ming Dynasty were found in Zhongxian county, also by the Yangtze River (CMCHI 2014), and Pengshui county [12], by the Wu River, a tributary of the Yangtze River. A few zinc smelting sites dated to the Qing Dynasty were also found in the lead-zinc deposits in counties of Youyang and Xiushan. In addition, several zinc smelting sites dated to the Qing Dynasty were found in counties of Huanjiang and Luocheng, Guangxi province (Huang *et al.* 2012; Huang and Liang 2012). In 2015-16, several zinc smelting sites were found during a field survey and two sites dated to the Qing Dynasty were excavated by Hunan Provincial Cultural Heritage and Archaeology Institute. Interestingly, there were rows of round hearth found in Guangxi and Hunan, which were

[12] From online news: http://www.cq.xinhuanet.com/view/2005-05/17/content_4249384.htm

used to roast sulphidic zinc ores. This indicates that sphalerite was also a type of zinc ore utilised at least in the Qing Dynasty in these two regions.
Analysing production remains from archaeological sites coupled with historical research will hopefully contribute to a comprehensive picture of zinc distillation technology in China – one that appreciates a diversity of technical solutions but also the sociocultural contexts in which they were embedded.

8.2 Zinc distillation, crucible smelting and technical ceramics

Crucibles are free standing, movable vessels used for high-temperature, metallurgical processes such as smelting, melting, cementation and assaying (Rehren 2003; Rehren and Bayley 2007; Zhou *et al.* 2016). Zinc distillation can be regarded as a crucible smelting process involving both reduction and condensation. There are other two important crucible smelting traditions in the metallurgical history of China, whose significance – let alone their relationship to zinc smelting – remains to be studied in detail: the crucible smelting of iron and lead.

Crucible iron smelting processes were practiced in central, east and northeast China, with Shanxi province being the largest production area throughout the nineteenth century and up to the 1950s (Wagner 1995; Wagner 2008, 38-46). Numerous Western, Chinese and Japanese ethnographic studies of the Chinese crucible iron smelting processes (cf. Wagner 2008, 42, and references therein) show that despite variations in details, the technology was generally consistent: the reduction of iron ores by coal in open tubular crucibles packed in coal-fuelled rectangular furnaces. Recently Liu Peifeng made comprehensive research on crucible iron smelting technology in Shanxi province through field surveys, historical literature, interviews and experiments (Liu 2014).

Crucible lead smelting was to use iron to displace lead from lead sulphide as iron has better chemical affinity to sulphur than lead. Such process was practiced in Linzhou, Gansu province, in the Qianlong period of the Qing Dynasty, as recorded in a memorial from the Gansu governor to the Emperor (Renmin University of China 1983, 382). It was also reported to be in operation in the 1950s to 1970s (Zhou *et al.* 2014a). Like zinc distillation and crucible iron smelting, this process also employed rectangular furnaces fuelled with coal. Recently great amounts of crucibles used for lead smelting have been found in several locations in north China, mostly dated to the Liao, Jin and Yuan periods (tenth to fourteenth centuries). Production remains from Quyang, Shanxi province, and Tongbai, Henan province, were analysed and comfirmed to be lead smelting with iron reduction process (Liu 2015b; Zhou *et al.* 2014a).

Crucible iron/lead smelting resembled the Chinese zinc distillation in many aspects. Their crucibles were typically tubular or cylindrical, with flat or pointed bases, and sometimes made of refractory kaolinitic clay from the coal measures. The reducing agents for zinc distillation and iron smelting were mainly coal, while metallic iron was used as reducing agents for crucible lead smelting. Their furnaces were rectangular fulled with coal, but those for iron/lead smelting were overall structurally simpler than the trough-shaped furnaces for zinc distillation. Sometimes alternate layers of iron smelting crucibles and coal were loaded into the furnaces, which would not be possible in zinc distillation furnaces, since lower temperatures had to be maintained around the condensers.

Without further evidence, it is difficult to assess what relationship may have existed between them. What we do know for certain is that they were three extraordinary metallurgical processes with no counterparts in other parts of the world. They used multiple crucibles and cheap coal as fuel, and may as well have exploited coal-associated fireclay to make crucibles. The three processes were organised in similar modes with a few craftspeople attending one furnace. In modern China, they were particularly suitable for small scale production with small capital investment in remote rural areas. At the same time, they could also facilitate mass production with a large number of furnaces agglomerated, as in some traditional zinc smelting processes that operated on a factory scale before the 1960s (see Section 2.3). This was found at the Ming sites in Fengdu and the Qing site in Shizhu presented in this research. Thus future work could include the analytical studies of production remains of crucible iron/lead smelting, which could broaden our knowledge of crucible smelting technology and also aid a better understanding of zinc distillation technology.

In addition to zinc distillation retorts and iron/lead smelting crucibles, other technical ceramics in China, including other types of crucibles, furnaces and moulds, are also worth scholarly attention. The importance of technical ceramics for the study of ancient metallurgy has been realised and explored by European scholars since the 1980s (Tite *et al.* 1985; Freestone and Tite 1986; Bayley 1992; Rehren 2003; Rehren and Bayley 2007; Martinón-Torres *et al.* 2008). A series of analytical methods has proved to be of great use in a number of case studies and also in this research. The most frequently employed instrument is SEM-EDS. The analyses of cross sections of technical ceramics attached with metallurgical residues by SEM-EDS coupled with other methods could not only help identify the choices of clay and temper, manufacturing methods and performance characteristics of technical ceramics, but also inform on the original charge and the metallurgical process within them, and the intended products. Little analytical work has been carried out on Chinese technical ceramics so far (Zhou *et al.* 2016; Liu 2015a). Further analyses following the established methodology will no doubt contribute to our knowledge of ancient Chinese metallurgy.

8.3 On the origins of zinc distillation in China

The problem of the origins of Chinese zinc distillation has attracted intense scholarly interest (see Section 1.2).

Previous scholars have tried to trace the earliest occurrence of Chinese zinc distillation technology by searching for the earliest historical and archaeological evidence. However, the problem is more complex and needs revisiting.

The origin of a particular technology refers to its earliest occurrence in time. The process of the origin can be divided into two stages. The first stage is 'invention', the creation or discovery of the original conception of a new idea, behaviour or thing; the second stage is 'innovation', the adoption and incorporation of what was invented within a society (Miller 2007, 180-181; Torrence and Van der Leeuw 1989; Fitzhugh 2001). When an invention is widely adopted, the technology becomes more easily identified in the archaeological and historical records. However, if there is a period during which the invention is not adopted or not widely adopted, the technology is less likely to be recognised. Previous discussion on the origins of zinc distillation has actually focused on the innovation stage, i.e. the moment when the technology had already been adopted. The production of zinc for making brass coins in Fengdu in the Ming Dynasty took place on a large scale, so it should be regarded as the innovation stage of this technology, rather than its invention stage. The degree of specialisation and standardisation in Ming zinc distillation revealed in this research seems at odds with the variability and trial and error that one might expect at the experimental stage associated with an invention. The invention of zinc distillation must have taken place earlier than its innovation stage. Although it is difficult to exactly trace when the technology was invented, how the original conception of distilling zinc occurred can be speculated.

It has been suggested by Zhou Weirong that during the cementation process, some metallic zinc could have been condensed if the upper parts of crucibles were cooler than the reaction zone (Zhou 1996). The craftspeople might have observed the accidental occurrence of the metal, realised its volatility and even intentionally experimented to obtain it. They would have discovered that the new metal could be used to make brass like *luganshi*, and that it was actually produced from *luganshi*. An early accidental discovery of the metal probably took place also in the lead/silver smelting furnaces charged with ores rich in zinc minerals. Occasionally droplets of metallic zinc would have been condensed in the flues of the furnaces. As a matter of fact, zinc droplets collected from German lead smelting furnaces did to some degree improve the understanding of this metal and also inspire its isolation in Europe (see Section 7.2.2). However, no archaeological or historical evidence of metallic zinc of this type has been found in the Chinese contexts.

Once zinc was realised as a volatile metal which could be alloyed with copper to make brass, the metal could be intentionally obtained by adapting distillation technology existing in other areas of metallurgy and beyond. China has a long tradition of distillation by ascending, which developed from the steaming methods of food cooking with pottery and bronze *yan* 甗 (a vessel made of two parts). Since the Eastern Han period, Chinese alchemists have mastered the distillation by ascending to extract volatile mercury from cinnabar (HgS) (Zhao 1984a). Distillation technology was also widely adopted to make floral water and alcohol (Zhao and Zhou 1998). In the processes employing distillation by ascending, the distillate was normally led off with a side-tube to a receiver. Sometimes no side-tube was added; instead a catch-bowl or a vessel was placed inside to collect the distillate. Distillation by ascending with a side-tube was a common practice in mercury production during the Ming Dynasty, as recorded in *Tian Gong Kai Wu* (Sun and Sun 1966, 280). However, it was not suitable for zinc distillation because zinc, condensing at higher temperatures (between 420 °C and 550 °C) than mercury, would solidify and oxidise when it entered the side-tube which was exposed to the air. Thus the catch-bowl way of distillation was adapted for the production of zinc with the pockets acting as catch-bowls.

The first occurrence of zinc distillation was probably related to alchemical practices. In *Tian Gong Kai Wu*, the process of zinc distillation was called *shenglian woyuan* 升炼倭铅 (Figure 1.4). *Shenglian*, literally 'sublimating and smelting', was a common term in alchemy and also used to describe the process of mercury distillation. Although no direct evidence of zinc distillation has been found in alchemical treatises so far, the process could have been practiced by the alchemists who operated the cementation brass making process. The first practices of zinc distillation must have taken place on a small scale, most likely in an individual retort, where the temperature and atmosphere could be well controlled. With the growing demand for the metal for coinage in the late Ming period, the process must have been adapted and perfected to fulfill the needs of large-scale production. The tradition of Chinese zinc distillation employing the multiple retorts fired in rectangular furnaces thus developed.

On the basis of the evidence available, it seems plausible that Chinese zinc distillation developed autonomously, probably through a combination of observation (e.g. zinc condensing in brass cementation crucibles and/or lead/silver smelting furnaces), experimentation, technological transfer (from distillation in other areas or other forms of crucible smelting) and adaption to mass production. The complexity of the origins of zinc distillation implies that we must cross narrow areas of specialisation and extend our research to technologies of cementation brass making, lead/silver smelting, distillation and crucible smelting in order to understand the development of the technology. Further discovery of early zinc production remains and brass cementation crucibles and their archaeometallurgical studies will hopefully contribute to a greater insight into the origins of the technology.

A further relevant point to discuss is how the development of zinc distillation technologies in China and Europe appear to represent different situations. European zinc distillation took a long period to complete until it finally replaced the cementation process in brass industry. This slow pace of innovation may have been due to the fact that, on the one hand, the traditional cementation method was a long-established tradition sufficient for brass making, and, on the other hand, the new method of zinc production was at first not very efficient and profitable, making it harder to compete with the imports of cheap Chinese zinc. In contrast, large-scale zinc production in China appears to emerge rather abruptly in the late Ming period, following a short period of the practice with the cementation process. Chinese zinc distillation was much more technically efficient than the cementation process and suitable for mass production, which fittingly met the demand for a cheap minting material in the late Ming period.

The two contrasting cases highlight the fact that no single theoretical model may be used to explain innovation processes and, more importantly, that technological innovation can only be explained by reference to other crafts and to the specific sociocultural and economic contexts that allow for both the discovery and the adoption of these techniques. The case studies presented in this research hopefully provide a starting reference for a study of the origins and transmission of zinc distillation that goes beyond a narrow focus on dates.

8.4 History, archaeology, science and ethnography

History, archaeology and science are regarded as three equally crucial approaches to ancient materials and technologies (Martinón-Torres 2008). For the study of Chinese zinc distillation technology, the ethnographic evidence of the traditional processes is also of great use. As discussed in Section 1.2, the historical records, the analyses of brass coins and the ethnographic documents showed problems and limitations in the interpretation of the ancient Chinese zinc distillation technology. However, the archaeological and archaeometric evidence of the production remains in Chongqing in the Ming and Qing periods, combined with those sources of evidence, has proved very useful in the reconstruction of the *chaînes opératoire* of the technology and organisation of zinc production within specific social contexts.

The technical treatises, *Tian Gong Kai Wu* in particular, provide invaluable information on the technology and on how zinc and its metallurgy were perceived by the contemporary people in the late Ming period. Although the treatise fails to correctly describe and illustrate the distillation principle of zinc smelting, other aspects of the technology documented, including ores, fuel and jars, have proved to be accurate. The treatise also shows that the contemporary people understood the extreme volatility of zinc and noticed the great zinc losses during smelting. In addition, relevant information from local gazetteers, administrative and geological documents helps contextualise the technology within its social and environmental contexts. This source of evidence also allows us to see the links between the production sites in Chongqing and the main consumption of zinc in official minting. It provides a wealth of information on the spatial organisation of production, state taxation and management, the political and economic environment, etc. With this information, we are in a better position to interpret the archaeological and archaeometric data. For example, the differences of the technology and choice of location at the two groups of sites in Chongqing were well explained by placing them within their specific historical contexts, which were mainly elicited from the literary evidence.

The ethnographic records of the traditional processes practiced in modern southwest China aid the interpretation of the archaeological data and the reconstruction of the *chaînes opératoire* and technological choices of the technology in Chongqing. Although they could not directly reflect the ancient technology, their holistic analysis contributes to a more diversified picture of the technological tradition of zinc distillation.

Archaeological and archaeometric evidence plays a crucial role in the study of Chinese zinc distillation technology. Previous studies concentrated on the analyses of brass coins and artefacts, but the use of zinc and cadmium levels to differentiate the cementation process and the speltering process has been questioned and remains in any case limited in terms of the technological information that we may obtain.

The study of zinc production remains is the most direct and reliable approach to the technology. The analyses of a variety of production remains with a number of scientific techniques in this research enable the reconstruction of the technology to be made in great detail. This reconstruction helps identify the inaccuracies of the description and illustration in *Tian Gong Kai Wu*. This further strengthens the role of archaeological and archaeometric research in promoting a more critical analysis of the historical evidence. The comparisons between the technologies reconstructed at the two groups of sites allow a comprehensive interpretation of the technological choices that characterised each of them in addition to providing systematic, high-resolution data for further comparisons in the future – a body of information unlikely to be attainable from historical records alone.

Overall, this research has provided a practical demonstration of the benefits of a critical integration of historical, archaeological, scientific and ethnographic data to approach Chinese zinc distillation technology. A number of questions have been answered; new questions have been posed too. In other cases, the cross-disciplinary approach has allowed us to propose new strategies to address old questions. All in all, it is hoped that this work has made a significant contribution to the field and paved the way for future studies of zinc distillation in China and beyond.

References

Aitchison, L., 1960. *A History of Metals, Volume 2*. London: Macdonald & Evans.

An Zhimin 安志敏, 1981. Some problems concerning Chinese early copper-based artefacts 中国早期铜器的几个问题. *Acta Archaeologica Sinica 考古学报* (3), 267-285.

Artioli, G., 2010. *Scientific Methods and Cultural Heritage: An Introduction to the Application of Materials Science to Archaeometry and Conservation Science*. Oxford: Oxford University Press.

Bachmann, H.-G., 1982. *The Identification of Slags from Archaeological Sites*. London: Institute of Archaeology. Institute of Archaeology Occasional Publication No.6.

Ball, D., 1995. *The Diana Adventure*. Kuala Lumpur: Malaysian Historical Salvors.

Bayley, J., 1984. Roman brass-making in Britain. *Historical Metallurgy* 18(1), 42-43.

Bayley, J., 1992. Metalworking ceramics. *Medieval Ceramics* 16, 3-10.

Bayley, J., 1998. The production of brass in antiquity with particular reference to Roman Britain. *In:* P. T. Craddock (ed.), *2000 Years of Zinc and Brass*. (2nd edition). London: British Museum Press, 7-26. British Museum Occasional Paper No.50.

Bayley, J., 2008. Medieval precious metal refining: archaeology and contemporary texts compared. *In:* M. Martinón-Torres and Th. Rehren (eds.), *Archaeology, History and Science: Intergrating Approaches to Ancient Materials*. Walnut Creek, California: Left Coast Press, 131-150.

Bonnin, A., 1924. *Tutenag and Paktong*. Oxford: Oxford University Press.

Bourgarit, D. and Bauchau, F., 2010. The ancient brass cementation processes revisited by extensive experimental simulation. *Journal of the Minerals, Metals and Materials Society* 62(3), 51-57.

Bowman, S., Cowell, M. R. and Cribb, J., 1989. Two thousand years of coinage in China: an analytical survey. *Historical Metallurgy* 23(1), 25-30.

Boyle, R., 1990. Essays of effluviums. *In:* M. Hunter and E. B. Davis (eds.), *The Works of Robert Boyle, Volume 7: Publications of 1672-1673*. London: Pickering & Chatto.

Browne, F., 1916. Early Chinese zinc. *Journal of the Royal Society of Arts* 64, 576.

Brownsword, R., 1988. A possible Romanesque object made from zinc. *Historical Metallurgy* 22(2), 102-103.

Caley, E. R., 1964. *Orichalcum and Related Ancient Alloys*. New York: American Numismatic Society.

Carsus, H. D., 1960. Historical background. *In:* C. H. Mathewson (ed.), *Zinc: The Science and Technology of the Metal, Its Alloys and Compounds*. New York: American Chemical Society, 1-8. American Chemistry Society Monograph 142.

CCLGSC, 1994. *Local Gazetteer of Shizhu County 石柱县志*. Chengdu 成都: Sichuan Lexicographical Publishing House 四川辞书出版社.

Chase, W. T., 1999. Zinc in Chinese Bronzes 中国青铜器中的锌. *Sciences of Conservation and Archaeology 文物保护与考古科学* 11(2), 57-64.

Chen Hailian 陈海连, 2013. Zinc transfer from China to Europe via trade, ca. 1600-1800: a transnational perspective. *Technikgeschichte* 80 (1), 71-94.

Chen Hailian 陈海连, 2014. Fueling the boom: coal as the primary source of energy for processing zinc in China and comparison with Europe, ca. 1720-1820. *Journal of the Economic and Social History of the Orient* 57 (1), 76-111.

Chen Hailian 陈海连 and Souza, G. B., forthcoming. Zinc in Ming and early Qing China, ca. 1400-1680s. *In:* U. Theobald and H.-U. Vogel (eds.), *Sino-Japanese Trade, Commodity Chains, Regional Economies and the Environment in China and East Asia (Monies, Markets, and Finance in China and East Asia conference proceedings volume in April 2010)*. Leiden: Brill.

Chen Zheng 陈正 and Zhang Guanhuan 章冠环, 1943. The mineralisation and transformation of Laochangping and Huagouping in Shizhu, Sichuan 四川石柱县老厂坪及花狗坪锌铅矿之成矿作用及其改变作用. *Geological Review 地质论评* 8(Z1), 67-84.

Childs, S. T., 1991. Style, technology, and iron smelting furnaces in Bantu-speaking Africa. *Journal of Anthropological Archaeology* 10(4), 332-359.

CMBCH and CMBI (eds.), 2001. *Archaeological Reports of Chongqing Reservoir in 1997 重庆库区考古报告集 1997 卷*. Beijing 北京: Science Press 科学出版社.

CMBCH and CMBI (eds.), 2003. *Archaeological Reports of Chongqing Reservoir in 1998 重庆库区考古报告集 1998 卷*. Beijing 北京: Science Press 科学出版社.

CMBCH and CMBI (eds.), 2006. Archaeological Reports of Chongqing Reservoir in 1999 重庆库区考古报告集 1999 卷. Beijing 北京: Science Press 科学出版社.

CMBCH and CMBI (eds.), 2007a. *Archaeological Reports of Chongqing Reservoir in 2000 重庆库区考古报告集 2000 卷*. Beijing 北京: Science Press 科学出版社.

CMBCH and CMBI (eds.), 2007b. *Archaeological Reports of Chongqing Reservoir in 2001 重庆库区考古报告集 2001 卷*. Beijing 北京: Science Press 科学出版社.

CMBCH and CMBI (eds.), 2010. *Archaeological Reports of Chongqing Reservoir in 2002 重庆库区考古报告集 2002 卷*. Beijing 北京: Science Press 科学出版社.

CMCHI, 2014. Preliminary excavation report on the zinc smelting site of Linjiangerdui, Zhongxian 忠县临江二队炼锌遗址发掘简报. *Southern Ethnology and Archaeology 南方民族考古* 10, 345-368.

CMCHI and FCBCH, 2013. A report on the excavation of the zinc smelting sites in Fengdu county between 2004 and 2005 重庆丰都炼锌遗址群 2004~2005 年发掘报告. *Jianghan Archaeology 江汉考古* (3), 70-86.

Cocks, E. J. and Walters, B., 1968. *A History of the Zinc Smelting Industry in Britain*. London: George G. Harrap.

Copper Development Association, 1961. *Brasses and Other Copper-Zinc Alloys*. London: Copper Development Association. C.D.A. Publication No.6.

Costin, C. L., 1991. Craft specialization: issues in defining, documenting, and explaining the organisation of production. *Archaeological Method and Theory* 3, 1-56.

Costin, C. L., 2000. The use of ethnoarchaeology for the archaeological study of ceramic production. *Journal of Archaeological Method and Theory* 7(4), 377-403.

Cotterill, C. H., 1950. *Industrial Plant Location: Its Application to Zinc Smelting*. St. Louis: American Zinc, Lead and Smelting Company.

Cowell, M. R., Cribb, J., Bowman, S. and Shashoua, Y., 1993. The Chinese cash: composition and production. *In:* M. M. Archibald and M. R. Cowell (eds.), *Metallurgy in Numismatics* 3. London: Royal Numismatic Society, 185-196.

Cowell, M. R., La Niece, S. and Rawson, J., 2003. A study of later Chinese metalwork. *In:* P. Jett, J. G. Douglas, B. McCarthy and J. Winter (eds.), *Scientific Research in the Field of Asian Art (Proceedings of the First Forbes Symposium at the Freer Gallery of Art)*. London: Archetype, 80-89.

Craddock, P. T., 1978. The composition of the copper alloys used by the Greek, Etruscan and Roman Civilisation 3: the origins of early use of brass. *Journal of Archaeological Science* 5(1), 1-16.

Craddock, P. T., 1985. A history of the distillation of metals. *Bulletin of the Metals Museum* 10, 3-25.

Craddock, P. T., 1995. *Early Metal Mining and Production*. Edinburgh: Edinburgh University Press.

Craddock, P. T., 1997. Low technology zinc mining in the south west of China. *Mining History: The Bulletin of the Peak District Mines Historical Society* 13(3), 41-51.

Craddock, P. T., 1998. Zinc in classical antiquity. *In:* P. T. Craddock (ed.), *2000 Years of Zinc and Brass*. (2nd edition). London: British Museum Press. British Museum Occasional Paper No.50.

Craddock, P. T., 2005. Enigmas of bidri. *Surface Engineering* 21(5/6), 333-339.

Craddock, P. T., 2009a. The origins and inspirations of zinc smelting. *Journal of Materials Science* 44(9), 2181-2191.

Craddock, P. T., 2009b. Perceptions and reality: the fall and rise of the Indian mining and metal industry. *In:* T. L. Kienlin and B. W. Roberts (eds.), *Metals and Societies: Studies in Honour of Barbara S. Ottaway*. Bonn: Habelt, 453-464.

Craddock, P. T., 2013. Two millennia of the sea-bourne metals trade with India. *Indian Journal of History of Science* 48 (1), 1-37.

Craddock, P. T. and Eckstein, K., 2003. Production of brass in antiquities by direct reduction. *In:* P. T. Craddock and J. Lang (eds.), *Mining and Metal Production through the Ages*. London: British Museum Press, 216-230.

Craddock, P. T., Freestone, I. C., Gurjar, L. K., Middleton, A. P. and Willies, L., 1989. The production of lead, silver and zinc in ancient India. *In:* A. Hauptmann, E. Pernicka and G. A. Wagner (eds.), *Old World Archaeometallurgy*. Bochum: Deutsches Bergbau-Museum, 51-69.

Craddock, P. T., Freestone, I. C., Gurjar, L. K., Middleton, A. P. and Willies, L., 1998. Zinc in India. *In:* P. T. Craddock (ed.), *2000 Years of Zinc and Brass*. (2nd edition). London: British Museum Press, 27-72. British Museum Occasional Paper No.50.

Craddock, P. T., Gurjar, L. K. and Hegde, K. T. M., 1983. Zinc production in medieval India. *World Archaeology* 15(2), 211-217.

Craddock, P. T., Gurjar, L. K. and Hegde, K. T. M., 1984. Ancient lead and zinc mining in Rajasthan, India. *World Archaeology* 16(2), 222-233.

Craddock, P. T. and Hook, D. R., 1997. The British Museum collection of metal ingots from dated wrecks. *In:* M. Redknap (ed.), *Artefacts from Wrecks: Dated Assemblages from the Late Middle Ages to the Industrial Revolution*. Oxford: Oxbow Books, 143-154.

Craddock, P. T. and Zhou Weirong, 2003. Traditional zinc production in modern China: survival and evolution. *In:* P. T. Craddock and J. Lang (eds.), *Mining and Metal Production Throughout the Ages*. London: British Museum Press, 267-292.

Cranmer-Byng, J. L., 1963. *An Embassy to China: Being the Journal Kept by Lord Macartney during his Embassy to the Emperor Ch'ien-lung, 1793-1794*. London: Longmans.

Dai Zhiqiang and Zhou Weirong, 1992. Studies of the alloy composition of more than two thousand years of Chinese coins (5th century B.C. - 20th century A.D.). *Historical Metallurgy* 26(1), 45-55.

David, N., 2001. Lost in the Third Hermeneutic? Theory and methodology, objects and representations in the ethnoarchaeology of African metallurgy. *Mediterranean Archaeology* 14, 49-72.

David, N. and Kramer, C., 2001. *Ethnoarchaeology in Action*. Cambridge: Cambridge University Press.

Day, J., 1973. *Bristol Brass: A History of the Industry*. Newton Abbot: David & Charles.

Day, J., 1988. The Bristol brass industry: furnace structures and their associated remains. Appendix I. Report on the examination of blue material from industrial remains at Warmley, by Freestone, I. C. *Historical Metallurgy* 22(1), 24-41.

Day, J., 1998. Brass and zinc in Europe from the Middle Ages until the nineteenth century. *In:* P. T. Craddock (ed.), *2000 Years of Zinc and Brass*. (2nd edition). London: British Museum Press, 133-158. British Museum Occasional Paper No.50.

de Ruette, M., 1995. From *conterfei* and *speauter* to zinc: the development of the understanding of the nature of zinc and brass in post-medieval Europe. *In:* D. R. Hook and D. R. M. Gaimster (eds.), *Trade and Discovery: The Scientific Study of Artefacts from Post-Medieval Europe and Beyond*. London: British Museum Press, 195-203. British Museum Occasional Paper No.109.

Diessel, C. F. K., 1992. *Coal-Bearing Depositional Systems*. Berlin, New York: Springer-Verlag.

Dungworth, D., 1995. *Iron Age and Roman Copper Alloys from Northern Britain*. Unpublished Ph.D. thesis, University of Durham.

Dungworth, D., 1996. Caley's 'zinc decline' reconsidered. *Numismatic Chronicle* 156, 228-234.

Dungworth, D., 1997. Roman copper alloys: analysis of artefacts from Northern Britain. *Journal of Archaeological Science* 24(10), 901-910.

Dungworth, D., 2003. *Scientific Examination of Glass and Glassworking materials from Silkstone, Yorkshire*. London: English Heritage. Centre for Archaeology Report 90/2003.

Dungworth, D. and White, H., 2007. Scientific examination of zinc-distillation remains from Warmley, Bristol. *Historical Metallurgy* 41(1), 77-83.

Dungworth, D. and Wilkes, R., 2010. *Taynton Brassmill, Newent, Gloucestershire: Early 17th-century Brass Manufacture*. Portsmouth: English Heritage. Reseach Department Report Series 28/2010.

Eramo, G., 2005. The melting furnace of the Derrière Sairoche glassworks (Court, Swiss Jura): heat-induced mineralogical transformations and their technological significance. *Archaeometry* 47(3), 571-592.

Fairbank, J. K. (ed.), 1978. *The Cambridge History of China, Volume 10: Late Ch'ing 1800-1911, Part 1.* Cambridge: Cambridge University Press.

Fan Liangming 范良明 and Liu Wenjun 刘文君, 1982. Hydrozincite in oxidising zone of lead-zinc deposits in Shizhu, Sichuan 四川省石柱铅锌矿床氧化带中的水锌矿. *Minerals and Rocks 矿物岩石* (2), 68-73.

Fan Xiaopan, Harbottle, G., Gao Qiang, ZhouWeirong, Gong Qiming, Wang Hua, Yu Xiaohan and Wang Changsui, 2012. Brass before bronze? Early copper-alloy metallurgy in China. *Journal of Analytical Atomic Spectrometry* 27(5), 821-826.

Fan Xiaopan 凡小盼, Huang Jie 黄洁, Zhao Ruiting 赵瑞廷, Zhou Weirong 周卫荣, Luo Wugan 罗武干 and Wang Changsui 王昌燧, 2010. Simulation experiments of Chinese early brass making by smelting mixed ores 中国早期黄铜混合矿冶炼工艺的模拟探索. *Relics from South 南方文物* (4), 143-148.

Farnsworth, M., Smith, C. S. and Rodda, J. L., 1949. Metallographic examination of a sample of metallic zinc from ancient Athens. *Hesperia (Supplement)* 8, 126-129.

Fitzhugh, B., 2001. Risk and invention in human technological evolution. *Journal of Anthropological Archaeology* 20, 125-167.

Freestone, I. C., 1995. Ceramic petrography. *American Journal of Archaeology* 99(1), 111-115.

Freestone, I. C., 2008. Pliny on Roman glassmaking. *In:* M. Martinón-Torres and Th. Rehren (eds.), *Archaeology, History and Science: Intergrating Approaches to Ancient Materials*. Walnut Creek, California: Left Coast Press, 77-100.

Freestone, I. C., Craddock, P. T., Gurjar, L. K., Hegde, K. T. M. and Paliawal, H. V., 1985a. Analytical approaches to the interpretation of medieval zinc-smelting debris from Zawar, Rajasthan. *Journal of Archaeologcial Chemistry* (3), 1-12.

Freestone, I. C., Craddock, P. T., Hegde, K. T. M., Hughes, M. J. and Paliawal, H. V., 1985b. Zinc production at Zawar, Rajasthan. *In:* P. T. Craddock and M. J. Hughes (eds.), *Furnaces and Smelting Technology in Antiquity*. London: British Museum Press, 229-245. British Museum Occasional Paper No.48.

Freestone, I. C., Middleton, A. P., Craddock, P. T., Gurjar, L. K. and Hook, D. R., 1991. Role of materials analysis in the reconstruction of early metal extraction technology: zinc and silver-lead smelting at Zawar, Rajasthan. *In:* P. B. Vandiver, J. Druzik and G. S. Wheeler (eds.), *Materials Issues in Art and Archaeology II*. Pittsburg: Materials Research Society, 617-626.

Freestone, I. C. and Tite, M. S., 1986. Refractories from the ancient and preindustrial world. *In:* W. D. Kingery (ed.), *High-technology Ceramics: Past, Present and Future. The Nature of Innovation and Change in Ceramic Technology.* Westerville (OH): The American Ceramic Society, 35-63. Ceramics and Civilization 3.

Gettens, R. J., 1969. *Freer Chinese Bronzes, Volume 2: Technical Studies*. Washington, D.C.: Smithsonian Institution Press.

Gilmour, B. and Worrall, E., 1995. Paktong: the trade in Chinese nickel brass to Europe. *In:* D. R. Hook and D. R. M. Gaimster (eds.), *Trade and Discovery: The Scientific Study of Artefacts from Post-Medieval Europe and Beyond.* London: British Museum Press, 259-282. British Museum Occasional Paper No.109.

Golas, P. J., 1999. *Science and Civilisation in China, Volume 5: Chemistry and Chemical Technology, Part 13: Mining*. Cambridge: Cambridge University Press.

Gong Yin 龚荫, 1992. *Chinese Tusi System 中国土司制度.* Kunming 昆明: Nationalities Publishing House of Yunnan 云南民族出版社.

Gu Zuyu 顾祖禹, 1993. *Essentials of Geography for Reading History 读史方舆纪要.* Shanghai 上海: Shanghai Old Archive Press 上海古籍出版社.

Haedecke, K., 1973. Gleichgewichtsverhaltnisse bei der Messingherstellung nach dem Galmeiverfahren. *Erzmetall* 26, 229-233.

Han Rubin 韩汝玢 and Ke Jun 柯俊, 1988. Reports on the identification of brass artefacts from the Jiangzhai I culture 姜寨第一期文化出土黄铜制品的鉴定报告. *In:* Banpo Museum 半坡博物馆 (ed.), *Excavation Reports of Neolithic Jiangzhai 姜寨－新石器时代遗址发掘报告.* Beijing 北京: Cultural Relics Press 文物出版社, 544-548.

Han Rubin 韩汝玢 and Ke Jun 柯俊, (eds.), 2007. *The History of Science and Technology in China: Mining and Metallurgy 中国科学技术史矿冶卷.* Beijing 北京: Science Press 科学出版社.

Han Rubin 韩汝玢, Sun Shuyun 孙淑云, Li Xiuhui 李秀辉 and Qian Wei 潜伟, 2002. Microstructures of ancient Chinese bronzes 中国古代铜器的显微组织. *Journal of University of Science and Technology Beijing 北京科技大学学报* 24(2), 219-230.

Hawthorne, J. G. and Smith, C. S., 1979. *Theophilus: On Divers Arts. The Foremost Medieval Treatise on Painting, Glassmaking and Metalwork.* New York: Dover Publications.

He Xin 何欣 and Hu Guoqiang 胡国强, 2012. Yongle-Xuande gilded Buddhist statues from the Palace Museum: discussion on the dating of early Ming Palace Buddhist statues 故宫博物院藏明永乐、宣德金铜佛造像－兼论明代早期宫廷造像分期. In Palace Museum 故宫博物院 (ed.), *Impacts of Yongle-Xuande Periods—The 2nd Symposium of Two Banks Palace Museum 永宣时代及其影响－两岸故宫第二届学术研讨会论文集.* Beijing 北京：Palace Museum Press 故宫出版社, 97-109.

Hegde, K. T. M., Craddock, P. T. and Sonawne, V. H., 1986. Zinc distillation in ancient India. *In:* J. S. Olin and M. J. Blackman (eds.), *Proceedings of the 24th International Archaeometry Symposium.* Washington, D.C.: Smithsonian Institution Press, 249-258.

Hein, A., Kilikoglou, V. and Kassianidou, V., 2007. Chemical and mineralogical examination of metallurgical ceramics from a Late Bronze Age copper smelting site in Cyprus. *Journal of Archaeological Science* 34(1), 141-154.

Henan Relics Group 河南省文化局文物工作队, 1962. *Gongxian Tieshenggou 巩县铁生沟.* Beijing 北京: Cultural Relics Press 文物出版社.

Highley, D. E., 1982. *Fireclay.* London: Her Majesty's Stationery Office.

Hommel, W., 1912. The origin of zinc smelting. *Engineering and Mining Journal* 93(24), 1185-1188.

Hopkins, D. W., 1954. *Physical Chemistry and Metal Extraction.* London: J. Garnet Miller.

HPCHAI, CCCHAI, CMBCH and FCBCH, 2007. Report on Zhangjiahe site in Fengdu 丰都张家河遗址发掘简

报. *In:* CMBCH and CMBI (eds.), *Archaeological Reports of Chongqing Reservoir in 2001 重庆库区考古报告集 2001 卷*. Beijing 北京: Science Press 科学出版社, 1698-1704.

Hu Wenlong 胡文龙 and Han Rubin 韩汝玢, 1984. Ancient Chinese zinc smelting technology seen from traditional zinc smelting 从传统法炼锌看我国古代炼锌技术. *Chemistry 化学通报* (7), 59-61.

Hua Jueming 华觉明, 1999. *Metal Technology in Ancient China—The Civilisation Made by Copper and Iron 中国古代金属技术—铜和铁造就的文明*. Zhengzhou 郑州: Elephant Press 大象出版社.

Huang Chao 黄超, 2016. *The Research on Chinese Paktong and Its Transmission to Europe during the 18th and 19th Centuries*. Aachen: Shaker Verlag.

Huang Chunhe 黄春和, 2003a. The gilded copper-based Tibetan Buddhist statues in the Ming Yongle and Xuande courts, Part 1 明代永乐宣德宫廷藏式金铜佛像 (上). *The Collectors 收藏家* (4), 20-25.

Huang Chunhe 黄春和, 2003b. The gilded copper-based Tibetan Buddhist statues in the Ming Yongle and Xuande courts, Part 2 明代永乐宣德宫廷藏式金铜佛像 (下). *The Collectors 收藏家* (5), 22-29.

Huang Guanghui 黄光辉 and Lang Chengshen 朗承诜, 1992. *Revised Local Gazetteer of Fengdu County in the Republic of China (1927) 民国重修丰都县志*. Chengdu 成都: Bashu Press 巴蜀书社. Chinese Local Gazetteer Collection: Prefectures and Counties in Sichuan Province, Volume 47 中国地方志集成四川府县志辑 47.

Huang Quansheng 黄全胜, Li Yanxiang 李延祥, Liang Fulin 梁福林 and Tan Jiale 谭家乐, 2012. Preliminary investigation of Hongshan smelting sites in Huanjiang, Guangxi 广西环江红山古代冶炼遗址初步考察. *China Mining 中国矿业* (6), 120-124.

Huang Quansheng 黄全胜 and Liang Xingquan 梁兴权, 2012. A preliminary investigation into ancient zinc smelting sites found in Luocheng county of Guangxi 广西罗城古代炼锌遗址群初步考察. *Journal of Guangxi University for Nationalities 广西民族大学学报* 34 (5), 140-145.

Huang, R., 1974. *Taxation and Governmental Finance in Sixteenth-Century Ming China*. Cambridge: Cambridge University Press.

Huang Tinggui 黄廷桂, 2008. *Comprehensive Gazetteer of Sichuan Province 四川通志*. Taipei 台北: Commercial Press 商务印书馆.

Huang Wei 黄维, 2006. *The Use of Coal as Fuel in Iron Smelting in the Song Dynasty—Evidence from the Examination of Iron Coins 从宋代铁钱探讨用煤炼铁*. Unpublished Master thesis, University of Science and Technology Beijing 北京科技大学.

IACASS, 1988. *Sanlihe in Jiaoxian 胶县三里河*. Beijing 北京: Cultural Relics Press 文物出版社.

IMRCAGS, 2007. *Instruction to the Map of Lead-zinc Deposits in China (1:5000,000) 中国铅锌矿矿产资源图 (1:500 万) 说明书*. Beijing 北京: Geological Press 地质出版社.

Ingalls, W. R., 1903. *The Metallurgy of Zinc and Cadmium*. London: The Engineering and Mining Journal.

Ingalls, W. R., 1936. History of the metallurgy of zinc. *Transactions of American Institute of Mining and Metallurgical Engineers, Volume 121: Metallurgy of Lead and Zinc*. New York: The American Institute of Mining and Metallurgical Engineers, 339-373.

Jackson, C. M. and Smedley, J. W., 2004. Medieval and post-medieval glass technology: melting characteristics of some glasses melted from vegetable ash and sand mixtures. *Glass Technology* 45(1), 36-42.

Jett, P. and Douglas, J. G., 1992. Chinese Buddhist bronzes in the Freer Gallery of Art: physical features and elemental composition. *Proceedings of the Materials Research Society Symposium* 267, 205-224.

Jettens, R. J. and Stout, G. L., 1966. *Painting Materials: A Short Encyclopaedia*. (2nd edition). New York: Dover Publication.

Juleff, G., 2009. Technology and evolution: a root and branch view of Asian iron from first-millennium BC Sri Lanka to Japanese steel. *World Archaeology* 41(4), 557-577.

Kearns, T., Martinón-Torres, M. and Rehren, Th., 2010. Metal to mould: alloy identification in experimental casting moulds using XRF. *Historical Metallurgy* 44(1), 48-58.

Kerr, R., 1990. *Later Chinese Bronzes*. London: Bamboo in association with the Victoria and Albert Museum.

Kerr, R. and Wood, N., 2004. *Science and Civilisation in China, Volume 5: Chemistry and Chemical Technology, Part 12: Ceramic Technology*. Cambridge: Cambridge University Press.

Kilikoglou, V., Vekinis, G., Maniatis, Y. and Day, P. M., 1998. Mechanical performance of quartz-tempered ceramics: Part I, Strength and toughness. *Archaeometry* 40(2), 261-279.

Killick, D., 1991. The relevance of recent African iron-smelting practice to reconstructions of Prehistoric smelting technology. *In:* P. D. Glumes (ed.), *Recent Trends in Archaeometallurgical Research (MASCA Research Papers in Science and Anthropology, Volume 8, Part 1)*. Philadelphia: MASCA, 47-54.

Killick, D., 2004. Social constructionist approaches to the study of technology. *World Archaeology* 36(4), 571-578.

Kühn, H., 2012. Zinc white. *In:* R. L. Feller (ed.), *Artists' Pigments: A Handbook of Their History and Characteristics, Volume 1*. London: Archetype, 169-186.

Lechtman, H., 1977. Style in technology—some early thoughts. *In:* H. Lechtman and R. S. Merrill (eds.), *Material Culture: Styles, Organization, and Dynamics of Technology*. St. Paul: West, 3-20.

Lee, S., Kim, Y. J. and Moon, H., 1999. Phase transformation sequence from kaolinite to mullite investigated by energy-filtering transmission electron microscope. *Journal of the American Ceramic Society* 82(10), 2841-2848.

Leeds, E. T., 1954. Zinc coins in medieval China. *Numismatic Chronicle* 14, 177-185.

Li Dadi 李大地, Bai Jiujiang 白九江 and Yuan Dongshan 袁东山, 2013. An analysis on the Zinc Smelting Archaeology 炼锌考古探析. *Jianghan Archaeology 江汉考古* (3), 113-122.

Li Hu 李虎, 2010. On the influence of changing the chieftain system to a new system of *Liuguan* on the economic development of the *Tujia* region in the Qing Dynasty 论清代改土归流对土家族地区经济开发的影响. *Journal of Chongqing Three Gorges University 重庆三峡学院学报* 26(5), 59-63.

Li Jiazhi 李家治 (ed.), 1998. *The History of Science and Technology in China: Ceramics 中国科学技术史陶瓷卷*. Beijing 北京: Science Press 科学出版社.

Li Jinyao 李进尧, Wu Xiaoyu 吴晓煜 and Lu Benshan 卢本珊, 2007. *Mining Engineering History of Ancient Metal and Coal Deposits in China 中国古代金属矿和煤矿开采工程技术史*. Taiyuan 太原: Shanxi Education Publishing House 山西教育出版社.

Li Liangpin 李良品, 2008b. Inspiration from the economic development of the *Tusi* chieftain period in the *Tujia* region along Wujiang valley 乌江流域土家族地区土司时期的经济发展及启示. *Journal of Hubei Institute for Nationalities (Philosophy and Social Sciences) 湖北民族学院学报 (哲学社会科学版)* 26(1), 1-7.

Li Mijia 李米佳, 2013. Non-destructive analyses on Xuan copper vessels from the Palace Museum 清宫旧藏宣铜器无损检测研究示例. *Wenwu Chunqiu 文物春秋* (2), 49-54.

Li Shizhen 李时珍, 1975. *Ben Cao Gang Mu, Volume 1 本草纲目第 1 册*. Beijing 北京: People's Medical Publishing House 人民卫生出版社.

Li Wenying 李文瑛 and Zhou Jinling 周金玲, 1998. Archaeological finds and studies of Yingpan cemetery 营盘墓地的考古发现与研究. *Xinjiang Cultural Relics 新疆文物* (1), 69-82.

Li Xiaocen 李晓岑, 2008a. Zinc smelting technology in Hezhang and Huize 赫章、会泽的炼锌技艺. *In:* Hua Jueming 华觉明, Li Xiaocen 李晓岑 and Tang Xuxiang 唐绪祥(eds.), *Mining, Smelting and Processing Skills of Metals 金属冶炼和加工工艺*. Zhengzhou 郑州: Elephant Press 大象出版社, 21-29.

Li Xiuhui 李秀辉, 2004. The metallographical studies of metal artefacts from Dongdajing and Qilangshan Xianbei cemetery, Inner Mongolia 内蒙古东大井和七郎山鲜卑墓地出土金属器物的金相学研究. *In:* Wei Jian 魏坚 (ed.), *The Discovery and Study of Xianbei Cemetery in the Inner Mongolia Region 内蒙古地区鲜卑墓葬的发现与研究*. Beijing 北京: Science Press 科学出版社, 319-327.

Li Xiuhui 李秀辉 and Han Rubin 韩汝玢, 1992. Studies of metal artefacts from Tubo cemetery, Dulan, Qinghai province 青海都兰吐蕃墓葬出土金属文物的研究. *Studies in the History of Natural Sciences 自然科学史研究* 11(3), 278-288.

Li Xiuhui 李秀辉 and Wu Kunyi 吴坤仪, 1994. The alloy compositions and material studies on an armillary sphere and an abridged armilla 浑仪、简仪合金成分及材质的研究. *Cultural Relics 文物* (10), 76-94.

Li Yao 李尧, 1944. Report on traditional zinc smelting in Yimen, Xikang 西康益门土法炼锌研究报告. *Mining and Smelting 矿冶* (4), 80-91.

Lin Meicun 林梅村, 1999. Studies on the introduction of *toushi* to China 鍮石入华考. *Archaeology and Cultural Relics 考古与文物* (2), 65-74.

Lin Yeqin 林业勤, 1987. *Ancient Chinese Casting Technologies 中国古代传统铸造技术*. Beijing 北京: Scientific and Technical Documentation Press 科学技术文献出版社.

Liu Damo 刘大谟 and Yang Shen 杨慎 (eds.), 1996. *Comprehensive Gazetteers of Sichuan Province (The Jiajing Period) 四川总志 (嘉靖)*. Beijing 北京: Bibliography Press 书目文献出版社. Beijing Library Ancient Precious Books Archives 42, History Department, Geography 北京图书馆古籍珍本丛刊 42 史部地理类.

Liu Guangding 刘广定, 1991. The history of zinc in China: reinvestigation on the existence of 'wo-chien' in the Five Dynasty period 中国用锌史研究：五代已知"倭铅"说重考. *Chinese Studies 汉学研究* 9(2), 213-221.

Liu Haifeng 刘海峰, 2015a. *Preliminary Research on Furnace Materials of Historical Ferrous Metallurgy in China 中国古代制铁炉壁材料初步研究*. Unpublished Ph.D. thesis, University of Science and Technology 北京科技大学.

Liu Haiwang, Chen Jianli, Li Yanxiang, Bao Wenbo, Wu Xiaohong, Han Ruibin, Sun Shuyun and Yuan Dongshan, 2007. Preliminary multidisciplinary study of the Miaobeihou zinc-smelting ruins at Yangliusi village, Fengdu county, Chongqing. *In:* S. La Niece, D. Hook and P. T. Craddock (eds.), *Metals and Mines: Studies in Archaeometallurgy*. London: Archetype in association with the British Museum, 170-178.

Liu Peifeng 刘培峰, 2014. *The Study of Crucible Smelting in Shanxi Province 山西传统坩埚炼铁技术研究*. Unpublished Ph.D. thesis, University of Science and Technology Beijing 北京科技大学.

Liu Siran, 2015b. *Gold and Silver Production in Imperial China: Technological Choices in Their Social-economic and Environmental Settings*. Unpublished Ph.D. thesis, University College London.

Lu Pengliang 陆鹏亮, 2008. A discriminative study of Xuande incense burners 宣炉辩疑. *Cultural Relics 文物* (7), 64-76.

Luo Chuandong 罗传栋, 1991. *The Shipping History of the Yangtze River 长江航运史*. Wuhan 武汉: China Communications Press 人民交通出版社.

Luo Wenhua 罗文华, 2010. The court and Tibet: the selection and introduction of manufacturing technology of Tibetan Buddhist copper-based ritual implements and sacrificial utensils 宫廷与西藏：乾隆时期藏传佛教铜法物加工技术之选择与引进. *In:* Palace Museum 故宫博物院 and Max-Planck-Gesellschaft 柏林马普学会科学史所 (eds.), *The Court and the Local: Technological Communications in the 17th and 18th Centuries 宫廷与地方—十七至十八世纪的技术交流*. Beijing 北京: Forbidden City Press 紫禁城出版社, 221-265.

Ma Qi 马琦, 2010a. The causes and background of the development of Guizhou *yuan* in the Qing Dynasty 论清代黔铅兴起的原因和背景. *Journal of Guizhou University (Social Sciences) 贵州大学学报 (社会科学版)* (3), 59-64.

Ma Qi 马琦, 2010b. The transport routes of Guizhou *yuan* during the Qing Dynasty 清代黔铅运输路线考. *Journal of Chinese Scocial and Economic History 中国社会经济史研究* (4), 39-49.

Ma Qi 马琦, 2011. Guizhou *yuan* output and sales volume during the Qing Dynasty—a re-examination of previous output estimate methods 清代黔铅的产量与销量—兼评以销量推算产量的方法. *Qing History Journal 清史研究* (1), 104-116.

Ma Yue 马越, 2008. *The Preliminary Research on the Smelting Technology of 'erhuo huangtong' and 'sihuo huangtong' in Ancient China 中国古代二火黄铜、四火黄铜冶炼技术初步研究*. Unpublished Master thesis, University of Science and Technology Beijing 北京科技大学.

Maniatis, Y. and Tite, M. S., 1975. Examination of ancient pottery using the scanning electron microscope. *Nature* 257(September 11), 122-123.

Maniatis, Y. and Tite, M. S., 1981. Technological examination of Neolithic-Bronze Age pottery from central and southeast Europe and from the Near East. *Journal of Archaeological Science* 8(1), 59-76.

Maréchal, J.-R., 1954. Petite histoire du laiton et du zinc. *Techniques et Civilisations* 16, 109-128.

Martinón-Torres, M., 2008. Why should archaeologists take history and science seriously? *In:* M. Martinón-Torres and Th. Rehren (eds.), *Archaeology, History and Science: Intergrating Approaches to Ancient Materials*. Walnut Creek, California: Left Coast Press.

Martinón-Torres, M., 2012. Inside Solomon's House: an archaeological study of the Old Ashmolean chymical laboratorory in Oxford. *AMBIX* 59(1), 22-48.

Martinón-Torres, M., Freestone, I. C., Hunt, A. and Rehren, Th., 2008. Mass-produced mullite crucibles in medieval Europe: manufacture and material properties. *Journal of American Ceramic Society* 91(6), 2071-2074.

Martinón-Torres, M. and Rehren, Th., 2002. Agricola and Zwickau: theory and practice of Renaissance brass production in SE Germany. *Historical Metallurgy* 36(2), 95-111.

Martinón-Torres, M. and Rehren, Th., 2009. Post-medieval crucible production and distribution: a study of materials and materialities. *Archaeometry* 51(1), 49-74.

Martinón-Torres, M., Rehren, Th. and Freestone, I. C., 2006. Mullite and the mystery of Hessian wares. *Nature* 444(23 November), 437-438.

Mei Jianjun 梅建军, 1990. Modern Chinese traditional zinc smelting 近代中国传统炼锌术. *China Historical Materials of Science and Technology 中国科技史料* 11(2), 33-37.

Mei Jianjun 梅建军, 1993. A comparison between the ancient zinc smelting techniques of India and China 印度和中国古代炼锌术的比较. *Studies in the History of Natural Sciences 自然科学史研究* 12(4), 360-367.

Mei Jianjun 梅建军, 1995. The history, metallurgy and spread of paktong. *Bulletin of the Metals Museum* 24, 43-55.

Mei Jianjun 梅建军 and Ke Jun 柯俊, 1989. Studies on the ancient Chinese paktong smelting technology 中国古代镍白铜冶炼技术的研究. *Studies in the History of Natural Sciences 自然科学史研究* 8(1), 67-77.

MHGBSIC, 1981. A preliminary study of Chinese early copper-based artefacts 中国早期铜器的初步研究. *Acta Archaeologica Sinica 考古学报* (3), 287-302.

Miller, H. M.-L., 2007. *Archaeological Approaches to Technology*. London: Elsevier/Academic Press.

MIP 冶金工业出版社, 1958. *Traditional Zinc Smelting 土法炼锌*. Beijing 北京: Metallurgical Industry Press 冶金工业出版社.

Misra, M. K., Ragland, K. W. and Baker, A. J., 1993. Wood ash composition as a function of furnace temperature. *Biomass and Bioenergy* 4(2), 103-116.

Moore-Bennet, A. J., 1915. The mineral area of Western China. *The Far Eastern Review* (November), 215-227.

Morgan, S. W. K., 1985. *Zinc and its Alloys and Compounds*. Chichester: E. Horwood.

Mosselman, M. E., 1825. The manufacture of zinc in Great Britain. *Annales des Mines* Tome X, unknown pages.

Needham, J., 1974. *Science and Civilisation in China, Volume 5: Chemistry and Chemical Technology, Part 2: Spagyrical Discovery and Invention: Magisteries of Gold and Immortality*. Cambridge: Cambridge University Press.

Newbury, B. D., Notis, M. R. and Newbury, D. E., 2005. Revisiting the zinc composition limit of cementation brass. *Historical Metallurgy* 39(2), 75-81.

Ohya, Y., Takahashi, Y., Murata, M., Nakagawa, Z. and Hamano, K., 1999. Acoustic emission from a porcelain body during cooling. *Journal of American Ceramic Society* 82(2), 445-448.

Olsen, S. L. (ed.), 1988. *Scanning Electron Microscopy in Archaeology*. Oxford: BAR. BAR International Series No.452.

Packe, C., 1689. *The Works of the Highly Experienced and Famous Chymist, John Rudolph Glauber*. London: Thomas Milbourn.

Palace Museum 故宫博物院(ed.), 2012. *Splendors from the Yongle and Xuande Reigns of China's Ming Dynasty 明永乐宣德文物图典*. Beijing 北京: Forbidden City Publishing House 紫禁城出版社.

Pan Jixing 潘吉星, 1990. *A Critical Biography of Song Yingxing 宋应星评传*. Nanjing 南京: Nanjing University Press 南京大学出版社.

Park, J.-S. and Voyakin, D., 2009. The key role of zinc, tin and lead in copper-base objects from medieval Talgar in Kazakhstan. *Journal of Archaeological Science* 36(3), 622-628.

Partington, J. R., 1961. *A History of Chemistry, Volume 2*. Macmillan: St. Martin's P.

Peng Furong 彭福荣 and Li Liangpin 李良品, 2009. *Studies of Tusi Culture in Shizhu 石柱土司文化研究*. Chongqing 重庆: Chongqing Publishing Group and Chongqing Press 重庆出版集团重庆出版社.

Peng Furong 彭福荣 and Wang Zhengyong 王正勇, 2007. The study of ancient transport in the *Tujia* region, Chongqing 重庆土家族地区古代交通研究. *Journal of Neijiang Teachers College 内江师范学院学报* 22(6), 102-107.

Peng Rongqiu 彭容秋, 2005. *Zinc Metallurgy 锌冶金*. Changsha 长沙: Central South University Press 中南大学出版社.

Peng Xinwei 彭信威, 1958. *The History of Currency in China 中国货币史*. Shanghai 上海: People Press 人民出版社.

Pinn, K., 1999. *Paktong: The Chinese Alloy in Europe, 1680-1820*. Suffolk: The Antique Collectors' Club.

Pollard, A. M., Batt, C. M., Stern, B. and Young, S. M. M., 2007. *Analytical Chemistry in Archaeology*. Cambridge: Cambridge University Press. Cambridge Manuals in Archaeology.

Pollard, A. M. and Heron, C., 1996. *Archaeological Chemistry, Chapter 6: The chemical study of metals—the European medieval and later brass industry*. London: Royal Society of Chemistry.

Prakash, O., 1984. *The Dutch Factories in India 1617-1623*. New Delhi: Munshiram Manoharlal.

Pryce, T. O., Pigott, V. C., Martinón-Torres, M. and Rehren, Th., 2010. Prehistoric copper production and technological reproduction in the Khao Wong Prachan Valley of Central Thailand. *Archaeological Anthropological Sciences* 2, 237-264.

Ran Yubing 冉玉炳 and Ran Hao 冉浩, 1991. Ramble on the present and the past of Laochangping 漫话老厂坪今昔. *Shizhu Cultural and Historical Documents 石柱文史资料* 13, 47-50.

Rapp, G., 2009. *Archaemineralogy*. London: Springer.

Reedy, C. L., 2008. *Thin-section Petrography of Stone and Ceramic Cultural Materials*. London: Archetype.

Rehder, J. E., 2000. *The Mastery and Uses of Fire in Antiquity*. Montreal: McGill-Queen's University Press.

Rehren, Th., 1996. A Roman zinc tablet from Bern, Switzerland: reconstruction of the manufacture. *In:* S. Demirci, A. M. Özer and G. D. Summers (eds.), Archaeometry 94: The Proceedings of the 29th International Symposium on Archaeometry. Ankara: Tübitak, 35-45.

Rehren, Th., 1999a. "The same, but different": a juxtaposition of Roman and medieval brass making in central Europe. *In:* S. M. M. Young, A. M. Pollard, P. Budd and R. A. Ixer (eds.), *Metals in Antiquity*. Oxford: BAR Publishing, 252-257.

Rehren, Th., 1999b. Small size, large scale: Roman brass making in Germania Inferior. *Journal of Archaeological Science* 26(8), 1083-1087.

Rehren, Th., 2003. Crucibles as reaction vessels in ancient metallurgy. *In:* P. T. Craddock and J. Lang (eds.), *Mining and Metal Production Through the Ages*. London: British Museum Press, 207-215.

Rehren, Th. and Bayley, J., 2007. Towards a functional and typological classification of crucibles. *In:* S. La Niece, D. Hook and P. T. Craddock (eds.), *Metals and Mines—Studies in Archaeometallurgy*. London: Archetype, 46-55.

Rehren, Th., Charlton, S. C., Humphris, J., Ige, A. and Veldhuijzen, H. A., 2007. Decisions set in slag—the human factor in African iron smelting. *In:* S. La Niece, D. Hook and P. T. Craddock (eds.), *Metals and Mines—Studies in Archaeometallurgy*. London: Archetype, 211-218.

Rehren, Th. and Martinón-Torres, M., 2008. Naturam ars imitata: European brassmaking between craft and science. *In:* M. Martinón-Torres and Th. Rehren (eds.), *Archaeology, History and Science: Integrating Approaches to Ancient Materials*. Walnut Creek, California: Left Coast Press, 167-188. Publications of the Institute of Archaeology, University College London.

Rehren, Th. and Papachristou, O., 2003. Similar like white and black: a comparison of steel-making crucibles from Central Asia and the Indian Subcontinent. *In:* T. Stollner, G. Korlin, G. Steffens and J. Cierny (eds.), *Man and Mining*. Bochum: Deutsches Bergbau-Museum, 393-404. *Der Anschnitt* Beiheft 16.

Rehren, Th. and Pernicka, E., 2008. Coins, artefacts and isotopes—archaeometallurgy and archaeometry. *Archaeometry* 50(2), 232-248.

Renmin University of China 中国人民大学清史研究所中国人民大学档案系中国政治制度史教研室 (ed.), 1983. *Mining Industry in the Qing Dynasty 清代的矿业*. Beijing 北京: Zhonghua Book Company 中华书局.

RGALHS, 2009. *Local Gazetteer of Shizhu 石柱厅志点校(乾隆《石柱厅志》道光《补辑石柱厅新志》)*. Chongqing 重庆: Xinrongming Press 重庆市欣荣鸣印务有限公司.

Rostoker, W. and Bronson, B., 1990. *Pre-Industrial Iron: Its Technology and Ethnology*. Philadelphia: Archeomaterials.

Schäfer, D., 2005. The congruence of knowledge and action: the Tiangong kaiwu and its author Song Yingxing. *In:* C. Moll-murata, J. Song and H. U. Vogel (eds.), *Chinese Handicraft Regulations of the Qing Dynasty: Theory and Application*. München: Ludicium, 35-60.

Schlanger, N., 2005. The chaîne opératoire. *In:* C. Renfrew and P. Bahn (eds.), *Archaeology: The Key Concepts*. London and New York: Routledge, 25-31.

Scott, D., 1991. *Metallography and Microstructure of Ancient and Historic Metals*. Singapore: The Getty Conservation Institute.

Shen Shixing 申时行 (ed.), 1989. *Ming Hui Dian 明会典 [1587]*. Zhonghua Book Company 中华书局: Beijing 北京.

Sichuan Geology Bureau 四川地质局, 1961. *Survey Report of Qiyao Mountain Coalfield in Sichuan Province 四川省齐跃山煤田地质普查报告*. Chongqing 重庆: Sichuan Geology Bureau 203 Geological Brigade 四川地质局 203 地质队.

Sillar, B. and Tite, M. S., 2000. The challenge of 'technological choices' for materials science approaches in archaeology. *Archaeometry* 42(1), 2-20.

Sisco, A. G. and Smith, C. S., 1951. *Lazarus Ercker's Treatise on Ores and Assaying, Translated from the German Edition of 1580*. Chicago: University of Chicago Press.

Song Yingxing 宋应星, 1637. *Tian Gong Kai Wu 天工开物*. Printed by Tu Shaokui 涂绍煃.

Souza, G. B., 1986. *The Survival of Empire: Portuguese Trade and Society in China and the South China Sea 1630-1754*. Cambridge: Cambridge University Press.

Souza, G. B., 1991. Ballast goods: Chinese maritime trade in zinc and sugar in the seventeenth and eighteenth centuries. *In:* R. Ptak and D. Rothermund (eds.), *Emporia, Commodities and Entrepreneurs in Asian Maritime Trade, c.1400-1750*. Stuttgart: Steiner Verlag, 291-315.

SPAI and CMBCH, 2007. Archaeological report of Puzihe site in Fengdu 丰都铺子河遗址考古发掘报告. *In:* CMBCH and CMBI (eds.), *Archaeological Reports of Chongqing Reservoir in 2001 重庆库区考古报告集 2001 卷*. Beijing 北京: Science Press 科学出版社, 1705-1770.

SPCHAI, 1998. Survey reports of Three Gorges reservoir region in Fengdu 丰都县三峡工程淹没区调查报告. *In:* SPCHAI (ed.), *Sichuan Archaeological Reports 四川考古报告集*. Beijing 北京: Cultural Relics Press 文物出版社, 281-349.

Stos-Gale, Z., 1993. The origin of metals from the Roman-period levels of a site in Southern Poland. *Journal of European Archaeology* 1(2), 101-130.

Sun, E.-T. Z. and Sun, S.-C., 1966. *T'ien-Kung K'ai-Wu: Chinese Technology in the Seventeenth Century*. University Park and London: Pennsylvania State University Press.

Sun Shanlun 孙善抡, 1945. Traditional zinc smelting during wartime 战时我国之土法炼锌. *Chemistry Information of Peking University 北大化讯* (8), 2-4.

Tan Qingxuan 谭清宣, 2009. The immigrant and the social impact after the management system reform in the *Tujia* region in the Qing Dynasty 清代改土归流后土家族地区的移民及其社会影响. *Chongqing Social Sciences 重庆社会科学* (5), 88-92.

Thompson, R. H., 1991. The archaeological purpose of ethnoarchaeology. *In:* W. A. Longacre (ed.), *Ceramic*

Ethnoarchaeology. Tucson: University of Arizona Press, 231-245.

Thondhlana, T. P. and Martinón-Torres, M., 2009. Small size, high value: composition and manufacture of second millennium AD copper-based beads from northern Zimbabwe. *Journal of African Archaeology* 7(1), 79-97.

Tite, M. S., Freestone, I. C., Meeks, N. D. and Craddock, P. T., 1985. The examination of refractory ceramics from metal-production and metalworking sites. *In:* P. Phillips (ed.), *The Archaeologist and the Laboratory*. London: Council for British Archaeology, 50-55.

Tite, M. S., Kilikoglou, V. and Vekinis, G., 2001. Review article: strength, toughness and thermal shock resistance of ancient ceramics, and their influence on technological choice. *Archaeometry* 43(3), 301-324.

Tite, M. S., Maniatis, Y., Meeks, N. D., Bimson, M., Hughes, M. J. and Leppard, S. C., 1982. Technological studies of ancient ceramics from the Near East, Aegean and Southeast Europe. *In:* T. A. Wertime and S. F. Wertime (eds.), *Early Pyrotechnology: The Evolution of the First Fire-Using Industries*. Washington, D.C.: Smithsonian Institution Press, 61-71.

Todd, M., 1993. Charterhouse on Mendip: An interim report on the survey and excavation 1993. *Proceedings of the Somerst Archaeological and Natural History Society* 137, 59-67.

Torrence, R. and Van der Leeuw, S. E., 1989. Introduction: what's new about innovation. *In:* S. E. Van der Leeuw and R. Torrence (eds.), *What's New? A Closer Look at the Process of Innovation*. London: Unwin Hyman, 1-15.

Twitchett, D. and Fairbank, J. K., 1988. *The Cambridge History of China, Volume 7: the Ming Dynasty 1368-1644, Part 1*. Cambridge: Cambridge University Press.

Twitchett, D. and Fairbank, J. K., 1998. *The Cambridge History of China, Volume 8: the Ming Dynasty 1368-1644, Part 2*. Cambridge: Cambridge University Press.

Tylecote, R. F., 1976. *A History of Metallurgy*. London: Metals Society.

Underhill, A. P., 2003. Investigating variation in organization of ceramic production: an ethnoarchaeological study in Guizhou, China. *Journal of Archaeological Method and Theory* 10(3), 203-275.

Vogel, H. U., 1987. Chinese central monetary policy, 1644-1800. *Later Imperial China* 8(2), 1-51.

Von Glahn, R., 1996. *Fountain of Fortune: Money and Monetary Policy in China 1000-1700*. London: University of California Press.

von Schroeder, U., 2001. *Buddhist Sculptures in Tibet*. Hong Kong: Visual Dharma.

Wagner, D. B., 1995. The traditional Chinese iron industry and its modern fate. *Chinese Science* 12, 138-161.

Wagner, D. B., 2007. Song Yingxing's illustrations of iron production. *In:* F. Bray, V. Dorofeeva-Lichtmann and G. Métailié (eds.), *Graphics and Text in the Production of Technical Knowledge in China: The Warp and the Weft*. Leiden and Boston: Brill, 615-632.

Wagner, D. B., 2008. *Science and Civilisation in China, Volume 5: Chemistry and Chemical Technology, Part 11: Ferrous Metallurgy*. Cambridge: Cambridge University Press.

Wan Ming 万明, 2005. The silverization and social transformation 白银货币化和中外变革. *In:* Wan Ming 万明 (ed.), *Problems and Studies of Social Transformation during the Late Ming Dynasty 晚明社会变迁问题与研究*. Beijing 北京: Commercial Press 商务印书馆, 143-246.

Wang Gang 王纲, 1990. The Management of copper and *yuan* production in Sichuan during the Qing Dynasty 论清代四川的铜铅生产管理. *Journal of Sichuan Normal University 四川师范大学学报* (4), 88-93.

Wang, H., Cowell, M. R., Cribb, J. and Bowman, S. (eds.), 2005. *Metallurgical Analysis of Chinese coins at the British Museum*. London: British Museum Press. British Museum Research Publication No.152.

Wang Jin 王琎, 1955. *Ancient Metal Chemistry and Alchemy in China 中国古代金属化学及金丹术*. Shanghai 上海: China Science Book Equipment Company 中国科学图书仪器公司.

Wang Wanzhi 王万知, 1991. Visits to Laochangping Mining Cave 老厂坪矿洞见闻. *Shizhu Cultural and History Documents 石柱文史资料* 13, 51-54.

Weber, A., 1929. *Theory of the Location of Industries*. Chicago: University of Chicago Press.

Welter, J.-M., 2003. The zinc content of brass: a chronological indicator? *Technc: La science aus service de l'histoire de l'art et des civilisations* 18, 27-36.

Wheler, A. S., 1922. Some Chinese metallurgical applicances. *Transactions of Institute of Mining and Metallurgy* 32, 256-283.

Wolf, S., 2002. Estimation of the production parameters of very large medieval bricks from St. Urban, Switzerland. *Archaeometry* 44(1), 37-65.

Wu Qijun 吴其浚, 2002a. *Dian Nan Kuang Chang Tu Lue* 滇南矿厂图略. *In:* Gu Tinglong 顾廷龙(ed.), *Xuxiu Siku Quanshu 续修四库全书*. Shanghai 上海: Shanghai Ancient Book Press 上海古籍出版社, 131-223.

Wu Xiaoyu 吴晓煜, 2002b. Ancient coal briquettes and coal balls 古代的煤饼与煤球. *Coal Enterprise Management 煤炭企业管理* (12), 60.

Xia Xiangrong 夏湘蓉, Li Zhongjun 李仲均 and Wang Genyuan 王根元, 1980. *The History of Mining Industry in Ancient China 中国古代矿业开发史*. Beijing 北京: Geology Press 地质出版社.

Xie Pengfei and Rehren, Th., 2009. Scientific analysis of lead-silver smelting slag from two sites in China. *In:* Mei Jianjun and Th. Rehren (eds.), *Metallurgy and Civilisation: Eurasia and beyond*. Beijing: Archetype, 177-183.

Xing Jizhu 邢继柱, 2011. Report on chemical compositions of gilded copper Buddhist statues from Ruibaoge 瑞宝阁藏金铜佛像金属元素分析报告. *Cultural Relics Worlds 文物天地* (6), 100-103.

Xu Li 许笠, 1986. Investigation of traditional zinc smelting technology in the Magu district of Hezhang County 贵州省赫章县妈姑地区传统炼锌工艺考察. *Studies in the History of Natural Sciences 自然科学史研究* 5(4), 361-369.

Xu Li, 1998. Traditional zinc-smelting technology in the Guma district of Hezhang County. *In:* P. T. Craddock (ed.), *2000 Years of Zinc and Brass*. (2nd edition). London: British Museum Press, 115-131. British Museum Occasional Paper No.50.

Yang Hongzhong 杨弘忠, Yi Dianwen 易点文, Zhou Qinglei 周庆磊 and Zhou Ruibo 周瑞波, 2008. Simple explanation of genesis of lead-zinc deposits in Southeast Chongqing 重庆渝东南地区铅锌矿矿床

成因浅析. *Journal of Sichuan Geology 四川地质学报* 28(3), 190-196.

Yih, T. D. and de Kreek, J., 1993. Preliminary data on the metallic composition of some S.E. Asian coinage as revealed by X ray fluorescence (XRF) analysis. *Seaby's Numismatic Circular* 101(1), 7-9.

Zhang Hongzhao 章鸿钊, 1923. The origin of the use of zinc in China 中国用锌的起源. *Science 科学* 8(3). Also in: Wang Jin 王琎, 1955. *Ancient Metal Chemistry and Alchemy in China 中国古代金属化学及金丹术*. Shanghai 上海: China Science Book Equipment Company 中国科学图书仪器公司, 21-28.

Zhang Hongzhao 章鸿钊, 1925. Further discussion on the origin of the use of zinc in China 再述中国用锌之起源. *Science 科学* 9(9), 1116-1125. Also in: Wang Jin 王琎, 1955. *Ancient Metal Chemistry and Alchemy in China 中国古代金属化学及金丹术*. Shanghai 上海: China Science Book Equipment Company 中国科学图书仪器公司, 29-38.

Zhang Jianwei 张剑葳, 2013. *Chinese Ancient Metal Architecture 中国古代金属建筑研究*. Shanghai 上海: Southeast University Press 东南大学出版社, 237-269.

Zhang Tingyu 张廷玉 (ed.), 1974. *Ming Shi, Volume 26 明史第26册*. Beijing 北京: Zhonghua Book Company 中华书局.

Zhang Zigao 张子高, 1964. *A Brief History of Chemistry in China 中国化学史稿*. Beijing 北京: Science Press 科学出版社.

Zhao Kuanghua 赵匡华, 1984a. The evolution and chemical achievement of ancient mercury distillation in China 中国古代"抽砂炼汞"的演进及其化学成就. *Studies in the History of Natural Sciences 自然科学史研究* 3(1), 11-23.

Zhao Kuanghua 赵匡华, 1984b. Further discussion on the origin of zinc in China 再探我国用锌起源. *Studies in the History of Natural Sciences 自然科学史研究* 5(4), 15-23.

Zhao Kuanghua 赵匡华 and Zhang Huizhen 张惠珍, 1987. Textual research and elucidation with simulated tests on various kinds of medicinal gold and silver in ancient Chinese alchemy 中国古代炼丹术中诸药金、药银的考释与模拟试验研究. *Studies in the History of Natural Sciences 自然科学史研究* 6(2), 105-122.

Zhao Kuanghua 赵匡华 and Zhou Jiahua 周嘉华, 1998. *The History of Science and Technology in China: Chemistry 中国科学技术史化学卷*. Beijing 北京: Science Press 科学出版社.

Zhao Kuanghua 赵匡华, Zhou Weirong 周卫荣, Guo Baozhang 郭保章, Xue Jie 薛婕 and Liu Junqi 刘俊祺, 1988. An analysis and discussion of the chemical composition of copper coins of the Ming Dynasty 明代铜钱化学成分剖析. *Studies in the History of Natural Sciences 自然科学史研究* 7(1), 54-65.

Zhao Yifeng 赵轶峰, 1986. On the historical origin and time characteristics of late Ming fiscal crisis 试论明末财政危机的历史根源及其时代特征. *Journal of Chinese Historical Studies 中国史研究* (4), 55-68.

Zhengzhou Museum 郑州市博物馆, 1978. Excavation report of the Han Dynasty iron smelting site in Guxing, Zhengzhou 郑州古荥镇汉代冶铁遗址发掘简报. *Cultural Relics 文物* (2), 28-43.

Zhou Weirong 周卫荣, 1990. A study of zinc in Xuande incense burners 关于宣德炉中的金属锌问题. *Studies in the History of Natural Sciences 自然科学史研究* 9(2), 161-164.

Zhou Weirong 周卫荣, 1992. A textual research of the ancient metal called 'water tin' "水锡"考辨. *Wenwu Chunqiu 文物春秋* (3), 57-61.

Zhou Weirong, 1993. A new study on the history of the use of zinc in China. *Bulletin of the Metals Museum* 19, 49-53.

Zhou Weirong, 1996. Chinese traditional zinc-smelting technology and the history of zinc production in China. *Bulletin of the Metals Museum* 25, 36-47.

Zhou Weirong, 2000. A new transliterational study of *toushi*. *Bulletin of the Metals Museum* 32, 65-72.

Zhou Weirong, 2001. The emergence and development of brass-smelting techniques in China. *Bulletin of the Metals Museum* 34, 87-98.

Zhou Weirong 周卫荣, 2002. A preliminary study of Chinese traditional coin-casting techniques 中国传统铸钱工艺初探. *In:* Zhou Weirong 周卫荣 and Dai Zhiqiang 戴志强(eds.), *Selected Paper on Numismatics and History of Metallurgy 钱币学与冶铸史论丛*. Beijing 北京: Zhonghua Book Company 中华书局, 199-233.

Zhou Weirong 周卫荣, 2004. *Studies on Alloy Compositions of Ancient Chinese Coins 中国古代钱币合金成分研究*. Beijing 北京: Zhonghua Book Company 中华书局.

Zhou Weirong, 2007a. The origin and invention of zinc-smelting technology in China. *In:* S. La Niece, D. Hook and P. T. Craddock (eds.), *Metals and Mines: Studies in Archaeometallurgy*. London: Archetype in association with the British Museum, 179-186.

Zhou Weirong 周卫荣, 2007b. The ternary evidence method and the researches of the history of science and technology and the researches of archaeometry 科技史与科技考古中的"三重证据法"及其实践. *In:* Chinese Academy of Cultural Heritage 中国文化遗产研究院 (ed.), *Scientific and Technological Research on Cultural Heritage 文物科技研究* (5). Beijing 北京: Science Press 科学出版社, 50-58.

Zhou Weirong 周卫荣, 2008. Theory and practice of the technical history—technology development and social demand 技术史的理论与实践—技术发展与社会需求. *Relics from South 南方文物* (3), 108-122.

Zhou Weirong and Fan Xiangxi, 1994. Application of zinc and cadmium for the dating and authenticity of metal relics in ancient China. *Bulletin of the Metals Museum* 22(2), 16-21.

Zhou Weirong 周卫荣 and Huang Wei 黄维, 2012. Alloying compositions of 'Eastern Treasures' Yongle-Xuande Buddhist statues and the interpretation "东方瑰宝"永宣佛像的合金成分及其解读. In Zhou Weirong 周卫荣 (ed.), *Numismatics and History of Metallurgy 钱币学与冶铸史* (2). Beijing 北京: Zhonghua Book Company 中华书局, 281-291.

Zhou Wenli 周文丽, Liu Siran 刘思然 and Chen Jianli 陈建立, 2016. Discovery and study of metallurgical crucibles in ancient China 中国古代冶金用坩埚的发现和研究. *Studies in the History of Natural Sciences 自然科学史研究* 35(3), 358-370.

Zhou Wenli 周文丽, Liu Siran 刘思然, Liu Haifeng 刘海峰 and Chen Jianli 陈建立, 2014a. A preliminary study of traditional Chinese crucible lead smelting technology 中国传统坩埚炼铅技术初探. *Studies in the History of Natural Sciences 自然科学史研究* 33(2), 201-215.

Zhou Wenli 周文丽, Martinón-Torres, M., Chen Jianli 陈建立 and Li Yanxiang 李延祥, 2014b. Not so efficient, but still distilled: the technology of Qing Dynasty zinc production at Dafengmen, Chongqing, southwest China. *Journal of Archaeological Science* 43(3), 278-288.

Zhou Wenli 周文丽, Martinón-Torres, M., Chen Jianli 陈建立, Liu Haiwang 刘海旺 and Li Yanxiang 李延祥, 2012. Distilling zinc for the Ming Dynasty: the technology of large scale zinc production in Fengdu, southwest China. *Journal of Archaeological Science* 39(4), 908-921.

Appendix 1 Chinese Dynasties

<table>
<tr><td colspan="2">The Shang Dynasty</td><td>sixteenth to eleventh century BC</td></tr>
<tr><td rowspan="3">The Zhou Dynasty</td><td>Western Zhou</td><td>eleventh century-771 BC</td></tr>
<tr><td rowspan="2">Eastern Zhou</td><td>Spring and Autumn: 770-475 BC</td></tr>
<tr><td>Warring States: 475-221 BC</td></tr>
<tr><td colspan="2">The Qin Dynasty</td><td>221-206 BC</td></tr>
<tr><td rowspan="2">The Han Dynasty</td><td>Western Han</td><td>202 BC-AD 9</td></tr>
<tr><td>Eastern Han</td><td>AD 25-220</td></tr>
<tr><td colspan="2">The Six Dynasties</td><td>AD 220-581</td></tr>
<tr><td colspan="2">The Sui Dynasty</td><td>AD 581-618</td></tr>
<tr><td colspan="2">The Tang Dynasty</td><td>AD 618-907</td></tr>
<tr><td colspan="2">The Five Dynasties</td><td>AD 907-960</td></tr>
<tr><td rowspan="2">The Song Dynasty</td><td>Northern Song</td><td>AD 960-1127</td></tr>
<tr><td>Southern Song</td><td>AD 1127-1279</td></tr>
<tr><td colspan="2">The Liao Dynasty</td><td>AD 907-1125</td></tr>
<tr><td colspan="2">The Jin Dynasty</td><td>AD 1115-1234</td></tr>
<tr><td colspan="2">The Yuan Dynasty</td><td>AD 1271-1368</td></tr>
<tr><td rowspan="11">The Ming Dynasty (AD 1368-1644)</td><td>Yongle 永乐</td><td>AD 1403-1424</td></tr>
<tr><td>Xuande 宣德</td><td>AD 1426-1435</td></tr>
<tr><td>Zhengtong 正统</td><td>AD 1436-1449</td></tr>
<tr><td>Chenghua 成化</td><td>AD 1465-1487</td></tr>
<tr><td>Hongzhi 弘治</td><td>AD 1488-1505</td></tr>
<tr><td>Jiajing 嘉靖</td><td>AD 1522-1566</td></tr>
<tr><td>Longqing 隆庆</td><td>AD 1567-1572</td></tr>
<tr><td>Wanli 万历</td><td>AD 1573-1620</td></tr>
<tr><td>Taichang 泰昌</td><td>AD 1620</td></tr>
<tr><td>Tianqi 天启</td><td>AD 1621-1627</td></tr>
<tr><td>Chongzhen 崇祯</td><td>AD 1628-1644</td></tr>
<tr><td rowspan="4">The Qing Dynasty (AD 1644-1911)</td><td>Kangxi 康熙</td><td>AD 1662-1722</td></tr>
<tr><td>Yongzheng 雍正</td><td>AD 1723-1735</td></tr>
<tr><td>Qianlong 乾隆</td><td>AD 1736-1795</td></tr>
<tr><td>Jiaqing 嘉庆</td><td>AD 1796-1820</td></tr>
<tr><td colspan="2">The Republic of China</td><td>AD 1912-1949</td></tr>
<tr><td colspan="2">The People's Republic of China</td><td>AD 1949-present</td></tr>
</table>

Appendix 2 Notes on Weights and Measures

Dan 担 (picol), *jin* 斤 (catty), *liang* 两 (tael), *qian* 钱 (mace), *fen* 分 and *li* 厘 are traditional Chinese units of weight.

1 picol = 133 1/3 pound = 59.7 kg = 0.0597 tonne
1 catty = 1/100 picol = 597 g
1 tael = 1/16 catty = 37.3 g
1 mace = 1/10 tael = 3.73 g
1 *fen* = 1/10 mace = 373 mg
1 *li* = 1/10 *fen* = 37.3 mg

Li 里, *chi* 尺, *cun* 寸 and *fen* 分 are traditional Chinese units of length.

1 *li* = 559.8 m
1 *chi* = 31.1 cm
1 *cun* = 1/10 *chi* = 3.11 cm
1 *fen* = 1/10 *cun* = 0.311 cm

Duo 斗 is a traditional Chinese unit of volume.

1 *dou* = 10.737 L

Appendix 3 Formulae of Minerals and Phases

Mineral/phase	Formula	Mineral/phase	Formula
albite	$NaAlSi_3O_8$	limonite	$FeO(OH)\cdot nH_2O$
anglesite	$PbSO_4$	malachite	$Cu_2CO_3(OH)_2$
anorthite	$CaAl_2Si_2O_8$	melilite	$Ca_2(Al, Mg, Fe)Si_2O_7$
barite	$BaSO_4$	monazite	$(Ce, La, Pr, Nd, Th, Y)PO_4$
calcite	$CaCO_3$	muscovite	$KAl_3Si_3O_{10}(OH)_2$
celsian, barium feldspar	$BaAl_2Si_2O_8$	mullite	$Al_6Si_2O_{13}$
cerussite	$PbCO_3$	olivine	$(Fe, Mg, Zn)_2SiO_4$, $(Zn, Fe)_2SiO_4$
chalcocite	Cu_2S	otavite	$CdCO_3$
chalcopyrite	$CuFeS_2$	plagioclase feldspar	$(Ca, Na)Al_{1-2}Si_{3-2}O_8$
cinnabar	HgS	potassium feldspar	$KAlSi_3O_8$
cristobalite	SiO_2	pyrite marcasite	FeS_2
dolomite	$CaMg(CO_3)_2$	pyroxene	$(Ca, Fe, Mg, Zn)_2Si_2O_6$
diopside	$MgCaSi_2O_6$	quartz	SiO_2
fluorapatite	$Ca_5(PO4)_3F$	rutile	TiO_2
fluorite	CaF_2	smithsonite	$ZnCO_3$
galena	PbS	sphalerite	ZnS
gahnite, zinc spinel	$ZnAl_2O_4$	spinel	$(Fe, Mg, Zn)Al_2O_4$, $(Zn, Fe)Al_2O_4$
goethite	FeO(OH)	willimite	Zn_2SiO_4
hematite	Fe_2O_3	wüstite	FeO
hemimorphite	$Zn_4Si_2O_7(OH)_2\cdot H_2O$	zircon	$ZrSiO_4$
hydrozincite	$Zn_5(CO_3)_2(OH)_6$	zinc ferrite	$ZnFe_2O_4$
kaolinite	$Al_2Si_2O_5(OH)_4$		

Appendix 4

Historical Records about Mining and Smelting in Fengdu and Shizhu

4.1 *Ming Shi* (History of the Ming Dynasty) [1739]

卷 312 四川土司二：
成化十六年，四川巡抚孙仁奏：“三月内盗三百人入石砫，杀宣抚马澄及隶卒二十余人，焚掠而去。以石砫地临酆都，互争银场相訐，有司不为区治，致相仇杀。”命责有司捕贼。仁奏：“石砫岁办铅课五千一百三十斤，正统后停之。邻境军民假以征课，乘机窃取，酿成祸阶。请除其课，闭其洞，仍移忠州临江巡检于酆都南宾里之姜池，以便防守。”从之。(Zhang 1974, 8059-8060)

Book 312, Sichuan *Tusi* 2:
In 1480, Sichuan provincial governor Sun Ren reported to the Emperor: 'In March, 300 interior robbers came to Shizhu, killed *Tusi* Ma Cheng and over 20 yamen runners, set fire, plundered and then left. This is because Shizhu and neighbouring Fengdu fought for areas producing silver, but officials did not punish either side, resulting in their hostility and vendetta between them.' The Emperor commanded officials to arrest robbers. Sun: 'Shizhu annually collected a lead tax of 5130 catties [c. 3 tonnes], which ceased after the Zhengtong period [AD 1436-1449]. Neighbouring soldiers and civilians substituted to levy, seized the opportunity to steal, finally leading to the disaster. Please exempt the tax, close the mines, and remove the patrol and inspection official of Linjiang, Zhongzhou to Jiangchi, Nanbinli, Fengdu to facilitate defence.' The Emperor followed Sun's request.

4.2 *Local Gazetteer of Shizhu* [1775]

物产志：
民间用铁，大山坪旧有数十厂，今以近厂木尽，次闭。铜、铅用以供铸。乾隆三十三年委官开采，隶本厅同知总理。厂官自华阳县典史金玉始，今则资州吏目刘璜。白沙岭白铅于三十五年报有成效，岁额三十余万斤解省局。其矿需煤火熔成。矿煤二峒相去百三十里，且羊肠鸟道，往返三日乃达。运邮不赀，初报时商家犹昂。同知德明详议，除课铅外，官买白铅百斤，给价银一两八钱半，任商买以盈补绌。厥后，黔铅贩川，商价渐减。近且减价亦不得售，炉户煎铅止供岁课及官买之额。煎铅百斤亏折本银至八九钱一两不等，渐积拖欠，官民交累，非别有变通之策，炉户将迁散一空矣。花椒园黑铅厂，前代取竭，入峒皆空。三十八年春，同知王萦绪详请封闭，而黑白二铅，其性相同，且白沙岭与花园椒相去仅十余里，故白铅矿内间有夹产黑铅，而为数无多，亦时有时无。厂民零星煎烧，铅斤分买银星，抽课随其多寡，岁无定额，亦议章程在案也。尖子山铜厂自三十三年试采至今，厂地屡易。岁官买铜数千斤，斤给价银九分，皆引矿所煎，未有成效。缘铜价官商皆昂，厂民岁日久拮据，犹竭力攻取。何时铜得大塘，以铜厂商人之利救白铅炉户之害，抽有余补不足，庶为白铅厂之变通之策耳！且白铅鎔于煤，黑铅与铜皆鎔于柴。厅山树木虽广，而流寓日多。刊树种包谷者，丁丁相闻。久而山童近地乏柴，又与白铅之困于煤无殊，亦恐贻将来之害矣！司事者，不得不早为计之。(RGALHS 2009, 94-95)

Records of Products:
The common people use iron. There were dozens of iron factories at Dashanping, but the factories are closed as trees nearby are all used up. Copper and lead (including black lead and white lead) are used for minting. In 1768, copper and lead ores were exploited by the government, managed by *Tongzhi* of Shizhu. At first the factory officer was Jing Yu, *Dianshi*[13] from Huayang county, and now is Liu Huang, an officer from Zizhou[14]. At Baishaling factory, white lead was finally mass-produced in 1770; over 300,000 catties [c. 179 tonnes] of white lead were sent to the Sichuan provincial mint annually. The ores of white lead need coal to smelt. There were 130 *li* [c. 73 km] between zinc mines and coal mines; the road was narrow, devious and steep and it took three days for a round trip. The transport cost was so high that at first the produced white lead was expensive commercially. *Tongzhi* suggested that besides the tax, the government bought white lead at a price of 1 tael 8.5 mace silver per 100 catties; the rest could be bought by merchants, so the surplus could cover the loss. Later, white lead from Guizhou was imported to Sichuan, so the commercial price of white lead decreased. Lately white lead cannot be sold even if the price of white lead produced locally is lowered. So the smelters only produce white lead for the tax and the amounts bought by the government. Smelting 100 catties of white lead resulted in a loss of 8 mace to 1 tael silver. Gradually both the government and the civilians are exhausted. If there is no alternative policy, all smelters will leave. At Huajiaoyuan black lead factory, the ores were exploited by previous generations and all the mines were empty. In spring 1773, *Tongzhi* Wang Yingxu asked to close the factory. Black lead and white lead have similar properties. There were only more than 10 *li* [c. 5.6 km] between Baishaling and Huajiayuan. So the ores of white lead occasionally produced black lead, but not too much. The factory workers sporadically smelted and the black lead could produce small amounts of silver; the tax depended on the output, so there were no fixed amounts of annual tax. This issue is also on record. Since Jianzishan copper factory had set up trial mining in 1768, the location of the factory often changed. Annually the government bought several thousand catties of copper and offered 9 *fen* silver for each catty of copper. The smelting of copper ores did not work well. Since both the governmental and commercial prices of copper were high, the workers still worked hard to obtain copper even if they ran short of capital. When copper is produced on a large scale, the copper factory could save the smelters of white lead. The profits made from copper smelting can cover the loss of white lead smelting, which is the alternative policy of white lead factory. In addition, white lead is smelted by coal, and black lead and copper are smelted from firewood. Although Shizhu is abundant in trees, trees have gradually been run short of. As with white lead which is restricted by coal, I am afraid that the shortage of trees will do harm to the production of black lead and copper. The officials must plan to solve this issue as soon as possible.

[13] *Dianshi* was a small officer in charge of the local police office and prison.
[14] Huayang county is in Chengdu, the capital city of Sichuan Province; Zizhou is in Zizhong county and Zhiyang county, Sichuan Province.

4.3 *Revised Local Gazetteer of Shizhu* [1843]

田赋志第二·税课：
铜：尖子山铜厂，乾隆三十三年试采。后解局验看，系红铜，不堪鼓铸；且前代采竭，久无成效，旋请关闭。铁：大山坪，旧有数厂。以近山木尽，封闭。黑铅：厂在花椒园，乾隆三十三年试采。后解局验看，系响镴，不堪鼓铸。亦经前代采竭，旋请封闭。白铅：厂在白沙岭，乾隆三十三年开采，报有成效，税额三十余万斤。四十年同知王萦绪修志时已称铅百斤亏折本银七、八、九钱、一两不等，渐积拖欠，官民交累。后因山水骤发，冲塌旧峒，屡易擂鼓台、冷水溪、黑田池、深溪沟、大富坪、鹿子坪、大火田等处，今厂又在中溪沟矣。旧白铅，以煤镕成，物有限而用无穷。即困于矿，又困于煤。在昔已然，于今为甚。厂之迁徙无常，雨因水患，余皆移炉就煤计也。四十八年，酌减税额，报解铅九万五千六百斤零。后复历年递增至道光十八年，报解铅至十二万七千五百余斤，同知荣庆禀请减额。十九年，报解铅六万余斤。盖厅地产铅，本属不多。从前历年尽采尽解定例，比较三届，有盈无绌，实则煎少报多，积欠巨亿。旧商既多亏欠，新商鲜克承充。前同知刘大经时，改商煎为官煎，为一时之权宜，然尚有铅也。今则峒老山空，任其事者均系采买他处，所领本银又于市价不敷，类皆捐廉填补。现拟禀请上宪，核实另办。(RGALHS 2009, 158-159)

Part 2, Records of Land Taxes, Taxation:
Copper: The Jianzishan copper factory set up trial mining in 1768. Later the produced copper was sent to provincial mints but identified to be red copper, which could not be used for minting. In addition, because copper mines exploited by previous generations were exhausted, the factory is asked to be closed. Iron: At Dashanping there used to be several factories, but they were closed as trees nearby were used up. Black lead: The Huajiaoyuan factory set up trial mining in 1768. Later the produced black lead was sent to provincial mints and identified to be *Xianglian*, which could not be used for minting. In addition, because black lead mines exploited by previous generations were exhausted, the factory is asked to be closed. White lead: The Baishaling factory was exploited since 1768 and produced white lead on a large scale with an annual tax of over 300,000 catties [c. 179 tonnes]. When *Tongzhi* Wang Yingxu compiled *Local Gazetteer of Shizhu* in 1775, it was said that producing 100 catties of white lead resulted in a loss of 7 mace to 1 tael silver. Gradually both the government and the civilians were exhausted. Later because the flood destroyed the old mine caves, the location of factory was changed many times (to Leigutai, Lengshuixi, Heishuichi, Shenxigou, Dafuping, Luziping and Dahuotian), and now the factory is at Zhongxigou. White lead was smelted with coal, which is limited but versatile. White lead is not only restricted by ores but also by coal. The situation was such before and is more serious now. The factory location was often changed because of the flood or moving the furnaces near coal mines. In 1783, the tax was reduced to only 95,600 catties [c. 57 tonnes]. Later it increased gradually to 127,500 catties [c. 76 tonnes] in 1838. *Tongzhi* Rongqing asked to reduce the tax. In 1839, the tax was about 60,000 catties [c. 36 tonnes]. This is because Shizhu did not produce too much white lead. The production of white lead was actually deep in debt. No merchants wanted to produce it any longer. Now the ore caves are empty, so the officers even bought white lead from the market to pay the tax. Please report to the Emperor and ask him to deal with this issue.

4.4 *Revised Local Gazetteer of Fengdu* [1927]

卷九食货志：
七曜山产铁间有铅矿，石梁河洗脚溪亦出铅矿。……煤矿：太平乡柏崖口产煤甚富，有厂六盐井沟厂三剑风垭厂一，距城甚远，只供一乡之用。又距城三十里新建乡清泊荡有煤洞，七日运至邑城发卖，其船由龙河出口极便利，但中阻险滩，提载过载运价之贵常倍厂价，然所出甚旺足敷治城半年之用。又忠义乡梁磉墩尖子山产煤亦旺，皆距城窵远无出河者。……七曜山产煤亦富。银铅矿：崇德乡老厂坪相传唐宋年间广产银铅，厂基尚存，故名老厂坪。明清两代皆有巨商集股开办，惟矿沙混杂炼之，若有若无，今已停止。(Huang and Lang 1992, 595)

Volume 9, Record of Food and Commodities:
The Qiyao Mountain produces iron and sometimes lead ores, and the Shiliang River and Xijiaoxi produce lead ores as well. … Coal deposits: there were rich coal deposits at Baiyakou, Taiping township. Several coal factories provided coal only for the township. In addition, there were coal caves at Qingbaidang, Xinjian township, 30 *li* [c. 17 km] away from the capital town of the Fengdu county. It was easy to transport coal by the Dragon River to the capital town, which took seven days. As there were dangerous rapids and shoals in the river, the prices of coal doubled. But the coal was enough for the capital town to consume for half a year. Jianzishan, Liangsangdun, Zhongyi township, also produced plenty of coal, but it was far away from the capital town, so no coal was exported by river. … The Qiyao Mountain also produced abundant coal. Silver and lead ores: It is said that Laochangping[15], Chongde township, produced silver and lead during the Tang and Song Dynasties. It was named so because the factory foundation was still remaining. In the Ming and Qing Dynasties, wealthy merchants ran the factory with raised funds. But the ores were mixed with sand, so the smelting has ceased.

[15] At that time, the Laochangping region belonged to Chongde township, Fengdu county.

www.ingramcontent.com/pod-product-compliance
Lightning Source LLC
LaVergne TN
LVHW071215210926
848482LV00015B/819